Changes in the
Human-Monsoon System
of East Asia in the
Context of Global Change

Monsoon Asia Integrated Regional Study on Global Change - Vol 1

Changes in the
Human-Monsoon System
of East Asia in the
Context of Global Change

editors

Congbin Fu
Chinese Academy of Sciences, China

J R Freney
CSIRO, Plant Industry, Australia

J W B Stewart
Salt Spring Island, British Columbia, Canada

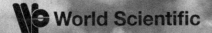

NEW JERSEY · LONDON · SINGAPORE · BEIJING · SHANGHAI · HONG KONG · TAIPEI · CHENNAI

Published by

World Scientific Publishing Co. Pte. Ltd.
5 Toh Tuck Link, Singapore 596224
USA office: 27 Warren Street, Suite 401-402, Hackensack, NJ 07601
UK office: 57 Shelton Street, Covent Garden, London WC2H 9HE

British Library Cataloguing-in-Publication Data
A catalogue record for this book is available from the British Library.

Monsoon Asia Integrated Regional Study on Global Change – Vol. 1
CHANGES IN THE HUMAN-MONSOON SYSTEM OF EAST ASIA IN THE CONTEXT OF GLOBAL CHANGE

Copyright © 2008 by World Scientific Publishing Co. Pte. Ltd.

All rights reserved. This book, or parts thereof, may not be reproduced in any form or by any means, electronic or mechanical, including photocopying, recording or any information storage and retrieval system now known or to be invented, without written permission from the Publisher.

For photocopying of material in this volume, please pay a copying fee through the Copyright Clearance Center, Inc., 222 Rosewood Drive, Danvers, MA 01923, USA. In this case permission to photocopy is not required from the publisher.

ISBN-13 978-981-283-241-2
ISBN-10 981-283-241-6

Typeset by Stallion Press
Email: enquiries@stallionpress.com

Printed in Singapore.

CONTENTS

Preface   xi

Acknowledgments   xiii

Introduction   1

Part I Variability of Monsoon   7

1. Thermal-Dynamical Effects of the Tibetan Plateau on
   the East Asian Monsoon   9

   Guoxiong Wu, Qiong Zhang, Anmin Duan and Jiangyu Mao
   E-mail: zhq@lasg.iap.ac.in

2. Paleo-Monsoon Variations in East Asia Reconstructed
   from Terrestrial Records   23

   Li Li and Zhisheng An
   E-mail: anzs@loess.llqg.ac.cn

3. Paleo-Monsoon Evolution and Variability Derived from
   Deep-Sea Sediments   39

   Pinxian Wang
   E-mail: pxwang@online.sh.cn

4. Late Quaternary Paleoclimate Simulations and Model
   Comparisons for the East Asian Monsoon   59

   Ge Yu, Sandy P. Harrison, Xing Chen and Yingqun Zheng
   E-mail: geyu33@hotmail.com

5. El Niño and the Southern Oscillation-Monsoon
   Interaction and Interannual Climate   75

   Chongyin Li and Ronghui Huang
   E-mail: lcy@lasg.iap.ac.cn

Part II   Atmospheric Composition   89

6. Recent Trends in Summer Precipitation in
   China and Mongolia   91

   Tetsuzo Yasunari, Nobuhiko Endo and Borjiginte Ailikun
   E-mail: yas.monsoon@nifty.com

7. Warming in East Asia as a Consequence of Increasing
   Greenhouse Gases   105

   Zongci Zhao, Akio Kitoh and Dong-Kyou Lee
   E-mail: zhaozc@cma.gov.cn

8. Climate Impacts of Atmospheric Sulfate and Black
   Carbon Aerosols   115

   Yun Qian, Qingyuan Song, Surabi Menon,
   Shaocai Yu, Shaw Liu, Guangyu Shi, Ruby Leung
   and Yunfeng Luo
   E-mail: Yun.Qian@pnl.gov

9. Long-Range Transport and Deposition of Dust Aerosols
   Over the Ocean, and Their Impacts on Climate   133

   Zifa Wang, Guangyu Shi and Young-Joon Kim
   E-mail: zifawang@mail.iap.ac.cn

Part III Land-Use Change   147

10. Land Use and Land Cover Change in East Asia and Its
    Potential Impacts on Monsoon Climate   149
    Congbin Fu and Shuyu Wang
    E-mail: wsy@tea.ac.cn

11. The Terrestrial Carbon Budget in East Asia:
    Human and Natural Impacts   163
    Hanqin Tian, Jiyuan Liu, Jerry Melillo,
    Mingliang Liu, David Kicklighter, Xiaodong Yan
    and Shufen Pan
    E-mail: tianhan@auburn.edu

12. Impacts of Agriculture on the Nitrogen Cycle   177
    Zhengqin Xiong, John Freney, Arvin Mosier,
    Zhaoliang Zhu, Youn Lee and Kazuyuki Yagi
    E-mail: John.Freney@csiro.au

13. Impacts of Land Use on Structure and Function
    of Terrestrial Ecosystems and Biodiversity   195
    Kanehiro Kitayama, Guangsheng Zhou
    and Keping Ma
    E-mail: kitayama@ecology.kyoto-u.ac.jp

14. Impact of Land Use in Semiarid Asia on Aridification
    and Desertification   211
    Dennis Ojima, Wen Jie Dong, Zhuguo Ma,
    Jeff Hicke and Tom Riley
    E-mail: ojima@heinzctr.org

Part IV   Marine/Coastal Systems   223

15. Transport of Materials Induced by Human Activities
    Along the Yellow and Yangtze Rivers   225
    Dunxin Hu, Chongguang Pang, Qingye Wang,
    Weijin Yan and Jiongxin Xu
    E-mail: dxhu@ms.qdio.ac.cn

16. Impacts of Global Warming on Living Resources
    in Ocean and Coastal Ecosystems   239
    Qisheng Tang and Ling Tong
    E-mail: tongling@ysfri.ac.cn

17. Sea-Level Changes and Vulnerability of the Coastal Zone   251
    Nobuo Mimura and Hiromune Yokoki
    E-mail: mimura@hcs.ibaraki.ac.jp

Part V   Driving Forces   265

18. Changes in Population Number, Composition,
    Distribution, and Consumption in East Asia   267
    Xizhe Peng and Xuehui Han
    E-mail: xzpeng@fudan.edu.cn

19. Application of Clean Technologies and New Policies to
    Reduce Emissions and the Impacts of Environmental
    Change on Human Health   279
    Kejun Jiang and Gang Wen
    E-mail: kjiang@eri.org.cn

20. Trends and Proactive Risk Management of Climate-
    Related Disasters   293
    Pei Jun Shi, Norio Okada, Joanne Linnerooth-Bayer
    and Yi Ge
    E-mail: spj@bnu.edu.cn

21. Assessment of Vulnerability and Adaptation to Climate
    Change in Western China   309
    Yongyuan Yin, Peng Gong and Yihui Ding
    E-mail: yyin@forestry.ubc.ca

22. Application of Asia–Pacific Integrated Model in East Asia   323
    Xiulian Hu and Kejun Jiang
    E-mail: huxl@eri.org.cn

23. Human Drivers of Change in the East Asian Monsoon System   335
    Karen Seto, Dennis Ojima, Qingyuan Song and Arvin Mosier
    E-mail: kseto@stanford.edu

List of Contributors   349

Index   359

PREFACE

Irreversible changes to regional biogeophysical processes and biogeochemistry, and terrestrial and marine ecosystem functioning brought about by increases in population, intensified land use, urbanization, industrialization and economic development may have global as well as regional consequences. Similarly, global changes have a significant impact on sustainable development at both regional and national levels.

The Earth System Science Partnership has recently identified Integrated Regional Studies as an important component of global change research. These studies are designed to contribute sound scientific understanding to support sustainable development at the regional level. They will assess the influence of regional processes on the Earth System functioning and vice-versa, and integrate the natural and social sciences. The regional studies must have relevance for people living in the regions and provide for the sustainable development of atmospheric, marine, terrestrial and human resources of the countries in the regions. The Integrated Regional Studies will include vulnerability analyses, identification of risks and the syndromes of environmental degradation which are crucial for sustainable development.

The key issues will include considerations of what the region will be like in ~50 years' time and the consequences of global changes for the region. The studies will consider (i) major demographic, socio-economic, and institutional drivers for change, including urbanization and industrialization, energy production and biomass burning, land use change and water resources harvesting, (ii) effects on regional and atmospheric composition, the regional water cycle and coastal systems, and local ecosystem structure and function, (iii) impacts on biogeochemical cycles and the physical climate system, and (iv) impacts of global and other feedback effects on the regional life support system, including food supply, water resources and health.

A regional approach to research is also important from an Earth system science perspective, because changes in regional biophysical, biogeochemical and anthropogenic components may produce considerable different consequences for the Earth system at the global scale. Regions are not closed systems and thus the linkages between regional changes and earth system are crucial. Regions may function as choke or switch points and small changes in a critical region may lead to profound changes in the ways in which the earth system operates.

The region of monsoon Asia has been identified as one of the priorities for an Integrated Regional Study by the Earth System Science Partnership. To provide the knowledge base of such a study, the global change System for Analysis, Research and Training (START) in collaboration with the Scientific Committee on Problems of the Environment (SCOPE)/ICSU has developed a rapid assessment project to evaluate the progress of global change research in the Asian region. The project systematically reviewed current knowledge regarding regional aspects of global change, in order to highlight gaps in knowledge and uncertainties, and define research priorities for the regional study. This book presents part of the results of that study. It provides a state-of-the-art summary of what we already know, and serves as a basis for identifying knowledge gaps that require study, including critical field experiments.

The book follows SCOPE's Rapid Assessment Project approach. Background chapters were prepared by experts from the region, then circulated within the author group and peer-reviewed in advance of a Dahlem-type workshop held in Hangzhou, China. During the workshop, groups examined crosscutting issues, and identified future research needs. This book is published as the first in a series of rapid assessments of global change research in Asia. The aim is to make sure that recent advances in the field of regional change are summarized, and their possible significance in understanding problems and potential solutions are discussed.

Congbin Fu, John R. Freney and John W. B. Stewart, Editors

ACKNOWLEDGMENTS

The editors and authors, on behalf of START and SCOPE, thank the International Council for Science (ICSU), United Nations Educational, Scientific and Cultural Organization (UNESCO), the Asia-Pacific Network for global change research (APN) and the Chinese Academy of Sciences for the financial support of this project. The assessment workshop was held in Hangzhou, China at the World Trade Center Grand Hotel. We are indebted to Yang Ying, Xie Li, Ailikun and the other staff of the Regional Center for Temperate East Asia (TEA) for their work to make the Hangzhou workshop a success.

INTRODUCTION

A systematic assessment of global change research in East Asia is presented in this book. The main aspects studied are monsoon variability, atmospheric composition, land use/land cover change, marine/coastal systems and the driving forces of change.

1. Variability of Monsoon

The East Asian winter and summer monsoons are the dominant atmospheric features determining the hydrologic conditions and natural vegetation cover for East Asia. As such, the system directly influences the climate experienced by about 57% of the world's people. Through our interconnected global economic system, interannual to interdecadal variations in the East Asian monsoons can affect the economies and societal well-being of people in the region and around the world. Consequently, understanding the potential for changes in the monsoons as a result of external factors, such as human-induced global warming, and internal factors, such as changes in land cover or the concentration of atmospheric aerosols, is important.

The nature of the East Asian monsoon, and how it resulted from the presence of the Tibetan Plateau is described in Chapter 1. Studies of climates of the past indicate that the monsoons did not develop until the region's land areas drifted into the current geography, thus creating a suitable land-sea thermal contrast under the seasonal cycle of solar radiation to excite the monsoon, and the Tibetan Plateau became elevated, thus positioning the monsoon over East Asia.

Further information about the variability of the East Asian monsoon over the last several million years is provided in Chapters 2 and 3. Of most importance is that the East Asian monsoon can vary in intensity and pattern, and that some of this variation results because of changes in heating patterns.

Model simulations and their comparisons with paleoenvironmental reconstructions indicate that models are capable of reproducing past variations of monsoon climate. This gives confidence that models will be useful for evaluating future changes of climate under the effects of changing greenhouse gas concentrations in the atmosphere and land use/cover patterns.

Connections between annual variations in the East Asian monsoon and the El Niño/Southern Oscillation are explored in Chapter 5. While these are not well defined, evidence suggests that the East Asian monsoon is affected, and there are indications that variations in the summer monsoon can have effects stretching even to the North Atlantic.

With the East Asian monsoon system being of such importance to the region and to the rest of the world, further research on the system is warranted.

2. Atmospheric Composition

Anthropogenic emissions over East Asia have increased, during the last several decades, in the same way as the global emissions. The surface air temperature over East Asia has increased by about 0.84°C per 100 years for 1900~1999 which is warmer than the increase in average globe temperature. This is interpreted to mean that human activities are most likely responsible for the warming in East Asia, especially for the last 50 years. Most projections suggest that the annual precipitation over East Asia could increase by about 23~151 mm for the period 2071~2100 relative to 1961~1990 and the East Asian winter monsoon might weaken and summer monsoon might strengthen, and more frequent floods and droughts might appear in some parts of East Asia (Chapter 6). Observed and projected results of the increase of greenhouse gases in recent years on regional aspects of global warming are discussed in Chapter 7.

It has been suggested that the changes in precipitation, and the tendency toward increased floods and droughts in certain regions (Chapter 6) are most likely caused by the large amounts, in the atmosphere, of black carbon particles which have been generated from industrial pollution, traffic, and burning of coal and biomass fuels. Simulations have been conducted to examine the effects of black carbon on the hydrologic cycle over China and the results indicated that black carbon affects regional climate by absorbing sunlight and heating the air, which rises, forming clouds and causes rain to fall in heavily polluted regions (Chapter 8).

Dust particles change optical depth, alter direct climatic forcing by scattering and absorbing solar short-wave radiation and earth's long-wave radiation. Because dust reacts with sulfur dioxide and nitric acid and becomes more hygroscopic, it may play a role in regional cloud formation. Cloud nucleation by dust particles produces indirect climatic effects through the alteration of cloud reflection and radiation by altering cloud microphysics and the production or suppression of rainfall (Chapter 9).

3. Land Use Change

Land cover can affect the impact of solar variation on climate by changing surface characteristics such as albedo, leaf area and surface roughness. Changes in management within land uses lead to changes in terrestrial carbon stocks and fluxes, and contribute to atmospheric carbon dioxide. East Asia has been transformed by land use change, and the environment has been subjected to stresses such as deforestation, desertification, and water and air pollution. The general theme in this section is that these changes have impacted regional climate. Because of the role it may play in future climate change and development, an accurate understanding of past and current land use practices and projections of future land use are provided in Chapter 10.

Land use change impacts the carbon balance of East Asian ecosystems (Chapter 11). Contemporary estimates of carbon exchange in Chinese forests indicate uptake rates of 0.05 to 0.1 Pg C yr^{-1}. Interannual estimates of net carbon exchange across ecosystems varies annually from net sinks to net sources with net carbon exchange typically related to precipitation. None of the inventory-based approaches provide estimates for all ecosystems in China, making it difficult to establish baselines for monitoring the effects of environmental change. Use of different nitrogen sources and increased fertilizer nitrogen to improve food supply, and the effects of changing agricultural systems and food consumption patterns on the nitrogen cycle are discussed in Chapter 12. More marginal land is now being used thereby decreasing nitrogen use efficiency. Environmental problems range from loss of soil nitrogen due to dust storms, to increased nitrate content of rivers.

The impacts of land use change on structure and function of terrestrial ecosystems and biodiversity of East Asia are discussed in Chapter 13. Using vascular plants for biodiversity studies the authors note that plant species information is variable across the region and that updates of species are needed to verify future environmental change. The dramatic recent changes in land use in Northern China and Mongolia due to increased population growth and political reforms of pastoral systems, and the effect this has had on vulnerable ecosystems in the region is described in Chapter 14.

4. Marine/Coastal Systems

East Asia is adjacent to some of the largest marginal seas in the world including the Yellow and East China Seas. These seas form the linkage between the world's largest continent and largest ocean, ventilate the deep oceans, receive land runoff, exchange with the open oceans, and sustain the life of millions of fishermen. Most of these seas are affected by the East Asian Monsoons as well as land use change.

Deforestation, cultivation, construction of dams, domestic and industrial uses of water, and use of fertilizers have affected water, sediment, and nutrient delivery to coastal zones with detrimental effects. The annual water flow at Lijin on the Yellow River decreased from 514 to $42 \times 10^8\,\mathrm{m^3\,yr^{-1}}$ between 1850 and 2002 and the sediment flux decreased from 13 to $0.5 \times 10^8\,\mathrm{t\,yr^{-1}}$ during that time. The water flux in the Yangtze River changed little between 1950 and 2002, but the sediment flux decreased by 40% between 1984 and 2000 due to the construction of dams and other practices. The nitrogen flux increased by a factor of two between 1970 and 2000 (Chapter 15).

Dams also affect the environment further downstream. Even the continental shelves could be affected as a result of reduced upwelling due to reduced freshwater outflow and buoyancy. Chapter 16 reports that decreases in all indices of species diversity, benthic community biomass, fish resources and recruitment of living marine resources are consistent with the notion that reduced freshwater outflow reduces the upwelling of nutrient rich subsurface waters from offshore.

Reduction in sediment outflow not only results in less food for the marine biota but it adds to the sea level rise problem. Local sea level rise depends not only on mean sea level rise, but also on sediment transport, land subsidence, waves and current flow patterns. On average, the mean sea level rise 10 to 20 cm during the past 100 years and may very well rise 9 to 88 cm more by the year 2100. These rates suggest that some slow growing reefs might be affected, as most corals do not survive when the water is too deep (Chapter 17).

5. Driving Forces

It is recognized that population trends, economic development, government, energy and environmental policies and technology advancement, will determine future greenhouse gas and sulfur dioxide emissions, which are the basic inputs for determining future change of the climate system with general circulation models. These major driving forces also provide the basis for setting socio-economic scenarios for the assessment of vulnerability, impacts and adaptation strategies, and policies to deal with climate change.

The population driving force, including current population, birth and death rates, factors influencing trends, correlations between economic and population growth rates, and population projection is described in Chapter 18. The importance of aging trends in the region, migration, urbanization, and trends to smaller household size are also discussed.

Application of clean technologies, and new energy and eco-city policies to reduce emissions is described in Chapter 19. For eco-city development, many Chinese cities have set up sustainability goals such as green land per capita, oxygen equivalent per capita, environmental quality, coordinating ability including energy, population, and industry increase elasticity, and impact ability of the city. The authors indicate that technology progress is a win-win option.

A different aspect of climate change, viz. climate-related disaster is presented in Chapter 20. Trends in losses (dollars & people) due to severe weather, institutional arrangements for managing risks, and case studies are presented to illustrate how management policies can be used to reduce risks. An integrated approach of vulnerabilities and adaptation assessment to climate change, along with methods and tools for measuring climate vulnerability are given in Chapter 21. A conceptual research framework which integrates climate change and socio-economic scenarios, vulnerability identification, sustainability indicator specification, adaptation option evaluation, and multi stakeholder participation is presented. A systems approach for designing a research framework to link modeling analysis with policy concerns for climate change is presented in Chapter 22. Implications of driving force changes for greenhouse gas emissions are presented.

The final chapter highlights the facts that East Asia is undergoing tremendous socioeconomic, political, and institutional changes and that the magnitude and rate of development in this region are unprecedented. It is emphasized that the continuous expansion of economies, demographic dynamics, globalization, policy reforms, and cultural and lifestyle changes will interact and affect the global Earth system. The authors point out that it is critical to understand the underlying human activities and social drivers of change in order to understand and predict changes in ecosystem dynamics, climate, atmospheric composition, and hydrology.

Part I

VARIABILITY OF MONSOON

Chapter 1

THERMAL-DYNAMICAL EFFECTS OF THE TIBETAN PLATEAU ON THE EAST ASIAN MONSOON

GUOXIONG WU, QIONG ZHANG, ANMIN DUAN and JIANGYU MAO

1. Introduction

The Tibetan Plateau (Qinghai-Xizang Plateau) extends over the latitude-longitude domain of 25–45°N, 70–105°E, with a size of about one quarter of the Chinese territory and a mean elevation of more than 4,000 m above sea level. Surface elevation changes rapidly across the boundaries of the Plateau, especially the southern boundary. Strong contrasts exist between the western and eastern parts of the Plateau in land surface features, vegetation and meteorological characteristics. At these altitudes the mass of the atmosphere over the surface is only 60% that of the sea level. Because of the lower densities, various radiative processes over the Plateau, particularly in the boundary layer, are quite distinct from those over lower-elevated regions. Therefore, the Tibetan Plateau exerts profound thermal and dynamical influences on the atmospheric circulation. This chapter reviews the recent progress of the thermal-dynamical effects of the Tibetan Plateau on the East Asian monsoon.

Before the 1950s, most papers concerned with the influence of large-scale topography on atmospheric circulation and climate focused on mechanical consequences. Queney (1948) summarized the former studies of air flow over mountains and identified three critical scales to distinguish different mountain waves using linearized equations. This work further proved that gravity, inertial gravity, or Rossby waves will be stimulated when an atmospheric current flows over mountains with various spatial scales. In the early 1950s, Bolin (1950) and Yeh (1950) suggested that the Tibetan Plateau divided the westerlies into two parts in winter and thus favors the generation of the Great Trough over East Asia. Koo (1951) argued that in winter the north and south fringe currents around the Tibetan Plateau always converge over the downstream region, resulting in the generation of the East Asian jet current. During the same period, numerical simulations based on a new atmospheric model by Chanery and Eliassen (1949) showed the role of large-scale topography on the formation of averaged troughs and ridges in westerlies.

Flohn (1957) and Yeh et al. (1958) found that the Tibetan Plateau is a source of atmospheric heat in summer. Since then studies have examined the temporal and spatial distribution of the heating and mechanical effects over the Tibetan Plateau and their impacts on weather and climate.

During the past decade, the theory of Ertel potential vorticity (Ertel 1942) has been applied to study the effects of the Tibetan Plateau on climate. Wu et al. (1997) employed newly released data and numerical simulation to examine the potential vorticity and temperature, and showed that the atmospheric motion driven by the Tibetan Plateau heating forms a huge air pump and impacts the variability of the Asian monsoon. It indicated that the Tibetan Plateau is not only a crucial heat source, but also a crucial negative vorticity source (Liu et al. 2001). The Rossby wave train stimulated by the vorticity source over the Tibetan Plateau creates an atmospheric circulation anomaly extending around the northern hemisphere (Wu et al. 2004).

2. Seasonal Transition of the Asian Monsoon

Yeh et al. (1958) pointed out that the atmospheric circulation of the northern hemisphere is characterized by an abrupt transition from winter to summer patterns. Mao et al. (2002a, b) defined the ridge-surface of the subtropical anticyclone belt as the westerly-easterly boundary surface, and investigated its temporal and spatial variations. They found that the Asian summer monsoon occurs first over the eastern Bay of Bengal (90–100°E). Such a seasonal transition is closely related to the earliest reversal of land-sea thermal contrast along the longitudes of the eastern Tibetan Plateau, due to the Tibetan Plateau heating during spring.

The westerly-easterly boundary is a good representation of the three-dimensional structure of the subtropical anticyclone. In the Asian monsoon area in winter, the westerly-easterly boundary tilts southward with increasing height. In summer it either tilts northward or is perpendicular to the surface, and the ridgelines of the subtropical anticyclone in the middle and lower troposphere are discontinuous. During the seasonal transition, the tilt of the westerly-easterly boundary changes from southward to northward. When the westerly-easterly boundary becomes perpendicular to the earth's surface or tilts northward, the meridional temperature gradient vanishes or becomes positive, implying the replacement of the winter monsoon by the summer monsoon. Therefore, Mao et al. (2002a) defined the "seasonal transition axis" as the vertical ridge axis that results from the switching of the tilt of the westerly-easterly boundary from southward to northward.

From the evolution of the westerly-easterly boundary and outgoing long wave radiation based on the climatology of 1968–1996, it is shown that prior to April 21, although the entire pattern is typical of winter, the ridgelines above 500 hPa are close together between 90° and 100°E, implying less meridional temperature gradient compared with other longitudes. Deep convection, resulting in outgoing long wave radiation of less than 215 W m^{-2}, exists only south of 5°N and east of 80°E. Essential changes in the westerly-easterly boundary start from May 1–5, during which the ridgelines in the middle-upper troposphere form two seasonal transition axis at 105°E and 90°E respectively, due to the northward (southward) migration of the upper (lower) ridgelines. The westerly-easterly boundary between two seasonal transition exhibits a northward tilt, indicating the establishment of the summer structure of subtropical anticyclone. In the lower troposphere, the 850 hPa ridgeline completely splits into two segments over the Bay of Bengal, where the monsoon trough forms.

With the northward tilting of the westerly-easterly boundary increasing, during May 6–10, the Bay of Bengal monsoon trough further deepens so that the 700 hPa ridgeline now splits. Meanwhile, deep convection develops all over the eastern Bay of Bengal and the Indochina Peninsula, indicating that the Asian summer monsoon first occurs over this region. Subsequently, the seasonal transition axis at 105°E and 90°E move eastward and westward respectively. Ten days later (16–20 May), the seasonal transition axis originally at 105°E reaches the northeastern South China Sea and the westernmost point of the 850 hPa ridgeline abruptly retreats eastward to 120°E; the 400 hPa ridgeline is also broken. At the same time, the south westerlies along the monsoon trough enter into the South China Sea, with deep convection thus becoming active across almost all of this region. When the seasonal transition axis at 95°E moves westward and reaches central and western India during the first ten days of June, the south Asian summer monsoon commences, accompanied by the occurrence of active convection over the southern Indian Peninsula and the eastern Arabian Sea.

These results support the conclusion of Wu and Zhang (1998) that the Asian summer monsoon onset is composed of three sequential stages, i.e., the Bay of Bengal, South China Sea and Indian monsoon onsets. Furthermore, the results also indicate that, each of the three stages closely corresponds to the change in the tilting of the westerly-easterly boundary and the reversal of its titling from southward to northward. This then can be considered as another characteristic associated with the Asian summer monsoon onset, in addition to a wind reversal at low-levels and an increase in precipitation.

The earliest breaking of the ridgelines over the Bay of Bengal, is closely related to the strong atmospheric warming over the Tibetan Plateau. The pressure-latitude cross sections along 90–100°E presented in Fig. 1.1 show the evolutions of the rate of local temperature change and warm temperature ridge during the Bay of Bengal monsoon onset (Mao et al. 2002c). The striking features seen from Fig. 1.1 are that the warm ridge between 200 and 500 hPa suddenly jumps from the south to the north of the geopotential height ridge around May 6–10. Prior to this period, the warm ridge is located between 5° and 10°N with a local temperature tendency less than $0.05\,\mathrm{K\,day^{-1}}$, while the tendency over the Tibetan Plateau to the north is more than $0.1\,\mathrm{K\,day^{-1}}$. The stronger warming to the north than to the south of the subtropical anticyclone ridge therefore contributes to shifting of the warm temperature ridgeline from the south of the subtropical anticyclone ridge before the monsoon onset to its north after the onset, resulting in the reversal of the meridional temperature gradient across the westerly-easterly boundary and the monsoon onset.

3. Summer Climate over Subtropical Asia

It is well established that a large-scale mountainous area can exert a significant influence on the atmospheric circulation through its mechanical and thermodynamical effects (e.g. Wu 1984; Rodwell and Hoskins 2001). However, the circulation in the summer subtropics seems to be more related to thermal forcing, and the formation mechanism is more complicated in that region than other latitudes (Hoskins 1987; Rodwell and Hoskins 2001).

3.1. *Summer heating and corresponding circulation*

In the lower layers over the Tibetan Plateau, vertical mixing, driven by a maximum heating rate of more than 10 K per day near the surface, is the

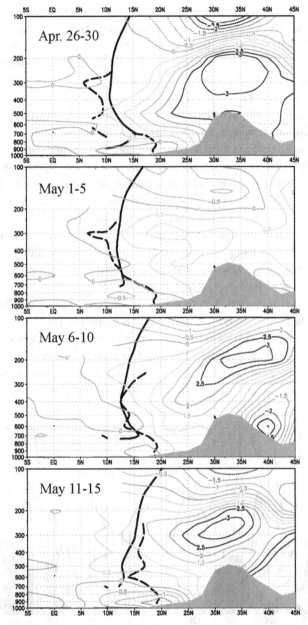

Fig. 1.1. Pressure-latitude cross sections (90°–100°E) of the local temperature trend (0.1 K day^{-1}) and warm temperature ridge during the Bay of Bengal monsoon onset (from April 25 to May 15). The thick solid line denotes the geopotential height ridge and the thick dashed line denotes a warm temperature ridge. Shading indicates orographic altitude.

primary component of total diabatic heating; sensible heating, however, decreases sharply to zero at about 500 hPa. Latent heating is only half the intensity of the sensible heating in the surface layer, but becomes dominant in the upper troposphere. Since the intensity of radiative cooling is not enough to balance either low-level sensible heating or mid- and high-level latent heating, the Tibetan Plateau creates a very strong atmospheric heat source, with the strongest heating in summer. At the West of the Tibetan Plateau, however, there is little condensation from latent heating at any altitude, and very little precipitation. The combination of sensible heating and radiative cooling, therefore, results in the appearance of a heat source in the lower troposphere, but a weaker heat sink above it. Over East China, both sensible heating and radiative cooling are much weaker, and deep convective latent heating (including intense precipitation) dominates. Therefore, the diabatic heating pattern over Asia is characterized by intense sensible heating and upper-layer latent heating over the Tibetan Plateau, deep convective latent heating to its east, as well as surface sensible heating and upper-level radiative cooling to the west.

In summertime, with the large and strong heat source over the Tibetan Plateau, a shallow cyclonic circulation will be generated in the ground layers and a deep anticyclonic circulation will form in the upper layers according to the thermal adaptation theory (Wu and Liu 2000). This vertical circulation structure can be found in Fig. 1.2 in which the July mean pressure-longitude cross-section of the meridional wind component and vertical velocity along

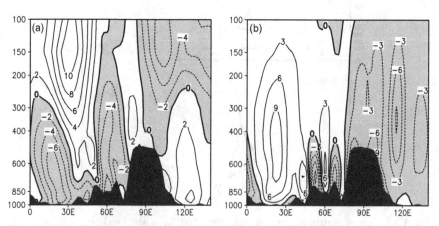

Fig. 1.2. Longitude-pressure section of July mean (a) meridional wind and (b) vertical velocity along 32.5°N. The contour intervals are $2\,\mathrm{m\,s^{-1}}$ in (a) and $1*10^2\,\mathrm{Pa\,s^{-1}}$ in (b). Shading indicates orographic altitude.

32.5°N are presented. Over the Tibetan Plateau between 75 and 105°E and the Iran Highland between 40 and 75°E, the meridional wind fields are characterized by southerly winds on their east and northerly winds on their west in the lower troposphere, and out-of-phase structure in the upper troposphere, corresponding to anticyclonic circulation in the upper troposphere and cyclonic circulation in the lower layers.

The two zero isolines near 60 and 90°E represent the bimodality of the South Asian High. The vertical structure of the circulation between 40 and 75°E and between 75 and 105°E is associated with the thermal forcing over the Iranian Highland (Zhang et al. 2002) and the Tibetan Plateau (Wu et al. 2004), respectively; whereas the lower tropospheric continental cyclonic circulation with northerly to the west of 40°E and southerly to the east of 105°E and the continental scale anticyclonic circulation aloft are forced by the configuration of the continental-scale diabatic heating (Wu and Liu 2003).

Following a simple vorticity balance argument, since the advection of relative vorticity in the summer subtropics is weak, the vertical shear of meridional wind over a heating source along subtropics should correspond to ascending motion on its east but descending motion on its west. This is the case presented in Fig. 1.2(b). Corresponding to the mountain crests at 50, 70, and 90°E, there exist three pairs of descent/ascent in the lower troposphere with ascent over the crests and their eastern sides. Such a circulation is embedded in the large-scale circulation that is forced by the Eurasian continental heating. Because the Tibetan Plateau together with the Iranian plateau are located at the central and eastern parts of the continent, and because the topography induced circulation is in phase with the continental scale circulation, the role of the Tibetan Plateau in thermal forcing is to intensify the southerly winds and ascend motion. This results in abundant rainfall over East Asia to its east, and the northerly winds and descending motion, result in dry climate to its west. In other words, the wet East Asian monsoon climate, and the dry, hot desert climate tend to be amplified by the Tibetan Plateau thermal forcing.

3.2. Influence of mountain waves

Figure 1.3 shows the mean precipitation distribution for July along 30°N and the corresponding pressure-longitude cross-section of potential temperature and vertical velocity. There are four ascending maxima with their centers located in southern (85°E) and southeastern (100°E) Tibetan

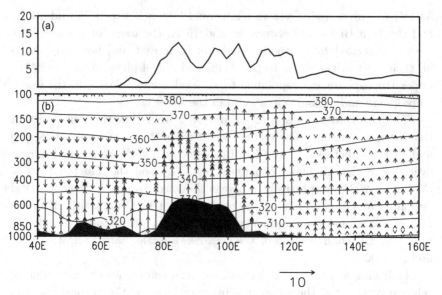

Fig. 1.3. July mean (a) precipitation (mm/day) and (b) vertical velocity–$(1 * 10^2 \times \text{Pa}\,\text{s}^{-1})$ and potential temperature (K) along 30°N.

Plateau, east China (115°E) and southern Japan (130°E). Each maxima lies just over the slope of a corresponding protuberant topography. Duan (2003) showed that topography does regulate the rainfall distribution to a certain degree even in summer.

4. Effects on Bimodality of the South Asian High in Summer

The South Asian High is a planetary scale high pressure system in the upper troposphere and lower stratosphere over the Tibetan Plateau and its surrounding area. A well-known character of this feature during summer is its east-west oscillation, which is represented as the longitudinal shifting of the high towards or away from the Tibetan Plateau (Tao and Zhu 1964). In subsequent work, Luo et al. (1982) further classified the South Asian High into the eastern, western and the belt-type patterns. They found that the occurrence of floods and drought over East China is closely related to the oscillating South Asian High circulation pattern.

The mechanisms responsible for the east-west oscillation of the South Asian High fall into two basic categories related to whether the thermal effect, or dynamical forcing of the Tibetan Plateau is more important. The proponents of thermal forcing as the key mechanism emphasize that it is the very large sensible heating of the Tibetan Plateau together with the effects of the latent heating over the eastern China plain, that results in the east-west oscillation (Liu et al. 1987). The proponents of dynamical forcing suggested that the adjustments of the nearby circulation induced by the Plateau cause a kind of forced oscillation that is different from the self-oscillation due to its thermal forcing. Some possible mechanisms based on dishpan experiments have also been suggested. Zhang and Yeh (1977) show that an obstacle can independently produce oscillation without any thermal effect. However, such an oscillation is confined to a small range, and the center of the South Asian High can hardly move out of the Tibetan Plateau. In contrast, a distinct large range of oscillation occurs when the external circulation effect is included into the dishpan experiment.

4.1. Bimodality of the South Asian high

Recently, the statistical results from the 40-yr monthly mean and 15-yr 5-day mean National Center for Environmental Prediction/National Center for Atmospheric Research reanalysis data reveal the main climate character of the South Asian High during summer to be the bimodality in its longitudinal location (Zhang et al. 2002). When the year-to-year distribution of the longitudinal location of the high in summer is statistically examined, as shown in Fig. 1.4(a), it is observed that its center possesses two preferable locations, the Tibetan Plateau to the east, and the Iranian Plateau to the west (and scarcely any occurrence near 70°–80°E). According to its preferable location, the South Asian High is classified into the Tibetan Mode and the Iranian Mode, which are shown in Fig. 1.4(b) and 1.4(c), respectively.

The diagnoses for the vertical structure of the Tibetan Mode and Iranian Mode indicate that the center of the South Asian High tends to stay over the warm air column (Zhang et al. 2002). Further examination revealed, (i) for the Tibetan mode, such a warming is mainly due to the diabatic heating over the Tibetan Plateau, and (ii) for the Iranian mode, besides the diabatic heating in the lower troposphere, the adiabatic heating associated with descent in the middle troposphere over the Iranian Plateau is more important. Therefore, the maintenance of the South Asian High over a certain region mostly depends on the thermal effect of the atmosphere over the region.

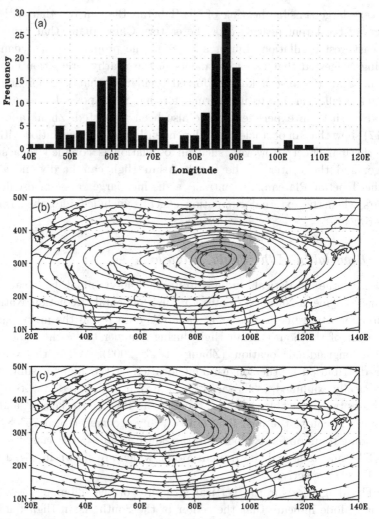

Fig. 1.4. (a) The longitude frequency distribution of the South Asia High major center during midsummer. Mean data for 900 days over 15 summers. (b) The 100 hPa streamline composite corresponding to the Tibetan Mode (77 cases) and (c) the corresponding Iranian Mode (62 cases). The shaded area denotes the Tibetan Plateau.

4.2. Bimodality and the climate anomaly

As indicated by Tao and Zhu (1964), the significant development and contraction of the 500 hPa subtropical anticyclone over the western Pacific corresponds with the east-west oscillation of the South Asian High.

Zhang et al. (2002) showed that the activities of the 100 hPa South Asian High bimodality and the 500 hPa subtropical anticyclone over the western Pacific is similar to the viewpoint of Tao and Zhu (1964). That is, following an eastward shifting of the 100 hPa South Asian High, the 500 hPa subtropical anticyclone behaves as extending westward (and vice versa). Furthermore, it is noticed that the 500 hPa Iranian subtropical high also has a similar development and contraction behavior, following the shifting of the 100 hPa South Asian High. Corresponding to the bimodality of the South Asian High, and the development-contraction of the 500 hPa subtropical high, the climate anomalies over the Asian area exhibit distinctly different patterns. In the Tibetan mode, due to the westward extension of the 500 hPa subtropical high over the western Pacific, more precipitation is observed along the northwest part of the subtropical high. Therefore, enhanced precipitation occurs over the Yangtze and Yellow river valleys, and northwest China. The Indian monsoon precipitation also increases, while precipitation over north China, south China, the south part of the Tibetan Plateau and the Bay of Bengal decreases. In the Iranian mode, the western Pacific subtropical high withdraws eastward, and the precipitation is distributed from the southwest China to north China. Several significant precipitation centers are observed over Taiwan, Beibu Gulf and the low reaches of the Yellow river valley. Less precipitation is found over northwest China. The precipitation over the north part of the Indian peninsula increases while that over the southern part decreases due to the eastward extension of the subtropical Iranian high and weakening of the Indian low.

5. Discussion

Yeh et al. (1957) and Flohn (1957) first identified the role of the heating of the Tibetan Plateau. Since then, many studies have been devoted to understanding how this functions and the consequent weather and climate impacts. It has become well established that the Tibetan Plateau is a heat sink in winter and a heat source in summer. The Tibetan Plateau is also a strong source of negative vorticity in summer, which, via energy dispersion, affects the anomaly of the atmospheric circulation at least in the northern hemisphere. The strong descending air column in winter (and ascending air column in summer) over the Tibetan Plateau works as a large air pump, and regulates the annual cycle of large scale atmospheric circulation and monsoon climate over Asia, Africa and Australia. Furthermore, due to its larger "memory" of the underlying surface, the Tibetan Plateau heating in late winter and spring can exert both simultaneous and delayed impacts on the weather and climate anomalies during summer in the surrounding areas, and therefore can be used as a prediction indicator.

However, understanding is primarily qualitative in nature. Due to the lack of knowledge about the land- air- sea-interaction in this area and how the Tibetan Plateau and its heating affect local and global climate remains unclear, and quantitative climate prediction is still not possible. To overcome this limitation, a coordinated research effort is needed, including:

(1) Expand analyses from summer to other months.

Most of the existing research has been focused on the impacts of the Tibetan Plateau on circulation and climate during the summer months. To achieve a coherent understanding, we need to understand how cooling of the Tibetan Plateau in winter affects weather and climate, and how the annual variation of the Tibetan Plateau heating is associated with the seasonal evolution of the atmospheric circulation.

(2) Analyze additional mechanisms related to the Asian monsoon onset.

The climate system is a nonlinear, dissipative and open system, and the timing and location of the Asian monsoon onset is influenced by many factors, not just the Tibetan Plateau heating. Other factors are persistent external forcing and low frequency oscillations caused by atmospheric motions such as the Madden-Julian Oscillation and the two- to three-week oscillation.

(3) Quantify the heating feature over the Tibetan Plateau.

Although reanalysis data can be used to determine the temporal and spatial distribution of the heating, the results are model dependent. Observation of vertical profiles of heating over the Tibetan Plateau and its neighbor will enable study of the circulation formation and its impacts on regional as well as global climate.

(4) Quantify the roles of various physical processes associated with the interactions between Tibetan Plateau and the atmosphere.

To achieve this, more field observations and numerical modeling experiments are required.

(5) Determine the response of the Tibetan Plateau thermal status to global change.

Most research has concentrated on how the Tibetan Plateau thermal status affects the global climate. However, the changes in global climate will in turn influence the thermal status over the Tibetan Plateau and its contrast with conditions over the Indian Ocean and Western Pacific, and eventually exert different forcing on global climate.

These and other related issues will require significant attention in order to be able to more completely understand how climate fluctuations and global climate change can affect the behavior and pattern of the East Asian monsoon.

Literature Cited

Bolin, B. 1950. On the influence of the Earth's orography on the general character of the westerlies. *Tellus* 2:184–195.

Charney, J. G., and A. Eliassen. 1949. A numerical method for predicting the perturbation of the middle latitude westerlies. *Tellus* 1:38–54.

Duan, A. M. 2003. *The influence of thermal and mechanical forcing of Tibetan plateau upon the climate patterns in East Asia.* Ph.D. Thesis, Institute of Atmospheric Physics, Chinese Academy of Sciences.

Ertel, H. 1942. A new hydrodynamical vorticity equation. *Meteorologische Zeitschrift* 59:33–49.

Flohn, H. 1957. Large-scale aspects of the "summer monsoon" in South and East Asia. *Journal of the Meteorological Society Japan* 75:180–186.

Hoskins, B. J. 1987. Diagnosis of forced and free variability in the atmosphere. Pp. 57–73 in *Atmospheric and oceanic variability*, edited by H. Cattle. London: Royal Meteorological Society.

Koo, C. C. 1951. On the importance of the dynamical influence of Tibetan Plateau on the circulation over East Asia. *Scientia Sinica* 2:283–303 (in Chinese).

Liu, F. M., and S. H. Wei. 1987. The east-west oscillation of the 100 h Pa South Asia High and its forecasting. Pp. 111–117 in *The effects of the Tibetan Plateau on weather of China during summer*. Beijing: Science Press (in Chinese).

Liu, Y. M., G. X. Wu, H. Liu, and P. Liu. 2001. Condensation heating of the Asian summer monsoon and the subtropical anticyclone in the Eastern Hemisphere. *Climate Dynamics* 17:327–338.

Luo, S. W., Z. A. Qian, and Q. Q. Wang. 1982. The study for 100 mb South Asian High and its association with climate of East China. *Plateau Meteorology* 1: 1–10 (in Chinese).

Mao, J. Y., G. X. Wu, and Y. M. Liu. 2002a. Study on model variation of subtropical high and its mechanism during seasonal transition. Part I: Climatological features of subtropical high structure. *Acta Meteorologica Sinica* 60:400–408 (in Chinese).

Mao, J. Y., G. X. Wu, and Y. M. Liu. 2002b. Study on model variation of subtropical high and its mechanism during seasonal transition. Part II: Seasonal transition index over Asian monsoon region. *Acta Meteorologica Sinica* 60:409–420 (in Chinese).

Mao, J. Y., G. X. Wu, and Y. M. Liu. 2002c. Study on model variation of subtropical high and its mechanism during seasonal transition. Part III: Thermodynamic diagnoses. *Acta Meteorologica Sinica* 60:647–659 (in Chinese).

Queney, P. 1948. The problem of air flow over mountains: A summary of theoretical studies. *Bulletin of American Meteorological Society* 29:16–29.

Rodwell, M. J., and B. J. Hoskins. 2001. Subtropical anticyclones and summer monsoons. *Journal of Climate* 14:3192–3211.

Tao, S. Y., and F. K. Zhu. 1964. The variation of 100 mb circulation over South Asia in summer and its association with march and withdraw of West Pacific Subtropical High. *Acta Meteorologica Sinica* 34:385–395 (in Chinese).

Wu, G. X. 1984. The nonlinear response of the atmosphere to large-scale mechanical and thermal forcing. *Journal of Atmospheric Sciences.* 41:2456–2476.

Wu, G. X., and Y. M. Liu. 2000. Thermal adaptation, overshooting, dispersion, and subtropical high. Part I: Thermal adaptation and overshooting. *Chinese Journal of Atmospheric Sciences* 24:433–436.

Wu, G. X., and Y. M. Liu. 2003. Summertime quadruplet heating pattern in the subtropics and the associated atmospheric circulation. *Geophysical Research Letters* 30:1201, doi:10.1029/2002GL016209.

Wu, G. X., and Y. S. Zhang. 1998. Tibetan plateau forcing and the timing of the monsoon onset over South Asia and the South China Sea. *Monthly Weather Review* 126:913–927.

Wu, G. X., Y. M. Liu, and J. Y. Mao. 2004. Adaptation of the atmospheric circulation to thermal forcing over the Tibetan Plateau. Pp. 92–114 in *Observation, theory and modeling of atmospheric variability*, edited by X. Zhu. New Jersey: World Scientific Press.

Wu, G. X., W. P. Li, H. Guo, H. Liu, J. S. Xue, and Z. Z. Wang. 1997. The sensible heat drive air-pump over the Tibetan Plateau and the Asian summer monsoon. Pp. 116–126 in *Collection in the memory of Dr. Zhao Jiuzhang*, edited by T. C. Yeh. Beijing: Chinese Science Press.

Yeh, T. C. 1950. The circulation of the high troposphere over China in the winter of 1945–1946. *Tellus* 2:173–183.

Yeh, T. C., S. W. Luo, and P. C. Chu. 1957. The wind structure and heat balance in the lower troposphere over Tibetan Plateau and its surrounding. *Acta Meteorologica Sinica* 28:108–121 (in Chinese).

Yeh, T. C., S. Y. Tao, and M. C. Li. 1958. The abrupt change of circulation over Northern Hemisphere during June and October. *Acta Meteorologica Sinica* 29:249–263 (in Chinese).

Zhang, K. S., and T. Z. Yeh. 1977. Simulation of the movements of the Tibetan High and the application on summer prediction. *Scientia Sinica* 4:360–368 (in Chinese).

Zhang, Q., G. X. Wu, and Y. F. Qian. 2002. The bimodality of the 100 hPa South Asia High and its relationship to the climate anomaly over East Asia in summer. *Journal of Meteorological Society of Japan* 80:733–744.

Chapter 2

PALEO-MONSOON VARIATIONS IN EAST ASIA RECONSTRUCTED FROM TERRESTRIAL RECORDS

LI LI and ZHISHENG AN

1. Introduction

The East Asian monsoon is an integral part of the global circulation system. During the winter season of the Northern Hemisphere, a persistent high pressure system develops as a result of the cold air mass in the high latitudes of Eurasia. This generates southward surface flow along the east of the Tibetan Plateau forming the strongest winter monsoon in the world. This monsoonal flow commonly penetrates as far south as the South China Sea and can even generate cross–equatorial flow. In summer, air masses originating in the low latitude ocean and cross-equator flow increase the intensity of the East Asian summer monsoon, which reaches up to the boundary of China and Mongolia. This is the highest latitude summer monsoon in the world providing important monsoon-associated precipitation for the populous regions of East Asia. The East Asian monsoon regime is not only responsible for a distinct seasonal climate (Chen et al. 1991), but is also an important subsystem of the global climatic system.

For more than 50 years, many climatologists, geologists and geographers have studied the East Asian monsoon. In the early 1950s, monsoon studies were mainly regionally focused. It was not until the late 1980s that Chinese Quaternary scientists acknowledged the importance of monsoon research and appreciated the importance of the East Asian monsoon in regulating the deposition of loess-paleosol sequences (An et al. 1991a). During the last decade of the 20th century, various types of paleo-monsoon records have been presented. These provide the framework for understanding the dynamics of the evolution of the East Asian monsoon over the last 130,000 years. This has led to the hypothesis that East Asian monsoon evolution is a principal and direct factor that has controlled environmental changes in East Asia (An et al. 1991b).

Because of the strong feedbacks with climate in other regions of the globe, research on the variations in the East Asian monsoon of necessity involves global climatic and environmental changes. In this chapter, we review the evolution of the East Asian monsoon, its relationship to the uplift of the Tibetan Plateau and the glaciation of the North Hemisphere, and summarize the studies on the variability of the East Asian monsoon and its connection with global climate change. Studies on past behavior have been conducted to provide the context for evaluating and projecting future changes in climate and overall sustainability for the region.

2. History

Aeolian deposits on the Chinese Loess Plateau provide valuable records of the evolution of the East Asian monsoon. Geological, biological and chemical evidence derived from loess-paleosol sequences for the last 2.5 million years (Kukla and An 1989) indicate that loess was deposited primarily from dust transported by the northern winter monsoon. Particle-size, aeolian dust flux, and detrital quartz grain size distributions can all be used as proxy indices of the winter monsoon (An et al. 1991b; Xiao et al. 1995). In contrast, the paleosols interleaved between loess layers were formed by soil forming processes under conditions where the summer monsoon dominated. Magnetic susceptibility, organic carbon content, stable carbon isotope ratio, chemical weathering indices and carbonate content of paleosols can be used as proxies of the summer monsoon strength (An et al. 1991b; Gu et al. 1997). The red clay sequence underlying the loess-paleosol sequence consists of inter-layered light-red to reddish-yellow silty loess and light-red to brownish-red paleosols. These cover a surface of variable relief and age. Using the grain-size distribution, the silt fraction, and other chemical and physical characteristics of the red-clay sequence, these sediments have been shown to be ancient loess-paleosol aeolian deposits (An et al. 1999). Similar to the loess-paleosol sequences, the magnetic susceptibility and chemical weathering index of the red clay can be used as a proxy index of the intensity of the summer monsoon whereas particle-size and aeolian dust flux can be used as an index of the intensity of the winter monsoon (An et al. 2001).

The loess or loess like-soil sequences found in Qinan county (Gansu province) are about 230 m in thickness. Paleomagnetic measurements and fossil evidence show that these sequences were deposited between 22 and 6.2 million years ago (Guo et al. 2002). These old aeolian deposits in the loess plateau suggest that large source areas of aeolian dust, and an energetic transport system must have existed in the interior of Asia during the early Miocene epoch. Regional tectonic changes and ongoing global cooling are probably linked to changes in aridity and circulation in Asia (Guo et al. 2002). Several aeolian deposit profiles found in the loess plateau, however, such as Lingtai, Zhaojiachuan, Chaona and Jiaxian indicate that the aeolian dust started to be deposited in a large region around 8 million years ago (An et al. 2001). This suggests that the East Asian monsoon regime that characterizes the winter/summer monsoon commenced at least around this time.

The records of two North Pacific sites which accumulated wind-blown dust from Asia, show a major dust peak about 7–8 million years ago (Rea et al. 1998). Likewise, substantial changes in the oxygen isotope composition of soil carbonates in Pakistan also occurred around 8–9 million years ago (Quade et al. 1989) (Fig. 2.1). These records imply changes in vegetation from $C3$ (forests) to $C4$ (grasses) in Pakistan beginning about 8 million years ago (Cerling et al. 1997), and a change from mixed needle-leaf and broad-leaf forests to grassland vegetation along the northeastern margin of the Tibetan plateau about 8.5 million years ago (Ma et al. 1998). An et al. (2001) suggested that these widely distributed observations could be interpreted as signaling an environmental response to a major phase of the Himalaya-Tibetan plateau uplift about 8–9 million years ago.

The stratigraphy and time series of magnetic susceptibility (Sun et al. 1998) (thin line) and $>19\,\mu m$ grain-size fraction (thick line) for the loess-paleosol-red clay sequence in Zhaojiachuan (35°53′N, 107°58′E), which is an excellent continuous record of the late Cenozoic, are shown in Fig. 2.1.

As discussed earlier, the magnetic susceptibility and grain size fraction can be used as proxy indices of the intensity of the summer and winter monsoons, respectively. It can be seen in Fig. 2.1 that for the period from 3.6 to about 6 million years ago, there was considerable variability of these indices, but relatively small trends compared to the subsequent period. The period from about 2.6 to 3.6 million years ago contains the most-sustained and simultaneous intensification of both summer and winter monsoons on the Loess Plateau. As expressed in An et al. (2001), this simultaneous intensification of both summer and winter monsoons is difficult to explain,

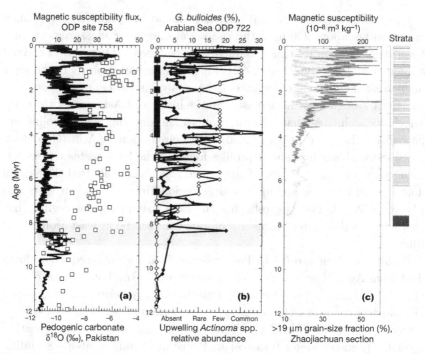

Fig. 2.1. Terrestrial and marine records from the Chinese Loess plateau and the southern margin of Asia. (a) The magnetic susceptibility flux (solid line) reflects the sea-level-mediated terrigenous flux to the Bay of Bengal. The $\delta^{18}O$ of soil carbonates (open squares) in Pakistan reflects increased aridity or a change in the precipitation source about 8–9 million years ago. (b) The abundance of upwelling planktonic foraminifer G. bulloides (filled diamonds) and radiolarian Actinoma spp. (open diamonds) at a site in the Arabian Sea. Intervals of poor carbonate preservation are indicated by filled squares on the time axis. (c) The stratigraphy and time series of magnetic susceptibility (thin line) and >19 μm grain-size fraction (thick line) from the Zhaojiachuan section on the Loess plateau (An et al. 2001).

because the rapid increase in the ice volume during this same period (as inferred from the marine oxygen isotope record (Shackleton et al. 1995)) implies a shift of the climate towards more glacial conditions. Based on climate-model simulations of glacial conditions, a weakening of the summer monsoon and a strengthening of the winter monsoon would be expected (Prell and Kutzbach 1992). Therefore, the simultaneous strengthening of both summer and winter monsoons on the Loess plateau might be attributed to additional, incremental plateau uplift or expansion (An et al. 2001). The geologic evidence along the east and north margin of the plateau (Zheng et al. 2000) and modeling results (An et al. 2001) support this idea.

The changes observed around 2.6 million years ago are significant for the evolution of the East Asian monsoon. After 2.6 million years ago, the red clay sediments shifted to loess-paleosol in the aeolian deposition region of the loess plateau. In these regions, loess deposition dominated during the glacial periods, while paleosol formation dominated during the interglacials. These profiles indicated that the east Asian summer monsoon became more variable, whereas the east Asian winter monsoon continued to be strong, and even intensified, as did the aeolian flux to the North Pacific (Fig. 2.2), indicating sustained or even greater central Asian aridity (Sun and An 2002; An et al. 2001). It has been suggested that the onset of major Northern Hemisphere glaciation occurring ~ 2.6 million years ago strongly influenced, and was perhaps influenced by the development of the Asian monsoons.

The aeolian sequences in Qinan indicate that the aridity of interior Asia started as early as 22 million years ago (Guo et al. 2002). From measurements of the dry bulk density and $CaCO_3$ content of the loess-paleosol-red clay sequences in the Lingtai profile, Sun and An (2002) reconstructed the aeolian flux in the Chinese Loess Plateau. Comparison of aeolian flux variation between the Lingtai profile and the sites in the North Pacific (Rea et al. 1998) Fig. 2.2 reveals significant wet-dry variability in addition to a gradual drying trend in the dust source regions in interior Asia (Sun and An 2002).

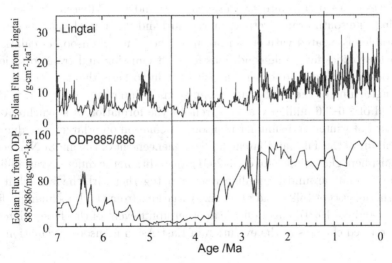

Fig. 2.2. Comparison of the aeolian fluxes between the Lingtai profile in the Loess Plateau and a site in the North Pacific ocean (Sun et al. 2002).

An increase of aeolian fluxes from both continental and pelagic aeolian sediments were also seen between 2.6 and 3.6 million years ago, indicating a sharp drying of the dust source regions. This might be attributed to tectonic uplift of the Tibetan Plateau, which reduced moisture input to the interior of Asia. The average value and variability of aeolian flux were higher after the latter time than before, and this may be related to the Quaternary climatic fluctuations on the glacial-interglacial cycle after the commencement of major Northern Hemisphere glaciations (Sun and An 2002).

The development of aridity in interior Asia is linked to the tectonic uplift of the Tibetan Plateau. Higher elevations result in blockage of moisture input derived from the Indian Ocean. In addition, the air masses which were transported over the high Plateau lost additional moisture as a result of adiabatic cooling during transport aloft. When these air masses subsequently sank and warmed adiabatically along the northern Plateau during the summer season, they produced under saturated air (Chen et al. 1991). The increased aridity during this period may also be related to a decrease in evaporation of ocean water resulting from lower sea surface temperatures during the Quaternary (Ruddiman 1997). The uplift of the Tibetan Plateau led not only to the drying of interior Asia and an expansion of deserts, but also to a seasonal change of the Northern Hemisphere westerlies (Ye and Gao 1979). In the winter half of the year, the westerlies divide into two branches around the Tibetan Plateau. In spring, the south branch retreats northward and the north branch becomes unstable, associated with enhancement of the winter monsoon circulation, which provides favorable conditions for dust emission and transportation. All of these factors are consistent with the hypothesis that a synchronous strengthening of the East Asian summer and winter monsoons during the period of 2.6–3.6 million years ago reflects the formation of a synchronous pattern of winter and summer monsoon circulation on a large scale due to the uplift of the Tibetan Plateau and commencement of the major Northern Hemisphere glaciations (An et al. 2001). In addition, the different variability of aeolian flux around 1.2 million years ago, together with the large shift of magnetic susceptibility and grain size variations from the Luochuan profile (Xiao and An 1999), may indicate another uplift event of the Tibet Plateau and related changes in the evolution of East Asian monsoon circulation.

3. Past Monsoon Variability

Multi-monsoon climate cycles, characterized by alternating dominance of warm-humid and dry-cold conditions, are recorded in the loess-paleosol sequences in central China. These correlate well with the glacial cycles recorded in deep sea sediments (Ding et al. 1994). Analysis of magnetic susceptibility and grain size records of aeolian sequences in the Chinese Loess Plateau shows that the East Asian monsoon variation exhibits strong orbital periodicities of 100, 41 and 23 thousand years (Ding et al. 1994). The 100,000-year cycle is quite prominent during the last 800,000 years. Lake Baikal also has a strong 100,000 year eccentricity cycle (Williams et al. 1997). This may imply that evolution of the East Asian and Lake Baikal monsoons is driven by the nonlinear rhythm of the ocean and ice sheets, together with orbitally induced changes in insolation. High-resolution loess records as well as lacustrine and speleotherm evidence also indicate that monsoonal variations show sub-orbital millennial fluctuations. This agrees well with high-frequency Late Quaternary climatic oscillations shown in Greenland ice cores and the North Atlantic deep-sea records (Wang et al. 2001).

The study of grain size data in the Luochuan (35°45′N, 109°25′E) loess sequence (Porter and An 1995) revealed that the last glacial loess at Luochuan contains a signature of 6 cold events, comparable to those preserved in the Greenland ice cores and to the Heinrich (iceberg rafting) events in the North Atlantic records (Fig. 2.3). This indicated that cold air activity in the high latitudes of the North Atlantic region had strongly influenced the East Asian winter monsoon through its effects on the westerly winds and associated pressure systems. Successive studies of loess grain size composition have revealed the fingerprint of North Atlantic cold events in the last glaciation loess (Ding et al. 1997). The chemical weathering index, which relates to the degree of pedogenesis and thus to variations of summer monsoon rain, has also indicated that there were 6 events with a low weathering index in loess profiles for the last four glacial loess periods (Guo et al. 1996). Recently, high-precision uranium-series dating and high-resolution oxygen isotope composition measurements of five stalagmites from Hulu cave near Nanjing, China, present evidence (Fig. 2.3) supporting the idea that millennial scale events first identified in Greenland are hemispheric or wider in extent (Wang et al. 2001). They also confirm that the East Asian monsoon was not a stable system during

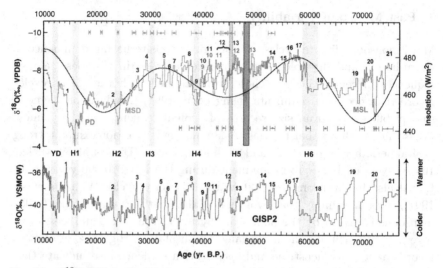

Fig. 2.3. δ^{18}O of Hulu Cave stalagmites (top graph) and Greenland Ice (bottom graph) and insolation at 33°N averaged over the months of June, July, and August (curved line) versus time. Numbers indicate GISs and correlated events at Hulu Cave. The Younger Dryas (YD) and Heinrich (H) events are depicted by vertical bars (Wang et al. 2001.)

the Last Glaciation. In addition, the chemical composition of dust from the source area (represented by aluminium) has also preserved Heinrich event signals (Zhang et al. 1997) that can be correlated with calcium analyses in a Greenland ice core (GRIP members 1993). These results show that dust from the Asian interior has contributed to that found in the Greenland ice core (Biscaye et al. 1997).

Based on the quartz fraction in the last interglacial paleosol ($S1$) at Luochuan and Xian, An and Porter (1997) found evidence of nine distinct dust epochs between 130,000 and 70,000 years ago. Six of these dust events during the period of 110,000 and 70,000 years ago correlate well with the 6 cold events observed in the North Atlantic (McManus et al. 1994), suggesting that there may be high frequency oscillations of the winter monsoon in East Asia during isotope stage 5 that are coupled to conditions in the North Atlantic. The micromorphology and carbonate content of the interglaciation soil at Beiyuan also suggest significant variability of summer monsoonal rainfall during the late stage 5 (Fang et al. 1996), and an instability of the East Asian summer monsoon. Contrary to these findings, analyses of stage 5 soil ($S1$) in other loess/paleosol sections indicate stable climatic conditions (Chen et al. 1999). Chen et al. (2000) interpreted the lack of marked oscillations in the particle size of the last interglacial paleosol (S1) in the western part of the loess plateau as an indication of stable winter monsoons. Recently, two stalagmites from Dongge Cave (Guizhou Province) contained an excellent record of the East Asian monsoon climate during Stage 5. Although smaller in range than that of the last glacial period, the $\delta^{18}O$ range of the stalagmites ($\sim 1\%$) is still significant, amounting to about half the amplitude of typical last glacial period events. This evidence strongly supports the idea of stage 5 monsoon instability (Yuan et al. 2004).

Shi et al. (1992) proposed that the strengthened summer monsoon activity indicated in various regions 35,000 to 25,000 years ago was caused by increased insolation resulting from orbital changes. These observations can be correlated with sea surface temperature changes observed in the South China Sea (Wang et al. 1995). However, they differ from the results of (Martinson et al. (1987) in that they show that the global ice volume during late Stage 3 was significantly greater than that during early Stage 3. Considering that late Stage 3 and the Last Glacial Maximum were periods of desert development in the interior of Australia (Wasson 1986) which resulted from strengthened southeasterly trade winds (de Deckker et al. 1991), we suggest that the East Asian summer monsoon may also have strengthened as a result of intensification and northward shift of the Australian high.

The Younger Dryas event was an abrupt cold climate period which occurred during the last deglaciation of the Pleistocene. It lasted from 11,600 to 12,900 years ago and seems to have manifested itself globally in different ways (Zhou et al. 2001). In East Asian areas, the climatic records from lacustrine and loess/paleosol deposits in the desert/loess boundary of interior China (Zhou et al. 1996), the pollen concentration of peat sediments from Dingnan Basin in Southern China (Xiao et al. 1998) indicate that a wet phase occurred during the Younger Dryas. This is similar to reports from sequences in the Sahel and the equatorial region of Africa. Based on evidence from these sites, together with other published data, Zhou et al. (2001) postulated that precipitation during the Younger Dryas was indicative of a low-latitude driving force superimposed on the high-latitude cold background. The variation of precipitation in East Asia may have been caused by an interaction between cold air advection and summer moisture transport from the tropical Pacific Ocean.

Two records of recent climate change show evidence of very high frequency climate behavior. A 500-year record of recent dust-fall shows that dust-fall events occurred on average 3.7 times per 10 years during cold periods, and 2.1 times per 10 years during warm periods (Zhang 1982). A similar 1,000-year record shows that there have been 5 high-frequency dust-fall periods during the past 1,000 years, each lasting from decades to around a century in length, reflecting frequent oscillations of the East Asian winter monsoon (Zhang 1984).

The Siberian high and the polar front shifted southward during the Little Ice Age significantly strengthening the East Asian winter monsoon circulation (Yoshino 1978). Variation of warm season temperature reconstructed by a 2,650-year stalagmite record reveals cyclic rapid warming on a centennial-scale (Tan et al. 2003). Tree-ring records in the far northwest margin of the East Asian Summer Monsoon show that this region is quite sensitive to monsoonal precipitation variation. These records clearly show variation on decadal and inter-decadal scales over the past 160 years. Synchronous variations on decadal-scale are observed in monsoonal rainfall records in Korea, where the precipitation is also affected by the East Asian summer monsoon (Liu et al. 2003). All of this evidence points to the high-frequency (10–100-year scale) oscillating nature of the East Asian monsoon system.

Besides temporal variations, the East Asian monsoon climate also shows significant spatial variability. Sun et al. (1996) examined the migration of the summer monsoon front (represented by the 250 mm rainfall boundary) for the last 130,000 years by reconstructing the precipitation variations reflected by magnetic susceptibility of loess and paleosols. During the last interglacial, the summer monsoon front reached the desert area, northwest of the Loess Plateau, while during the Last Glacial Maximum, the front extended only to a line from Xian to Qinling at the southern margin of the plateau. Stratigraphic investigations together with radiocarbon and some thermoluminescence dating of 10 profiles in the desert/loess transition zone in northern China indicate latitudinal shifts of the southern desert margin over the last 20,000 years (Zhou et al. 2002). During the last glacial maximum the desert margin was at its most southerly position (38°N), whereas it moved northward to 41°N during the early Holocene. On the basis of geological data and numerical modeling, An et al. (2000) suggested that the Holocene optimum, as defined by peak East Asian summer monsoon precipitation, was asynchronous in central and eastern China, reaching a maximum at different times in different regions, e.g. ∼8,000–10,000 years ago in northeastern China, 7,000–8,000 years ago in north-central and northern east-central China, ∼5,000–7,000 years ago in the middle and lower reaches of the Yangtze River, and ~3,000 years ago in southern China. At present, the desert margin is again close to its most southerly position (38°21′N) (Fig. 2.4), and is not consistent with the northern boundary position (higher than 41°N) of modern summer monsoon activity. This situation cannot be explained from natural climate models. Zhou et al. (2002) suggest that it could have been caused by human activity over the last 3,000 years.

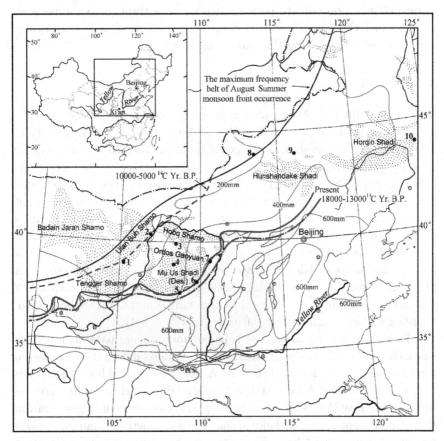

Fig. 2.4. Reconstruction of the desert-loess boundary positions. The inset map indicates the main study area. The shaded area indicates the Loess Plateau, while deserts are indicated by stipple. Boundary positions during the last glacial maximum (dashed line) and the early Holocene (dot and dashed line), and their relationship with the present day boundary (solid line) is indicated. Numbers 1 to 10 indicate the study locations in the paper by Zhou et al. (2002). Present mean annual precipitation is shown. (Zhou et al. 2002.)

In summary, since the commencement of the East Asian Monsoon, especially subsequent to the development of Quaternary glacial cycles in the northern Hemisphere, the East Asian monsoon climate has undergone substantial change in time as well as frequency. During the last interglacial-glacial cycle, both the East Asian winter and summer monsoons show high-frequency oscillations and associated abrupt climatic events of millennial scale. The magnitude of climatic oscillations during the last interglacial may be smaller than that during the last glacial. This variable and unstable nature of monsoon climate persists in East Asia through to the present, during which time there have been frequent incursions of cold air and great variability of precipitation (Fu and Zheng 1998).

4. Discussion

Variations in the East Asian paleo-monsoon reflect interactions between the global atmosphere, ocean, land and ice systems. In addition, variability in solar radiation, topographical changes of the East Asian continent have also produced significant changes. The evolution of the East Asian monsoon is closely coupled to the uplift history of the Tibetan Plateau, and to glacial development in the northern Hemisphere. Changes in ocean-continent configuration, abnormal temperature of the ocean water and desertification of the land also feedback on the evolution of the East Asian monsoon. As indicated in previous studies, the East Asian monsoon exhibits frequent climatic oscillations with 1,000 years scale or less, which are connected with other climate change mechanisms, such as cold air activity in high north latitude, changes in tropical ocean surface temperature, atmosphere-ocean coupling between high and low latitudes and solar activity.

Therefore, two research areas in paleo-monsoon and paleoenvironment research in China are suggested. First, understanding the causes of inception and the evolution of the East Asian monsoon and how this couples to aridity in interior Asia. Significant information on the coupling process between the East Asian monsoon and the arid environment in Asia, within the framework of orbital and global change, can be derived from the various terrestrial deposits in China. Understanding the evolution and interaction of these two systems is highly desirable, and we need to pay more attention to the extraordinary contribution of the uplifting of the Tibetan Plateau to these systems.

Second, study of high-resolution records in loess, lacustrine sediments, ice core and cave deposits and tree rings could yield a greater understanding of high frequency climate change during the last glacial and deglacial periods, and for the Holocene. With more accurate chronologies and comparisons with historical documents for the last 2,000 years, we may be able to produce a rigorous reconstruction of the history of the East Asian Monsoon during this time. Finally, detailed understanding of past environmental change in East Asia may lead to better prediction of future climate change.

Literature Cited

An, Z. S., J. E. Kutzbach, W. Prell, and S. C. Porter. 2001. Evolution of Asian monsoons and phased uplift of the Himalaya-Tibetan plateau since Late Miocene times. *Nature* 411:62–66.
An, Z. S., G. Kukla, S. C. Porter, and J. L. Xiao. 1991a. Late Quaternary dust flow on the Chinese Loess Plateau. *Catena* 18:125–132.
An, Z. S., S. M. Wang, X. H. Wu, M. Y. Chen, D. H. Sun, X. M. Liu, F. B. Wang, L. Li, Y. B. Sun, W. J. Zhou, J. Zhou, X. D. Liu, H. Y. Lu, Y. X. Zhang, G. R. Dong, and X. K. Qiang. 1999. Eolian evidence from the Chinese Loess Plateau: The onset of the late Cenozoic Great Glaciation in the Northern Hemisphere and Qinghai-Xizang Plateau uplift forcing. *Science in China* D 42:258–271.
An, Z. S., and S. C. Porter. 1997. Millennial-scale climatic oscillations during the last interglaciation in Central China. *Geology* 25:603–606.
An, Z. S., S. C. Porter, J. E. Kutzbach, X. H. Wu, S. M. Wang, X. D. Liu, X. Q. Liu, and W. J. Zhou. 2000. Asynchronous Holocene Optimum of the East Asian monsoon. *Quarternary Science Reviews* 19:743–762.
An, Z. S., X. H. Wu, P. X. Wang, S. M. Wang, G. R. Dong, X. J. Sun, D. E. Zhang, Y. Z. Lu, S. H. Zheng, and S. L. Zhao. 1991b. Paleomonsoon of China over the last 130,000 years. *Science in China*, Series B 34:1007–1024.
Biscaye, P. E., F. E. Grousset, M. Revel, S. Van der Gasst, G. A. Zielinski, and A. Vaars. 1997. Asian provenance of Last Glacial Maximum dust in GISP2 ice core, Summit, Greenland. *Journal of Geophysical Research* 102:26765–26781.
Cerling, T. E., J. M. Harris, B. J. MacFadden, M. G. Leakey, J. Quade, V. Eisenmann, and J. R. Ehleringer. 1997. Global vegetation change through the Miocene/Pliocene boundary. *Nature* 389:153–158.
Chen F. H., J. Bloemendal, Z. D. Feng, J. M. Wang, E. Parker, and Z. T. Guo. 1999. East Asian monsoon variations during the last interglacial: Evidence from the northwestern margin of the Chinese Loess Plateau. *Quaternary Science Reviews* 18:1127–1135.
Chen, F. H., Z. D. Feng, and J. W. Zhang. 2000. Loess particle size data indicative of stable winter monsoons during the last interglacial in the western part of the Chinese Loess Plateau. *Catena* 39:112–121.
Chen, L. X., Q. G. Zhu, H. B. Luo, J. H. He, M. Dong, and Z. Q. Feng. 1991. The East Asian monsoon. Beijing: Meteorological Press (in Chinese).

De Deckker, P. T. Corrège, and J. Head. 1991. Late Pleistocene record of cyclic eolian activity from tropical Australia suggesting the Younger Dryas is not an unusual climatic event. *Geology* 19:602–605.

Ding, Z. L., Z. W. Yu, N. W. Rutter, and T. S. Liu. 1994. Towards an orbital time scale for Chinese loess deposits. *Quaternary Science Reviews* 13:39–70.

Ding, Z. L., N. W. Rutter, T. S. Liu, J. M. Sun, J. Z. Ren, D. Rokosh, and S. F. Xiong. 1997. Correlation of Dansgaard-Oeschger cycles between Greenland ice and Chinese loess. *Palaeoclimates* 4:1–11.

Fang, X. M., X. R. Dai, J. J. Li, J. X. Cao, D. H. Guan, Y. P. He, J. L. Wang, and J. P. Wang. 1996. The mutability and instability of the Eastern monsoon evolution — evidence derived from the last interglacial soil. *Science in China*, Series D 26:154–160 (in Chinese).

Fu, C. B., and Zheng, Z. M. 1998. Monsoon regions: The highest rate of precipitation changes observed from global data. *Chinese Science Bulletin* 43:662–666.

GRIP (Greenland Ice Core Project) members. 1993. Climate instability during the last interglacial period recorded in the GRIP ice core. *Nature* 364:203–207.

Gu, Z. Y., D. Lal, T. S. Liu, Z. T. Guo, J. Southon, and M. W Caffee. 1997. Weathering histories of Chinese loess deposits based on Uranium and Thorium series nuclides and cosmogenic ^{10}Be. *Geochimica et Cosmochemica Acta* 61:5221–5231.

Guo Z. T., W. F. Ruddiman, Q. Z. Hao, H. B Wu, Y. S. Qiao, R. X. Zhu, S. Z. Peng, J. J. Wei, B. Y. Yuan, and T. S. Liu. 2002. Onset of Asian desertification by 22 M yr ago inferred from loess deposits in China. *Nature* 416:159–163.

Guo, Z., T. Liu, J. Guiot, N. Wu, H. Lu, J. Han, J. Liu, and Z. Gu. 1996. High frequency pulses of East Asia monsoon climate in the last two glaciations: Link with the North Atlantic. *Climate Dynamics* 12:701–709.

Kukla, G., and Z. S. An. 1989. Loess stratigraphy in central China. *Palaeogeography, Palaeoclimatology, Palaeoecology* 72:203–225.

Liu Y., W. K. Park, Q. F. Cai, J. W. Seo, and H. S. Jung. 2003. Monsoonal precipitation variation in the East Asia since A.D. 1840-Tree-ring evidences from China and Korea. *Science in China* Series D 46:1031–1039.

Ma, Y. Z., J. J. Li, and X. M. Fang. 1998. Pollen assemblage in 30.6-5.0 Ma red beds of Linxia region and climate evolution. *Chinese Science Bulletin* 43:301–304.

Martinson, D. G., N. G. Pisias, J. D. Hays, J. Imbrie. T. C. Moore, and N. J. Shackleton. 1987. Age dating and the orbital theory of the ice ages: Development of a high-resolution 0 to 300,000 year chronostratigraphy. *Quaternary Research* 27:1–29.

McManus, J. F., G. C. Bond, W. S. Broecker, S. Johnsen, L. Labeyrie, and S. Higgins. 1994. High-resolution climate records from the North Atlantic during the last interglacial. *Nature* 371:326–327.

Porter, S. C., and Z. S. An. 1995. Correlation between climate events in the North Atlantic and China during the last glaciation. *Nature* 375:305–308.

Prell, W. L., and J. E. Kutzbach. 1992. Sensitivity of the Indian monsoon to forcing parameters and implications for its evolution. *Nature* 360:647–652.
Quade, J., T. E. Cerling, and J. R. Bowman. 1989. Development of Asian monsoon revealed by marked ecological shift in the latest Miocene of northern Pakistan. *Nature* 342:163–166.
Rea, D. K., H. Snoeckx, and L. H. Joseph. 1998. Late Cenozoic eolian deposition in the North Pacific: Asia drying, Tibetan uplift, and cooling of the Northern hemisphere. *Paleoceanography* 13:215–224.
Ruddiman, W. F. 1997. *Tectonic uplift and climate changes*. New York: Plenum Press.
Shackleton, N. J., M. A. Hall, and D. Pate. 1995. Pliocene stable isotope stratigraphy of site 846. Pp. 337–355 in *Proceedings of the ocean drilling program, scientific results 138*, edited by N. G. Pisias, L. A. Mayer, T. R. Janecek, A. Palmer-Julson, and T. H van Andel.
Shi, Y. F., Z. Y. Kong, S. M. Wang, L. Y. Tang, F. B. Wang, T. D. Yao, X. T. Zhao, P. Y. Zhang, and S. H. Shi. 1992. The climatic fluctuation and major events during the Holocene Optimum in China. *Science in China*, Series B 12:1300–1308 (in Chinese).
Sun, Y. B., and Z. S. An. 2002. History and variability of Asian interior aridity recorded by eolian flux in the Chinese Loess Plateau during the past 7 Ma. *Science in China*, Series D 45:420–429.
Sun, D. H., Z. S. An, J. Shaw, J. Bloemendal, and Sun. 1998. Magnetostratigraphy and palaeoclimatic significance of Late Tertiary aeolian sequences in the Chinese Loess Plateau. *Geophysical Journal International* 134:207–212.
Sun, D. H., X. H. Wu, and T. S. Liu. 1996. Evolution of the summer monsoon regime over the Loess Plateau of the last 150 ka. *Science in China*, Series D 39:503–511.
Tan, M., T. S. Liu, J. Z. Hou, X. G. Qin, H. C. Zhang, and T. Y. Li. 2003. Cyclic rapid warming on centennial-scale revealed by a 2650-year stalagmite record of warm season temperature. *Geophysical Research Letters* 30:1617–1626.
Wang Y. J., H. Cheng, R. L. Edwards, Z. S. An, J. Y. Wu, C. C. Shen, and J. A. Dorale. 2001. A high-resolution absolute-dated late Pleistocene monsoon record from Hulu Cave, China. *Science* 294:2345–2348.
Wang, P. X., L. J. Wang, Y. H. Bian, and Z. M. Jian. 1995. Late Quaternary paleoceanography of the South China Sea: Surface circulation and carbonate cycles. *Marine Geology* 127:145–165.
Wasson, R. J. 1986. Geomorphology and Quaternary history of the Australian continental dune fields. *Geographical Review of Japan*, Series B 59:55–67.
Williams, D. F., J. Peck, E. B. Karabanov, A. A. Prokopenko, V. Kravchinsky, J. King, and M. I. Kuzmin. 1997. Lake Baikal record of continental climate response to orbital insolation during the past 5 million years. *Science* 278:1114–1117.
Xiao, J. L., and Z. S. An. 1999. Three large shifts in East Asian monsoon circulation indicated by loess-paleosol sequences in China and late

Cenozoic deposits in Japan. *Paleogeography, Paleoclimatology, Paleoecology* 154:179–189.
Xiao, J. L., S. C. Porter, Z. S. An, H. Kumai, and S. Yoshikawa. 1995. Grain size of quartz as an indicator of winter monsoon strength on the Loess Plateau of central China during the last 130,000 yr. *Quaternary Research* 43:22–29.
Xiao, J. Y., J. Wang, Z. S. An, X. H. Wu, and W. J. Zhou. 1998. Evidence for Younger Dryas event in the eastern part of Nanjing region. *Acta Botanica Sinica* 40:1079–1082 (in Chinese with English abstract).
Ye, D. Z., and Y. X. Gao. 1979. *Meteorology of Qinghai-Xizang (Tibet) Plateau.* Beijing: Science Press (in Chinese).
Yoshino, M. M. 1978. *Climate change and food production.* Tokyo: University of Tokyo.
Yuan, D. X., H. Cheng, R. L. Edwards, C. A. Dykoski, M. J. Kelly, M. L. Zhang, J. M. Qing, Y. S. Lin, Y. J. Wang, J. Y. Wu, J. A. Dorale, Z. S. An, and Y. J. Cai. 2004. Timing, duration, and transitions of the Last Interglacial Asian Monsoon. *Science* 304:575–578.
Zhang, D. E. 1982. Analysis of dust rain in the historic times of China. *Chinese Science Bulletin* 27:294–297 (in Chinese).
Zhang, D. E. 1984. Synoptic climatic studies of dust fall in China since the historic time. *Scientia Sinica* Series B 27:825–836.
Zhang, X. Y., R. Arimoto, and Z. S. An. 1997. Dust emission from Chinese desert sources linked to variations in atmospheric circulation. *Journal of Geophysical Research* 102 (D23):28041–28047.
Zheng, H., C. Powell, Z. An, J. Zhou, and G. Dong. 2000. Pliocene uplift of the northern Tibetan Plateau. *Geology* 28:715–718.
Zhou, W. J., J. Dodson, M. J. Head, B. S. Li, Y. J. Hou, X. F. Lu, D. J. Donahue, and A. J. T. Jull. 2002. Environmental variability within the Chinese desert-loess transition zone over the last 20000 years. *The Holocene* 12:117–122.
Zhou, W. J., D. J. Donahue, S. C. Porter, A. J. T. Jull, X. Q. Li, M. Stuiver, Z. S. An, E. Matsumoto, and G. R. Dong. 1996. Variability of monsoon climate in East Asia at the end of the Last Glaciation. *Quaternary Research* 46:219–229.
Zhou, W. J., M. J. Head, Z. S. An, P. De Deckker, Z. Y. Liu, X. D. Liu, X. F. Lu, D. Donahue, A. J. T. Jull, and J. W. Beck. 2001. Terrestrial evidence for a spatial structure of tropical-polar interconnections during the Younger Dryas episode. *Earth and Planetary Science Letters* 191:231–239.

Chapter 3

PALEO-MONSOON EVOLUTION AND VARIABILITY DERIVED FROM DEEP-SEA SEDIMENTS

PINXIAN WANG

1. Introduction

At the beginning, reconstruction of paleo-sea surface temperature based on a census of planktonic foraminifera was used to extract a paleo-monsoon signal from deep-sea sediments of the South China Sea (Wang and Wang 1990), and compared with the results from numerical modeling of monsoon-driven seasonal surface circulation patterns (Wang and Li 1995). On-board cruises and deep-sea coring in the South China Sea, and post-cruise studies have yielded a long sequence of East Asian monsoon records over the last 30 million years and enabled high-resolution studies for the Quaternary (Wang et al. 2003). In this chapter, recent progress in paleo-monsoon studies in deep-sea sediments from East Asia at three time scales: tectonic ($=10^6$ yrs), orbital (10^{4-5} yrs), and sub-orbital ($=10^3$ yrs) is reviewed.

2. Monsoon Evolution at Tectonic Time Scales

Long-term monsoon records in deep-sea sediments have been developed from Deep Sea Drilling Project/Ocean Drilling Program cruises to the Indian Ocean, the Mediterranean Sea, and the South China Sea. Therefore, all our knowledge on the Pre-Quaternary evolution of the East Asian monsoon in the marine realm comes from Ocean Drilling Program Leg 184 in the South China Sea (see http://www-odp.tamu.edu/sched and http://www.intermargins.org/maps/drill).

2.1. Land–sea distribution and initiation of the Asian monsoon system

Dealing with the long-term evolution, the first question to answer is how long the East Asian monsoon history can be traced back. So far, the loess-paleosol profile spanning the last 2.6 million years provides the most intact record of the East Asian monsoon. With the study of the underlying Red Clay, the history of the monsoon has been extended to 7–8 million years (An et al. 2001), but the recent discovery of the Neogene (23–5 million years) loess-paleosol profile in the western Loess Plateau implied that the initiation of the East Asian monsoon should be no later than 22 million years ago (Guo et al. 2002).

The early beginning of the monsoon circulation is supported by pollen and lithological data in China. According to the compiled data, a broad arid belt stretched across China in the Paleogene (23–65 million years ago) from west to east, but retreated to the northwest by the end of the Oligocene (~23 million years ago), indicating a transition from a planetary to a monsoonal system in atmospheric circulation over the region (Sun and Wang 2005).

Although the Ocean Drilling Program Leg 184 provides no pollen constraint on the age of the monsoon system due to the practical absence of pollen in the Miocene (23–5 million years) deposits, carbon isotope evidence may fill this gap. As revealed by stable isotope analysis of black carbon from a 30 million year old deep-sea sequence at Ocean Drilling Program site 1148, C4 plants appeared first as a component of land vegetation in East Asia during the early Miocene, about 20 million years ago. Since C4 photosynthesis is commonly associated with hot, dry environments with warm-season precipitation in a low atmospheric pCO_2 background, the secular changes in terrestrial vegetation have been ascribed to the climate evolution toward a monsoon circulation system, especially to summer monsoon intensification (Jia et al. 2003), in good agreement with the loess and pollen records in China.

In terms of tectonic–climate links, the contrast in atmospheric circulation system between the Paleogene and Neogene is, as believed, related to the changes in land–sea distribution patterns. In the Eocene, about 50 million years ago, Europe was separated from Asia by an epicontinental sea, and the Asian continent without India and Europe was much smaller and "slimmer" than it is today (Wang 2004). General Circulation Model simulation results show a zonal distribution of high precipitation in the tropics and much more arid conditions in east China, corresponding well with the paleo-environmental records (Chen et al. 2000). The role of sea–land distribution was also shown by simulation with Paratethys, an epicontinental sea that existed in Eurasia 30 million years ago. Its progressive recession during the Miocene resulted in continentalization of the Asian interior and enhancement of the monsoon circulation (Ramstein et al. 1997).

Generally, the monsoon circulation must have existed through geological history whenever the tropics were occupied by land and sea, but monsoon intensity would have varied greatly with tectonically induced geographic and topographic changes. Since the monsoon is caused by the land–sea contrast in heating rate, ideal conditions for development of the monsoon system appear when all the land masses on the Earth are assembled into one major continent, with only one complementary ocean. This was the case about 200–250 million years ago, when all continents assembled into the super-continent Pangaea and a "megamonsoon" developed on the megacontinent (Kutzbach and Gallimore 1989). During the late Cenozoic, the continents started gathering again, giving rise to the immense African–Asian–Australian monsoon system evident today. This may explain, at least partly, the Paleogene/Neogene transition from the planetary to monsoonal circulation in East Asia.

2.2. Uplift of plateau and stepwise development of monsoons

Plateau uplift is the most widely discussed tectonic factor responsible for the Asian monsoon system. General Circulation Modeling experiments indicate that strong monsoons can be induced by solar forcing only when the elevation of Tibet–Himalaya is at least half that of today (Prell and Kutzbach 1992). The main problem in testing this uplift-climate relationship is the poor constraints of the Tibetan Plateau elevation history. It is now generally accepted that the Himalayan–Tibetan plateau uplift has been a stepwise process. Geological data and computer modeling were used to support the hypothesis of intensified uplift of the Tibetan Plateau around 8 million years ago, causing enhanced aridity in the Asian interior and onset of the Indian and east Asian monsoons (An et al. 2001).

Although no remarkable change in sediment accumulation around 8 million years ago was found in the Ocean Drilling Program Leg 184 in the South China Sea (Wang et al. 2000), there is some indication from micropaleontology. If the planktonic foraminifer *Neogloboquadrina dutertrei* is used as an indicator of the East Asian monsoon and enhanced productivity in the South China Sea, its abrupt increase in frequency on the northern slope at 7.6 million years and, again, at 3.2 million years (Fig. 3.1(B); Wang et al. 2003) corresponds well with the Indian monsoon records. The strengthening of the Indian monsoon at 8 million years broadly correlated with the initiation of the Red Clay aeolian deposition on the Loess Plateau; the new results from the South China Sea confirm a significant enhancement of the Asian monsoon system around 8 million years ago.

Further development of monsoons at 3.2–2.0 million years was also manifested by the increase of magnetic susceptibility and grain size of

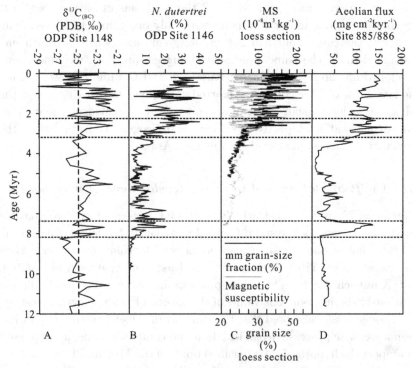

Fig. 3.1. Paleo-monsoon records in the South China Sea during the last 12 million years and comparison with records in the Loess Plateau and Pacific. (A) $d^{13}C$ of black carbon from Ocean Drilling Program Site 1148 (Jia et al. 2003); (B) *Neogloboquadrina dutertrei* % from Site 1146 (Wang et al. 2003); (C) Magnetic susceptibility (gray line) and >19 μm grain size from the Loess Plateau (An et al. 2001); (D) Dust flux (mg cm^{-2} thousand yr^{-1}), Sites 885/886, central Pacific (Rea et al. 1998). Dotted lines mark the time intervals of enhancement of the East Asian monsoons around 8 and 3 million years.

aeolian dust on the Loess Plateau (Fig. 3.1(C)). From marine records, both the events around 8 and 3 million years can be seen from $d^{13}C$ of black carbon at the Ocean Drilling Program site 1148, South China Sea (Fig. 3.1(A)), and from aeolian flux at the site in the northern Pacific (Fig. 3.1(D)).

Noteworthy are the different trends in $d^{18}O$ records of the benthic vs planktonic foraminifera in the South China Sea over the 2.0–3.2 million year period. Both at the northern and southern sites, the benthic foraminiferal $d^{18}O$ gradually became positive, whereas the planktonic foraminiferal $d^{18}O$ remained stable over this period. The benthic $d^{18}O$ changes recorded the growth of a boreal ice-sheet, and the stable planktonic $d^{18}O$ is probably caused by the decrease of sea surface salinity which counteracted the effects of glacial ice volume. If the salinity decrease is interpreted as intensification of monsoon precipitation, the development of the East Asian monsoon system must have been coupled with the Northern Hemisphere ice cap (Tian et al. 2004).

In summary, both the deep-sea and terrestrial data suggest that the East Asian monsoon system was established before the Oligocene/Miocene boundary, most probably around 25 million years ago when the reorganization of distribution pattern of climate and vegetation took place in China (Sun and Wang in press). Since then, the monsoon system has experienced significant variations, including enhancement of aridity and monsoon intensity around 3 and 8 million years ago, as shown above.

3. Monsoon Response to Glacial Cycles

Similar to those at tectonic time scales, the South China Sea has been the focus of East Asian monsoon studies over orbital timescales. The first question is the proxy used for monsoon reconstruction.

3.1. *Use of monsoon proxies*

The first attempt, as mentioned above, was to reconstruct paleo-sea surface temperature changes in the northern South China Sea, and the results showed that the seasonal contrast in temperature in this region was 3–4°C higher in the last glacial maximum as compared to the Holocene (Wang and Wang 1990). This enhanced seasonality was interpreted as evidence for intensification of the winter monsoon, a conclusion later confirmed by Wang et al. (1999a). In an extensive study of the *Sonne* cores, Wang et al. (1999a) used carbon isotope of planktonic foraminifera to indicate monsoon-induced upwelling, and foraminifer- and alkenone-based temperature records as indicators of winter-monsoon strength. The 250,000 year long records indicated intensified winter and weakened summer monsoons during glacial periods; interglacial periods were interpreted to have weakened winter and strengthened summer monsoons (Wang et al. 1999a).

However, there are two groups of monsoon proxies: those related to monsoon winds and those associated with monsoon-induced precipitation (Wang et al. in press). Both approaches have been using pollen and spores in deep-sea sediment as a paleo-monsoon proxy. For example, Morley and Heusser (1997) interpreted *Cryptomeria japonica* pollen as a proxy for summer-monsoon strength in the North Pacific of Japan because the source area of this plant is Japan and China, and its occurrence in the ocean is indicative of wind direction and strength. On the other hand, deep-sea pollen is indicative of the vegetation changes in the surrounding lands. Working on pollen from surface sediments and cores from the South China Sea, Sun et al. (2003) interpreted the influx of tree pollen as a proxy for winter-monsoon strength, as tree pollen is mainly transported from mainland China by the northwestern winter monsoon wind and monsoon-forced surface currents. Meanwhile, they also used the percentage of fern spores as a proxy for summer-monsoon strength and the percentage of herbs as its inverse proxy (Fig. 3.2(E)) because ferns grow under humid conditions and herbs show the opposite. The results again indicate strengthened winter monsoons during glacials and enhanced summer monsoons during interglacials.

The lithology of pelagic sediments in the marginal seas are highly sensitive to glacial cycles; for example the color reflectance is determined by the input of terrigenous clay and organic carbon (Fig. 3.2(B)). Changes in grain size of terrigenous sediments, in clay mineral composition, and in element ratios have also been used as East Asian monsoon proxies. In the northern South China Sea, the median size of siliciclastics was used to distinguish aeolian and fluvial sediments, and the fluctuations in their relative abundance were interpreted to indicate predominance of winter- vs summer-monsoon strength. In general, these proxies indicate enhanced winter monsoons during glacials and enhanced summer monsoons during interglacials (Wang et al. 1999a). Among clay minerals deposited in the modern South China Sea, smectite basically originates from the south, whereas illite and chlorite come from the north. Because different clay

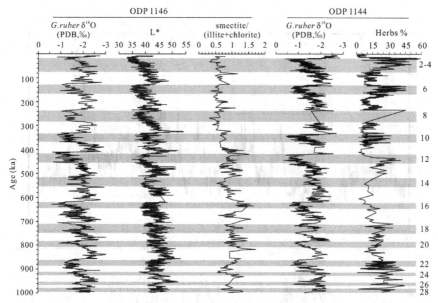

Fig. 3.2. Variations of monsoon proxies in the northern South China Sea in the last one million years. Ocean Drilling Program Site 1146: (A) planktonic $d^{18}O$, (B) lightness of sediment, (C) ratio of smectite/(illite+chlorite) (Liu et al. 2003). Site 1144: (D) planktonic $d^{18}O$ (Buehring et al. in press), (E) herbs% (percentages in the total pollen sum) (Sun et al. 2003).

minerals are transported by different monsoon-driven surface currents, the ratio of smectite/(illite+chlorite) was therefore used to estimate the relative intensity of the winter vs summer monsoon over the past 2 million years at Site 1146, northern South China Sea (Fig. 3.2(C)). The results again display clear monsoon variations in glacial cycles (Liu et al. 2003).

Many other proxies have been used to reconstruct the history of the East Asian monsoon during glacial cycles. Generally, it is always better to adopt a multiproxy approach in estimating paleo-monsoon intensity than to rely on a single fossil or geochemistry (Wang et al. in press). To estimate upwelling in the South China Sea, for example, a number of proxies can be used in combination, including those of paleo-sea surface temperature, paleo-productivity (benthic foraminifer flux, organic carbon flux, infauna/epifauna ratio in benthic foraminiferal fauna), and thermocline depth based on planktonic foraminifera (Fig. 3.3; Jian et al. 2001; Huang et al. 2003).

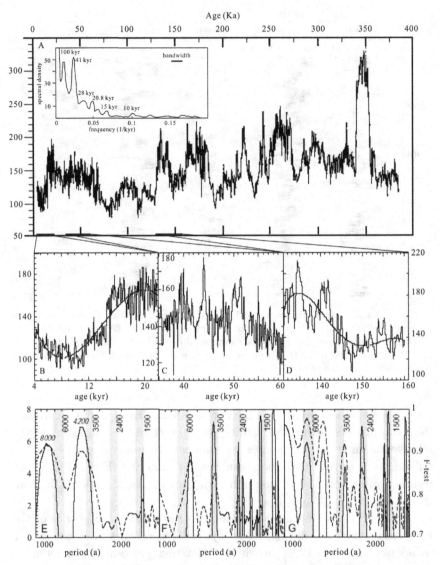

Fig. 3.3. Primary productivity recorded in Sulu Sea Core MD97-2141, showing cyclicities at orbital (upper panel) and millennial (lower panel) time scales. (A) A 400 thousand year record based on nanofossils, with results of spectrum analysis in the upper left inset (Beaufort et al. 2003); (B–D) Enlarged presentation of primary productivity for three time intervals: (B) 4–22 thousand years, (C) 35–60 thousand years, (D) 130–160 thousand years. (E–G) Sub-orbital cycles revealed by the multitaper method spectral analysis for the above three time windows. Line amplitude (dashed line) and F test (estimate of confidence; solid line) are plotted versus period. The shaded areas indicate significant frequency bands (de Garidal-Thoron et al. 2001).

3.2. Geographic differences in monsoon records

All the above discussions show enhancement of summer monsoons during interglacials and winter monsoons during glacial times, but this does not mean that the temporal variations of monsoon proxies are uniform in the region. On the contrary, monsoon records display remarkable geographic differences. In the South China Sea, the prevailing monsoon system coupled with the Ekman effect results in winter upwelling in the eastern part of the Philippines and summer upwelling in the western part of Vietnam. Based on down-core variations in proxies of productivity and thermocline depth, it was found that upwelling intensified of Vietnam during interglacials and of the Philippines during glacial times (Huang et al. 2003). There is nothing unusual, therefore, if geochemical data from the southwest South China Sea indicate increased productivity during interglacials (Wei et al. 2003a), because opal content and diatom abundance give the same conclusion and imply enhanced summer monsoon and upwelling (Wang and Abelmann 2002).

Another aspect is precipitation. Monsoon precipitation provides freshwater input to the South China Sea, rendering the salinity of the surface water highly sensitive to the intensity of the summer monsoon. Because the oxygen isotope record of planktonic foraminifera contains a local salinity signal, the difference in $d^{18}O$ records from the South China Sea from those from elsewhere can be considered as a proxy of summer monsoon intensity. Unlike the South China Sea, the Sulu Sea has no large river runoff and is less sensitive to summer monsoon precipitation. A comparison between the $d^{18}O$ records of *Globigerinoides ruber* over the last 0.8 million years from the southern South China Sea and the Sulu Sea shows that the former are always lighter than the latter, and the difference is always increased during the interglacial and reduced during glacial (Wei et al. 2003b). Similar results are derived from a comparison between the late Quaternary $d^{18}O$ records of *G. ruber* from the southern South China Sea and the Ontong Java Plateau of the open Pacific (Tian et al. 2004). The amplified planktonic $d^{18}O$ signals of the South China Sea indicate an intensified East Asian summer monsoon during interglacials.

3.3. Orbital forcing of monsoon variations

While the early studies on the East Asian monsoon in marine records have been based primarily on visual inspection, recent studies started to pay attention to statistical evaluations of variations at different orbital bands and of phase relationships among the proxies themselves and relative to insolation and glacial boundary conditions. It turns out that the East Asian monsoon is not only passively responding to ice-volume changes in the Quaternary, but is also directly driven by the orbital forcing. In other words, the summer monsoon may vary directly in response to insolation changes in lower latitudes.

For example, spectra of two 220,000 year long records from the South China Sea indicate that winter-monsoon upwelling varies at 100,000 and 41,000 year periods of eccentricity and obliquity in orbital forcing, whereas the 23,000 year precession is the main period for the summer-monsoon upwelling (Jian et al. 2001). Time series analyses of other monsoon-proxy records from the South China Sea also reveal variations at the three orbital frequency bands (e.g., Liu et al. 2003). In the Sulu Sea, a 400,000 year record of East Asian monsoon was established using nannoplankton *Florisphaera profunda* as an inverse proxy for primary productivity driven by winter monsoon (Fig. 3.3(A)) (Beaufort et al. 2003). The winter monsoon was found to have strong concentrations of variance at 100,000 and 41,000 years with considerably smaller variance in the 23,000 year precession band (see insert in Fig. 3.3(A)). Cross-spectrum analyses show only small phase differences between winter monsoon strength and ice maximum, suggesting ice-volume control of the variations of the winter monsoon (Fig. 4 in Beaufort et al. 2003).

While glacial boundary conditions represent a primary factor in driving winter monsoon strength, there are different factors responsible for the summer monsoon. Because the intensity of chemical weathering depends on precipitation, a weathering index is often used to estimate summer monsoon. Thus, K/Si ratios from Site 1145, in the northern South China Sea, were successfully employed to reconstruct the history of chemical weathering, which in turn depends on summer monsoon precipitation. Clear precession forcing was found for the summer monsoon for the analyzed time interval of 2.5–3.2 million years (Wehausen and Brumsack 2002).

In the Sulu Sea, Beaufort et al. (2003) used micro-charcoal as an inverse proxy of summer monsoon precipitation, and the record exhibits complex mechanisms attributed to the competing influence of the long-term El Nino Southern Ocillation-like forcing and the glacial/interglacial cycles, resulting in a frequency spectrum with power around 30,000 and 19,000 years (Fig. 5 in Beaufort et al. 2003). The complicated relationship between the East Asian monsoon climate and the glacial conditions can further be illustrated with pollen records from the South China Sea. Although the pollen-based vegetation changes are in phase with the global ice-volume variations, a detailed comparison shows that the pollen assemblages changed earlier, at the glacial–interglacial transition (about 13,000 years ago), than the ice-volume indicated by the oxygen isotope record, implying that mid-low latitude climate warming preceded high latitude ice sheet retreat (Luo et al. 2005).

In summary, monsoon variability is sensitive to any process that alters interhemispheric pressure gradient (winds) or the availability and transport of moisture.

4. Monsoon Variations at Millennial and Decadal Scales

In paleo-monsoon research, efforts have generally been focused on understanding links between monsoon intensity and large-scale boundary conditions such as orbital forcing and changes in global ice-volume. However, ice cores from Greenland have revealed records of climate change on far shorter time scales as well, showing pronounced climate variability with abrupt amplitude shifts that occur on millennia to decades and shorter time scales (Schulz et al. 1999). Subsequent to these findings, a number of marine sediment archives have been recovered, including varved sediments and coral heads with annual resolution, or sediment cores where sedimentation rates are sufficiently high for resolving the long-term monsoon history on decades and shorter time spans.

4.1. Millennial-scale variations

Based on high-resolution records from the northern South China Sea, Wang et al. (1999a, c) found millennial to centennial scale spells of cold and dry conditions and spells of warm and humid conditions from marine isotope stage 3 to the Holocene, comparable to Dansgaard–Oscheger events during the glacial and to subharmonics of the 1,500-year cycles in the Holocene (Wang et al. 1999b). The late deglaciation has recorded a series of millennial reoccurrences of century-scale changes in the East Asian monsoon climate, and the phase relationship between events of monsoon climate and ice sheet suggests a faster monsoon response and slower ice sheet response to the insolation change (Wang et al. 1999c). Recently, high-resolution records such as millennial-scale variations in nitrogen isotope composition in the South China Sea during marine isotope stage 3, corresponding to Dansgaard–Oscheger events, are being published (Higginson et al. 2003).

In the East China Sea, analyses of planktonic foraminifera and their oxygen isotopes revealed Heinrich-like events during the last glaciation and last deglaciation (Li et al. 2001), and ~1,500,000 year cyclicity in the Holocene (Jian et al. 2000). These cyclic events were attributed to changes of the Kuroshio Current. Further work is needed to find out whether the changes are related to those of the East Asian monsoon or reflect changes on a broader scale.

In the Sulu Sea the oxygen isotopes of planktonic foraminifera show important rapid oscillations that appear synchronous with some Heinrich events (Linsley 1996). Recently, detailed studies of Core MD97-2141 indicate that the marine isotope stage 3 millennial $d^{18}O$ events in the Sulu Sea were primarily the result of salinity changes, which in turn are related to the East Asian monsoon, and the authors argue that suborbital variability on 4,000–10,000 year timescales are linked to those in the North Atlantic (Oppo et al. 2003). The coccolith-based paleo-productivity record from the same core also displays frequencies (1.5, 2.3, 3.4–4.2 and ~6 thousand years), some of which are near those of the Dansgaard–Oeschger cycles, indicating a teleconnection of the East Asian winter monsoon with the Greenland climate; but the pervasive 1,500 years cyclicity is not driven by high-latitude forcing (Fig. 3.3(B–D); de Garidel-Thoron et al. 2001).

In the Sea of Japan, the millennial-scale fluctuations of environment are exhibited in the form of centimeter- to decimeter-scale alterations of the dark and light layers. There is a suggestion that the alternations are correlated to Dansgaard–Oscheger cycles with dark layers corresponding to interstadials and light layers to stadials, and the millennial-scale variations were ascribed to humidity changes in East Asia related to the East Asian monsoon (Fig. 3.4) (Tada 2004). The notion was supported by the analyses of aeolian dusts in the Sea of Japan sediments, with enhanced accumulation rate in the light layers when winter monsoon strengthened and decreased accumulation rate in the dark layers when summer monsoon prevailed (Irino and Tada 2002).

4.2. Centennial- and decadal-scale variations

The lack of laminated deep-sea sediments in East Asian waters with high sedimentation rates precludes the possibility of studying the East Asian monsoon on decadal or centennial scales. However, such opportunities are provided by coral heads. Massive corals with annual resolution produce ideal records of decadal time scales. To date, only a few corals have been

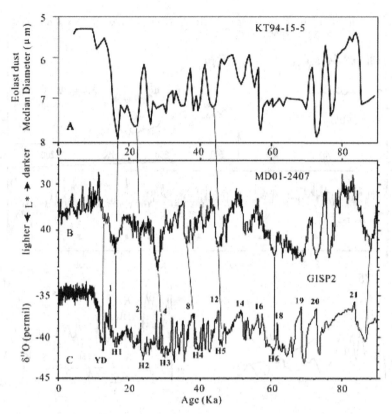

Fig. 3.4. Millennial-scale variations in lithology recorded in the Sea of Japan over the last 0.9 million years. (A) Aeolian dust median diameter (mm) from Core KT94-15-5, NE Sea of Japan; (B) color reflectance L* from Core MD01-2407, Oki-Ridge; (C) oxygen isotope of GISP 2 (modified from Tada 2004).

used to derive a climate proxy in the East Asian region, mostly focused on El Nino events. Recently, a detailed ecological and skeletal Sr/Ca study on a *Goniopora* reef profile near the northern South China Sea coast revealed nine abrupt cooling events during the period 7,000 to 7,500 years ago, each resulting in sudden coral mortality in winter (Yu et al. 2004). The abrupt cooling is interpreted as bursts of the winter monsoon.

Hemipelagic sediments with an unusually high depositional rate may also provide opportunities for centennial and even decadal scale studies. Thus, the sedimentation rate of the Holocene section from Core 17,940, northern South China Sea, was 40–85 cm per 1,000 years, and 1–2 cm sampling provides a time resolution of 15–25 years over the last 10,000 years.

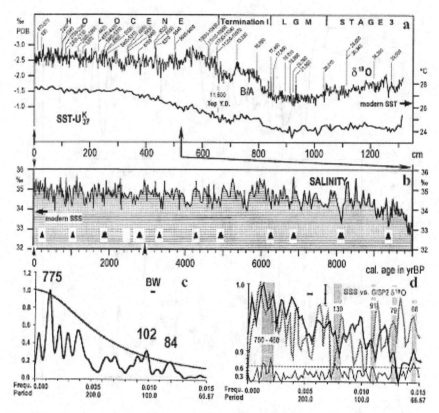

Fig. 3.5. Centennial-scale variations of East Asian monsoon over the last 10,000 years recorded in northern South China Sea Core 17940. (A) planktonic $d^{18}O$, U^{137} sea surface temperature (SST) and AMS C-14 datings of the 13 m long core, representing the last 39,000 years; (B) estimated paleo-sea surface salinity over the last 10,000 years; (C) frequency spectra of sea surface salinity over the last 10,000 years, numbers indicate periods in years, stippled curve shows the upper limit of red noise at 80% confidence level; (D) cross spectra (lower thin line) between sea surface salinity record (thick solid line) and the $d^{18}O$ record in the GISP2 ice core, revealing a significant coherency near periods of 77–79, 91, 130, 460, and 750 years (Wang et al. 1999b).

The high-resolution paleo-sea surface salinity data exhibit variations of the summer monsoon with periodicities of 84, 102, and ~775 years (Fig. 3.5), implying a climate forcing not only by oceanic circulation, but probably also by solar cycles (Wang et al. 1999b).

In summary, our understanding of the monsoon variability on decadal to millennial scales is still developing. It remains unclear whether the monsoon responds through teleconnections to high-latitude mechanisms, or plays a primary role in initiating and/or amplifying abrupt changes originating from the tropics.

5. Conclusions

Summarizing the paleo-monsoon studies in deep-sea records of East Asia, we found the following problems to be critical for further work:

1. *Proxies and observations*: Opinions diverge in terms of monsoon proxies, particularly for indirect indicators of monsoon directions or strength. Because most natural processes are driven by more than one factor, it is crucial to distinguish the monsoon from other factors, and for this purpose long-term modern observations are needed.

2. *Long high-resolution records*: To determine the driving mechanisms of monsoon variations on orbital and suborbital time scales, phase relationships are the key to test the origin of variations in monsoon and their possible driving processes. For this purpose, an appropriate time resolution is absolutely critical. Thus we need long, undisturbed, continuous, and large-diameter cores, and adaptation of precise dating techniques.

3. *Relationships*: Quite often when climate changes are found in records of East Asia, it is difficult to identify whether the changes were caused by variations in the monsoon, El Nino and the Southern Oscillation, or Kuroshio. Modeling results indicate that El Nino and the Southern Oscillation shutdowns may lead to a semi-precession cyclicity (11 thousand years) in the monsoon influenced sediment record (Clement et al. 2001). In general, little effort has been devoted to numerical modeling of paleo-monsoon variations in East Asia, and even less has been devoted to testing the modeling results.

4. *Synthetic approach*: East Asian monsoon records are extensively studied in loess, stalagmite, coral, lake and deep-sea sediments. However, the correlation between different kinds of records in most cases rely on visual inspection of the curves, whereas the causal mechanisms should be explored through a synthetic approach involving the variety of paleo-monsoon records.

Additional efforts are required to make further progress in two critical areas: (1) proxy development and evaluation, and (2) geographic coverage of paleo-monsoon records. Our new challenge is to incorporate the research results from the Indian and East Asian monsoon records, and to uncover the causative mechanism behind the monsoon evolution and variations. This will enable us to provide the scientific background for prediction of future climate change.

Literature Cited

An, Z., J. E. Kutzbach, W. L. Prell, and S. C. Porter. 2001. Evolution of Asian monsoons and phased uplift of the Himalaya–Tibetan plateau since Late Miocene times. *Nature* 411:62–66.

Beaufort, L., T. de Garidel-Thoron, B. K. Linsley, D. Oppo, and N. Buchet. 2003. Biomass burning and oceanic primary production estimates in the Sulu Sea are over the last 380 kyr and the East Asian monsoon dynamics. *Marine Geology* 201:53–65.

Bühring, C., Sarnthein, M., and Erlenkeuser, H. 2004. Toward a high-resolution stable isotope stratigraphy of the last 1.1 m.y., Site 1144, South China Sea. In: *Proceedings of the Ocean Drilling Program, Scientific Results*, Volume 184, edited by Prell, W. L., Wang, P., Blum, P., Rea, D. K., and Clemens, S. C., College Station, ODP, 1–29 [online].

Chen, L., J. Liu, X. Zhou, and P. Wang. 2000. Impact of uplift of Tibetan Plateau and change of land-ocean distribution on climate over Asia. *Acta Meteorologica Sinica* 14:459–474.

Clement, A. C., M. A. Cane, and R. Seager. 2001. An orbitally driven tropical source for abrupt climate change. *Journal of Climate Change* 14:2369–2375.

de Garidel-Thoron, T., L. Beaufort, B. K. Linsley, and S. Dannenmann. 2001. Millennial-scale dynamics of the East Asian winter monsoon during the last 200,000 years. *Paleoceanography* 16:491–502.

Guo, Z. T., W. F. Ruddiman, Q. Z. Hao, H. B. Wu, Y. S. Qiao, R. X. Zhu, S. Z. Peng, J. J. Wei, B. Y. Yuan, and T. S. Liu. 2002. Onset of Asian desertification by 22 M yr ago inferred from loess deposits in China. *Nature* 416:159–163.

Higginson, M., J. R. Maxwell, and M. A. Altabet. 2003. Nitrogen isotope and chlorine paleoproductivity records from the Northern South China Sea: Remote vs local forcing of millennia- and orbital-scale variability. *Marine Geology* 201:223–250.

Huang, B., Z. Jian, X. Cheng, and P. Wang. 2003. Foraminiferal responses to upwelling variations in the South China Sea over the last 220,000 years. *Marine Micropaleontology* 47:1–15.

Irino, T., and R. Tada. 2002. High-resolution reconstruction of variation in Aeolian dust (Kosa) deposition at ODP site 797, the Japan Sea, during the last 200 ka. *Global and Planetary Change* 35:143–156.

Jia, G., P. Peng, Q. Zhao, and Z. Jian. 2003. Changes in terrestrial ecosystem since 30 Ma in East Asia: Stable isotope evidence from black carbon in the South China Sea. *Geology* 31:1093–1096.

Jian, Z., B. Huang, W. Khunt, and H.-L. Lin. 2001. Late Quaternary upwelling intensity and East Asian monsoon forcing in the South China Sea. *Quaternary Research* 55:363–370.

Jian, Z., P. Wang, Y. Saito, J. Wang, U. Pflaumann, T. Oba, and X. Cheng. 2000. Holocene variability of the Kuroshio current in the Okinawa trough, northwestern Pacific Ocean. *Earth and Planetary Science Letters* 184:305–319.

Kutzbach, J. E., and R. G. Gallimore. 1989. Pangaean climates: Megamonsoons of the megacontinent. *Journal of Geophysical Research* 94:3341–3357.

Li, T., Z. Liu, M. A. Hall, S. Berne, Y. Saito, S. Cang, and S. Z. Cheng. 2001. Heinrich event imprints in the "Okinawa Trough" evidence from oxygen isotope and planktonic foraminifera. *Palaeogeography, Palaeoclimatology, Palaeoecology* 176:133–146.

Linsley, B. K. 1996. Oxygen-isotope record of sea-level and climate variations in the Sulu Sea over the past 150,000 years. *Nature* 380:234–237.

Liu, Z. F., A. Trentesaux, S. Clemens, S. C. Colin, and P. Wang. 2003. Clay mineral assemblages in the northern South China Sea: Implications for East Asian monsoon evolution over the past 2 million years. *Marine Geology* 20:133–146.

Luo, Y., X. Sun, and Z. Jian. Environmental change during the penultimate glacial cycle: A high-resolution pollen record from ODP Site 1144, South China Sea. *Marine Micropaleontology* 54:107–123.

Morley, J. J., and L. E. Heusser. 1997. Role of orbital forcing in east Asian monsoon climates during the last 350 kyr: Evidence from terrestrial and marine climate proxies from core RC14-99. *Paleoceanography* 12:483–494.

Oppo, D. W., B. K. Linsley, Y. Rosenthal, S. Dannenmann, and L. Beaufort. 2003. Orbital and suborbital climate variability in the Sulu Sea, western tropical Pacific. *Geochemistry, Geophysics, Geosystems* 4, 1003, doi: 10.1029/2001GC000260.

Prell, W. L., and J. E. Kutzbach. 1992. Sensitivity of the Indian monsoon to forcing parameters and implications for its evolution. *Nature* 360:647–653.

Ramstein, G., F. Fluteau, J. Besseand, and S. Joussaume. 1997. Effect of orogeny, plate motion and land-sea distribution on Eurasian climate change over the past 30 million years. *Nature* 386:788–795.

Rea, D. K., H. Snoeckx, and L. H. Joseph. 1998. Late Cenozoic eolian deposition in the North Pacific: Asian drying, Tibetan uplift, and cooling of the northern hemisphere. *Paleoceanography* 13:215–224.

Schulz, M., W. H. Berger, M. Sarnthein, and P. Grootes. 1999. Amplitude variations of 1,470-year climate oscillations during the last 100,000 years linked to fluctuations of continental ice mass. *Geophysical Research Letters* 26:3385–3388.

Sun, X., and P. Wang. How old is the Asian monsoon system? Palaeobotanical constraints from China. *Palaeogeography, Palaeoclimatology, Palaeoecology* 222:181–222.

Sun, X., Y. Luo, F. Huang, J. Tian, and P. Wang. 2003. Deep-sea pollen from the South China Sea: Pleistocene indicators of East Asian monsoon. *Marine Geology* 201:97–118.

Tada, R. Onset, and evolution of millennial-scale variability in the Asian monsoon and its impact on paleoceanography of the Japan Sea. in *Continental-ocean* interactions in the East Asian marginal seas. Geophysical Monograph 149, edited by P. Clift, P. Wang, W. Kuhnt, and D. Hayes, 285–298.

Tian, J., P. Wang, and X. Cheng. 2004. Development of the East Asian monsoon and Northern Hemisphere glaciation: Oxygen isotope records from the South China Sea. *Quaternary Science Reviews* 23:2007–2016.

Wang, B., S. C. Clemens, and P. Liu. 2003a. Contrasting the Indian and East Asian monsoons: Implications on geologic timescales. *Marine Geology* 201:5–21.

Wang, L. J., and P. X. Wang. 1990. Late Quaternary paleo-oceanography of the South China Sea: Glacial-interglacial contrast in an enclosed basin. *Palaeoceanography* 5:77–90.

Wang, L., M. Sarnthein, H. Erlenkeuser, J. Grimalt, P. Grootes, S. Heilig, E. Ivanova, M. Kienast, C. Pelejero, and U. Pflaumann. 1999a. East Asian monsoon climate during the late Pleistocene: High-resolution sediment records from the South China Sea. *Marine Geology* 156:245–284.

Wang, L., M. Sarnthein, H. Erlenkeuser, P. M. Grootes, J. O. Grimalt, C. Pelejero, and G. Linck. 1999b. Holocene variations in Asian monsoon moisture: A bidecadal sediment record from the South China Sea. *Geophysical Research Letters* 26:2889–2892.

Wang, L., M. Sarnthein, H. Erlenkeuser, P. M. Grootes, and H. Erlenkeuser. 1999c. Millennial reoccurrence of century-scale abrupt events of East Asian monsoon: A possible heat conveyor for the global deglaciation. *Paleoceanography* 14:725–731.

Wang, P. 2004. Cenozoic deformation and the history of sea–land interactions in Asia. In *Continental-Ocean Interactions in the East Asian Marginal Sea*, Geophysical Monograph 149, edited by P. Clift, P. Wang, W. Kuhnt, and D. Hayes, 1–22.

Wang, P. X., and R. F. Li. 1995. Numerical simulation of surface circulation of South China Sea during the last glacation and its verification. *Chinese Science Bulletin* 40:1813–1817.

Wang, P., W. Prell, P. Blum et al. 2000. *Proceedings of Ocean Drilling Program, Initial Reports*, Volume 184. College Station: Ocean Drilling Program, 77p.

Wang, P., S. Clemens, L. Beaufort, P. Braconnot, G. Ganssen, Z. Jian, P. Kershaw, and M. Sarnthein. SCOR/IMAGES Working Group 113 SEAMONS. Evolution and Variability of the Asian Monsoon System" State of the Art and Outstanding Issues. *Quaternary Science Reviews*, in press.

Wang, P., Z. Jian, Q. Zhao, Q. Li, R. Wang, Z. Liu, G. Wu, L. Shao, J. Wang, B. Huang, D. Fang, J. Tian, J. Li, X. Li, G. Wei, X. Sun, Y. Luo, X. Su, S. Mai, and M. Chen. 2003. Evolution of the South China Sea and monsoon history revealed in deep-sea records. *Chinese Science Bulletin* 48:2549–2561.

Wang, R. J., and A. Abelmann. 2002. Radiolarian responses to paleoceanographic events of the southern South China Sea during the Pleistocene. *Marine Micropaleontology* 46:25–44.

Wehausen, R., and H. J. Brumsack. 2002. Astronomical forcing of the East Asian monsoon mirrored by the composition of Pliocene South China Sea sediments. *Earth and Planetary Science Letters* 201:621–636.

Wei, G., Y. Liu, X. Li, M. Chen, and W. Wei. 2003a. High-resolution elemental records from the South China Sea and their paleoproductivity implications. *Paleoceanography* 18:1054, doi: 10.1029/2002PA000826.

Wei, K.-Y., T.-C. Chiu, and Y.-G. Chen. 2003b. Toward establishing a maritime proxy record of the East Asian summer monsoons for the late Quaternary. *Marine Geology* 201:67–79.

Yu. K. F., J. X. Zhao, T. S. Liu, G. J. Wei, P. X. Wang, and K. D. Collerson. High-frequency winter cooling and coral mortality during the Holocene climate optimum. *Earth and Planetary Science Letters* 224:143–155.

Chapter 4

LATE QUATERNARY PALEOCLIMATE SIMULATIONS AND MODEL COMPARISONS FOR THE EAST ASIAN MONSOON

GE YU, SANDY P. HARRISON, XING CHEN and YINGQUN ZHENG

1. Introduction

The wealth of paleoenvironmental evidence from East Asia shows that dramatic changes in the strength and extent of the Asian monsoon occurred during the Quaternary, in response to changes in insolation, glaciation, land-surface conditions, human activities, etc. (An et al. 1990; Winkler and Wang 1993). Physically based models provide a powerful tool for attempting to understand these past changes in climate. The Paleoclimate Modeling Intercomparison Project (Joussaume and Taylor 1995) has examined how the Northern Hemisphere monsoons responded to changes in climate forcing at the Last Glacial Maximum (~21,000 years ago) and during the mid-Holocene (Joussaume et al. 1999), but the major focus for comparisons with paleoenvironmental data has been the African and North American monsoons (Joussaume et al. 1999). Japanese modelers have diagnosed the tropical and subtropical Pacific climate 6,000 years ago and the Last Glacial Maximum (Kitoh and Shigenori 2001, 2002), but there have been a few attempts to examine the realism of these simulations for the East Asian mainland.

The Paleoclimate Modeling Intercomparison Project intervals represent two climate extremes. The Last Glacial Maximum is a time when the Northern Hemisphere ice sheets were much larger than today, the oceans were colder and the concentrations of greenhouse gases in the atmosphere were low. During the mid-Holocene the climate forcings of the ice sheets, atmospheric CO_2, and ocean conditions were not significantly different from today, but the changes in the Earth's orbital parameters resulted in enhanced seasonal contrast in insolation forcing. One question that these simulations cannot address is what an enhancement of the seasonal contrast in insolation would lead to when combined with glacial conditions. There have been some attempts to look at this combination of forcings during the early Holocene (Hewitt and Mitchell 1998), but the Northern Hemisphere ice sheets were rather small at that time and their impact largely local. The period around 35,000 years ago during the last interstadial (~24–59,000 years ago) provides an opportunity to examine the joint impact of enhanced seasonal contrast in insolation forcing during an interval when the ice sheets were considerably larger than present or during the early Holocene (Shi and Yu 2003).

Analyses of the Paleoclimate Modeling Intercomparison Project simulations have shown that the atmospheric response to orbital forcing 6,000 years ago is generally not sufficient to produce the observed increase in precipitation in the African monsoon region (Coe and Harrison 2002) and this raises questions about whether the strengthening of the Asian monsoon can be correctly simulated as an atmospheric response to orbital forcing. Furthermore, the Paleoclimate Modeling Intercomparison Project simulations for 6,000 years ago produce less warming at high northern latitudes than indicated by vegetation data (Prentice et al. 2000). Changes in high latitudes have the potential to influence the strength of the Asian monsoon through changing the extent and duration of snow cover and sea ice, and hence the magnitude of mid-continental warming in the subsequent summer season (Chen et al. 2002). Feedbacks due to changes in vegetation cover have been shown to produce an enhancement of the African monsoon (Doherty et al. 2000) and enhanced warming in high northern latitudes (Ganopolski et al. 1998), and may therefore play an important role in explaining changes in the Asian monsoon region during the mid-Holocene.

Vegetation feedbacks could also have been important through physical conditions of land surface (e.g., albedo), CO_2 changes on transpiration, and soil moisture at the Last Glacial Maximum. The Paleoclimate Modeling Intercomparison Project simulations for 21,000 years ago tend to produce conditions warmer than observed, particularly in the high- and mid-latitudes (Kageyama et al. 2001). Incorporation of vegetation feedback, either through prescription of observed vegetation changes (Wyputta and McAvaney 2001) or through the simulation of vegetation changes (Levis et al. 1999) results in colder and drier conditions which are more consistent with climate reconstructions from the tropics (Farrera et al. 1999) and the replacement of forest with steppe and/or desert vegetation in the Mediterranean region (Peyron et al. 1998) and in China (Harrison et al. 2001). However, there has been no systematic analysis of the role of vegetation feedback on the Asian monsoon during the Last Glacial Maximum.

2. Data Synthesis

Quaternary climate records from East Asia (approximately 60°–150°E and 0°–60°N in the present study) document large changes in environmental conditions during this period. The spatial pattern of these environmental changes can be used to provide insights into the changes in atmospheric circulation regimes that drove the regional changes in the climate. The recognition of this fact has led to multiple attempts to synthesize and map the available paleoenvironmental data from East Asia. This kind of synthesis and interpretation is greatly aided by the existence of paleoenvironmental databases, which provide comprehensive, well-documented, and quality-controlled records from individual sites for continental-scale regions. There are three databases that provide such information for China: the Chinese Lake Status Database, the BIOME 6000 data sets for China, and the Chinese Loess database.

2.1. Chinese lake status database

Fluctuations in the water balance (precipitation minus evaporation) over the catchment of a lake are reflected in changes in lake area, level, and volume, and can thus be reconstructed from geomorphic and stratigraphical records from the lake basin (Harrison and Digerfeldt 1993). Although such changes may be influenced by local non-climatic factors, regionally synchronous changes in lake area, level or volume (collectively referred to as lake status) are generally a response to climate. Thus, regional changes in lake status have been used to reconstruct regional paleoclimates and the changes in atmospheric circulation patterns that gave rise to them (Harrison et al. 1996). Continental-scale reconstructions of changes in lake status have also been used as a benchmark to evaluate model simulations of changes in water balance (Coe and Harrison 2002).

The Global Lake Status Database is an effort to compile the geomorphic and biostratigraphical data for changes in lake status for individual lake basins, in order to document changes in regional water balance during the last 40,000 years. The Chinese Lake Status Database is a component of the Global Lake Status Database which contains 68 lakes (Yu et al. 2001a,b).

Lake status data from the Chinese Lake Status Database show that the conditions 6,000 years ago were wetter than today in northeastern and northern China and most areas of western China. The lake data suggest that southeastern China was drier than today. Lake status data for 21,000 years ago show that conditions were wetter than today in western China (west of 100°E) including the Tibetan Plateau and inland Xinjiang, but were drier in eastern China. Lake status data for 35,000 years ago show conditions wetter than today in most areas of central and western China.

2.2. BIOME 6000 data sets for China

The Paleovegetation Mapping Project (BIOME 6000: Prentice and Webb 1998) has developed global maps of paleovegetation patterns for the Last Glacial Maximum and the mid-Holocene. Broadscale vegetation types (biomes) are reconstructed from pollen or plant-macrofossil data using a standardized, objective method (biomization: Prentice et al. 1996) based on plant functional types. Plant taxa are first assigned to plant functional types, and then the set of plant functional types that can occur in each biome is specified. The allocation of pollen or plant-macrofossil assemblages to biomes is made on the basis of an affinity-score procedure which takes into account both the diversity and the abundance of taxa belonging to each plant functional types in the sample.

China was one of the first regions to be considered in the BIOME 6000 project and there have been several subsequent attempts to improve the vegetation reconstructions (Yu et al. 2000). A set of 840 modern pollen samples spanning all biomes and regions (Yu et al. 2001c) was tested to determine whether they match with modern vegetation distributions. The results show convincing agreement between reconstructed biomes and present natural vegetation types, both geographically and in terms of the elevation gradients in mountain regions of northeastern and southwestern China.

The pollen data show that forests in eastern China were systematically shifted northwards during the mid-Holocene: the northern limit of tropical forest was shifted ~100 km north, while the northern limit of broad-leaved evergreen forest was shifted 300–400 km north and the northern limit of temperate deciduous forest ~800 km further north relative to today. The northern limits of these forests are controlled by minimum winter temperatures and thus the northward extension of forest belts reflects warmer conditions in winter. Forests also extended further westwards into the continental interior, encroaching on non-forested areas. This shift reflects an increase in plant-available moisture, and presumably reflects increased precipitation during the growing season in the interior.

Non-forest vegetation (steppe and/or desert) extended to the coast of eastern China at the Last Glacial Maximum, extending into areas occupied by temperate deciduous forest today, ~20° longitude. Temperate deciduous forest is known from only one site on the exposed continental shelf. The extension of non-forest vegetation reflects drier conditions across the mid-latitudes. Further south, the temperate broad-leaved evergreen and warm mixed forests occurred further south than today, and tropical forests were not recorded in mainland China. This southward shift of the subtropical and tropical forests reflects colder winters. Colder winter conditions are also shown by the southward extension of taiga and cool mixed forests to ~44°N.

Pollen data from 35±3,000 years ago have been classified using the biomization procedure (Yu et al. 2002). According to these reconstructions cold mixed forests occurred much further south than today in eastern China. At the same time, these forests expanded westward into regions occupied today by desert and steppe in the continental interior. The southward displacement of this forest zone implies that winters were colder in eastern China, but its westward expansion suggests that this was accompanied by significantly wetter conditions during the growing season. Temperate forests were not represented 35,000 years ago, and warm-temperate forest appears to be confined to a similar area as today. This implies that conditions in southeastern China were not significantly colder than today.

2.3. Loess records

The Chinese Loess Plateau contains an extensive record of aeolian deposition through multiple glacial–interglacial cycles. Independent chronologies based on pedostratigraphy, magnetic susceptibility, radiocarbon, and luminescence dating have been developed for 79 sites from the Loess Plateau and the surrounding region (Kohfeld and Harrison 2003) and used to estimate aeolian mass accumulation rates (MAR) for marine isotope stages 1 through 5 (0–125,000 years ago).

In general, estimated MAR are highest at sites in the northwest of the Loess Plateau and lowest in sites in the southeast (Fig. 4.1(a), (b), (c)).

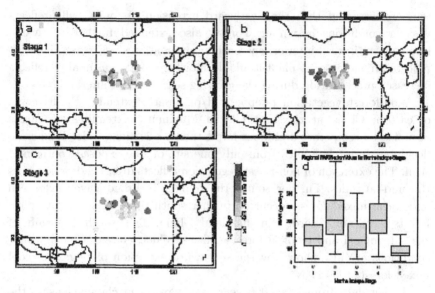

Fig. 4.1. Changes in mass accumulation rates (MAR) from Chinese loess records. The maps show estimates of MAR (g m^{-2} yr^{-1}) at individual sites for marine isotope stages 1 (a), 2 (b), and 3 (c) (d) show the regionally averaged MAR for marine isotope stages 1 through 5, where the upper and lower ends of the shaded box represent the 75 and 25 percentiles respectively.

Such a pattern is consistent with a dust source to the northwest of the Loess Plateau and the persistence of a broadly similar trajectory for dust transport during both glacials and interglacials. The results suggest that regional median MAR were high during the last glacial, with values of ~310 g m^{-2} yr^{-1} for marine isotope stages 2 and 4 and 154 g m^{-2} yr^{-1} during marine isotope stage 3 (Fig. 4.1(d)). These values are considerably higher than those during the last interglacial (marine isotope stage 5: 65 g m^{-2} yr^{-1}). The value for the last interstadial (marine isotope stage 4) is lower than during the Holocene (Stage 1: 171 g m^{-2} yr^{-1}), but this probably reflects an inflation of MAR during the Holocene as a result of human activity on the Plateau. Even given the inflated values, the results suggest that Holocene accumulation rates were lower than accumulation rates during full glacial conditions. The increased vigor of the dust cycle during the glacial is probably a result of the colder, drier, and windier glacial climate.

High resolution records of Loess accumulation during the Holocene show that accumulation rates were relatively high initially and fell to a minimum between ~7,000 and 3,000 years ago. These low accumulation rates during the mid-Holocene reflect an expansion of the Pacific monsoon resulting in wetter conditions on the Loess Plateau. The timing of the onset and cessation of reduced dust accumulation varies from site to site, and probably reflects different spatial patterns in the period of increased monsoonal rainfall (Kohfeld and Harrison 2003).

3. Paleoclimate Simulations

An atmospheric general circulation model, which produces a reasonable simulation of modern climate in fields of temperature, precipitation, and air pressure was used to explore climate changes in the Asian region at 6,000, 21,000, and 35,000 years ago (Wu et al. 1996; Liu and Wu 1997).

3.1. *6,000 years ago*

The simulated summer (June, July, August) temperature at that time, in response to orbital and CO_2 forcing was about 1°C warmer than present in the mid-latitudes (20°–40°N) of East Asia (Fig. 4.2(b)). Winter (December, January, February) temperature decreased by about 1°C over most of East Asia (Fig. 4.2(c)), with maximum cooling (>4°C) in regions north of 40°–60°N. The simulated summer warming enhanced the land–sea contrast and resulted in a strengthening of the Indian summer monsoon, which in turn resulted in an increase in the summer precipitation by $\sim 2\,\mathrm{mm\,day^{-1}}$ over the Indian Subcontinent (Fig. 4.2(e)). The expected strengthening of the Pacific monsoon is not apparent, and simulated conditions in southern and central China are drier than present although there is increased precipitation along the northern coast of China which extends up to Russia.

The simulated temperature changes reflect the changes in Northern Hemisphere insolation, which was ~5% greater in summer and 5% lesser in winter. This pattern of change is common to all atmosphere-only models, although the precise location of the area of summer warming (winter cooling) varies from model to model (Joussaume et al. 1999). In the Paleoclimate Modeling Intercomparison Project simulations, the average summer warming over central Asia (40°–150°E, 30°–60°N) varied from 0.6°C to 2.1°C and the average winter cooling varied from 1°C to 3°C. The enhancement of the Indian monsoon was also common to all simulations of the mid-Holocene by the atmospheric general circulation

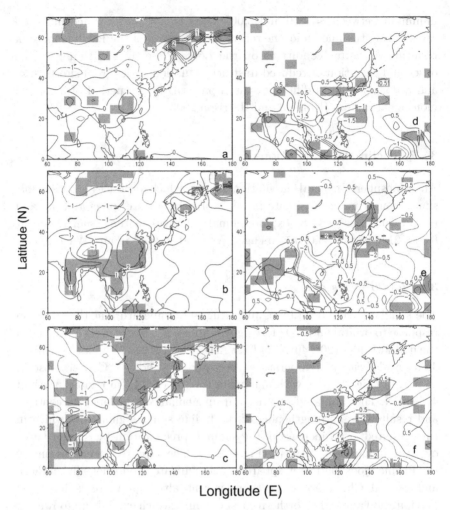

Fig. 4.2. Simulated changes in (a) annual, (b) summer, and (c) winter temperature, and (d) annual, (e) summer, and (f) winter precipitation, 6,000 years ago compared to present resulting from orbital forcing. Gray shaded areas are significantly different from present at the 95% confidence level.

model (Joussaume et al. 1999), although the magnitude of the increase in precipitation varied considerably.

The simulated prescription for 6,000 years ago, with vegetation feedback, produced warmer summers at high northern latitudes than result from orbital-forcing alone, and significantly warmer summers (>2°C warmer than present) in the mid-latitudes of East Asia, except for highlands of Tibet. The Indian monsoon is significantly enhanced, as is the area of increased summer precipitation along the east coast in the mid-to-high latitudes. There is increased precipitation along the northern coast of China, extending up into Russia and extending inland to near Lake Baikal. However, the reduction in precipitation over southern China is enhanced compared to the situation resulting from orbital-forcing alone.

Vegetation feedback, however, has an impact on the winter climate of East Asia. According to these simulations, vegetation feedback reversed the orbitally forced winter cooling and resulted in temperatures between 1–2°C warmer than present in the mid-latitudes of Asia. As a result, mean annual temperature is up to 1°C warmer than in the experiment on response to orbital and CO_2 forcing.

3.2. 21,000 years ago

The simulated annual temperatures over Eurasia at this time, in the experiment on atmospheric response to orbital and glacial forcing, were significantly lower than today (Fig. 4.3(a)). The cooling was comparatively small at low latitudes ($\sim -2°$ to $-4°C$), but increased northwards to reach >12°C in the ice-free polar areas of central Eurasia. Although the largest cooling occurred during the summer in the high northern latitudes (>20°C), this is not the case for the mid- to low-latitudes of East Asia where the summer cooling was $\sim 2°-4°C$ (Fig. 4.3(b)) but the winter cooling was >4°C inland, increasing steeply toward the east coast where it reached values of >8°C colder than present (Fig. 4.3(c)). Precipitation was decreased year-round over most of the region (Fig. 4.3(d)), although the largest decrease (>5 mm day^{-1}) compared to present occurred in the summer (Fig. 4.3(e)), reflecting a significant decrease in the strength of the Pacific summer monsoon. The decrease in winter precipitation, which was >2 mm day^{-1} in southeastern China, may reflect a strengthening of the winter monsoon that brings dry conditions to the region.

The general pattern of simulated changes in the climate is consistent with the atmospheric general circulation model simulations in the Paleoclimate Modeling Intercomparison Project (Kagayama et al. 2001). These all show a strong gradient in temperature, with comparatively small changes in the tropics (that was caused by the prescription of comparatively warm tropical oceans in the data sets; CLIMAP Members 1981), and larger changes in the northern polar region (caused by the downstream effects of the large Northern Hemisphere ice sheets). The Paleoclimate Modeling

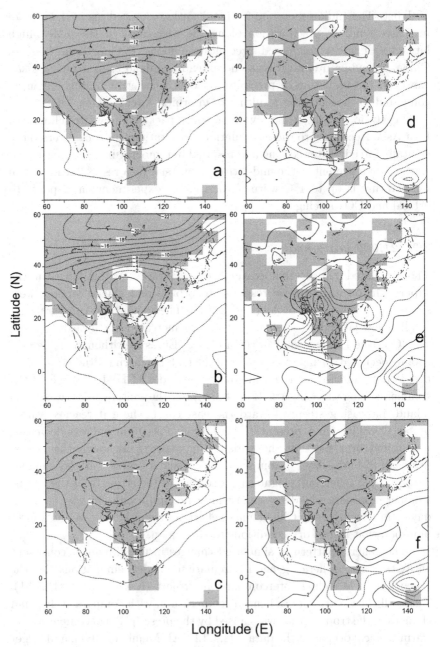

Fig. 4.3. Simulated changes in (a) annual, (b) summer, and (c) winter temperature, and (d) annual, (e) summer, and (f) winter precipitation, 21,000 years ago compared to present resulting from orbital and glacial forcing. Gray shaded areas as in Fig. 4.2.

Intercomparison Project simulations also show similar seasonal patterns, with larger cooling in summer than winter in the mid- to low-latitudes of South East Asia. They also show significantly drier conditions year-round, with a reduction in summer precipitation associated with suppression of the Pacific monsoon that is generally larger than the reduction in winter precipitation.

The prescription for the 21,000 years ago simulation with vegetation feedback resulted in significant cooling over much of the Eurasian continent compared to the simulations forced by the orbital and glacial boundary conditions. The impact of vegetation is most marked in winter, when there was a marked expansion of the area in eastern and central China with temperatures >10°C colder than today. The impact of vegetation is less pronounced in summer, with little change in the low latitudes although there is a tendency for the high northern latitudes to be slightly warmer than in the simulations forced by orbital and glacial boundary conditions. Vegetation feedback also has a large impact on precipitation, causing conditions considerably less arid than in the standard simulation for that time. In eastern China (east of 100°E), mean annual precipitation was reduced compared to present but only by $\sim$2 mm day^{-1} compared to >4 mm day^{-1} in response to orbital and glacial forcing. In western China (west of 100°E), mean annual precipitation was slightly increased compared to present.

3.3. 35,000 years ago

The increase in summer insolation in the 35,000 year atmospheric response to orbital forcing experiment resulted in increased summer temperature in the mid-latitudes (Fig. 4.4(b)), which in turn enhanced land–sea contrast and resulted in a strengthening of both India and the Pacific summer monsoons (Fig. 4.4(e)). The increase over Indian and the Tibetan Plateau was $\sim$2 mm day^{-1}, but smaller changes were registered over much of central China. The simulated increase in precipitation (up to 0.5 mm day^{-1} compared to present) on the northwestern margin of the Pacific monsoon zone was accompanied by a decrease in precipitation in the coastal regions of southeastern China.

In contrast to 6,000 years ago, the winters 35,000 years ago were warmer than present in the mid-latitudes (30°–50°N) of East Asia. Winter insolation at these latitudes was reduced by $\sim$9 W m^{-2} compared to today, so this warming cannot be a response to orbital forcing, but rather appears

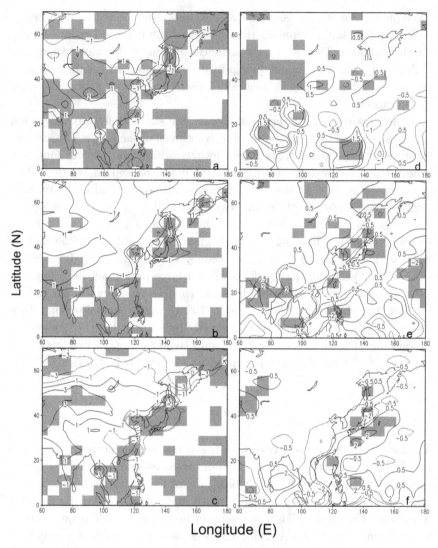

Fig. 4.4. Simulated changes in (a) annual, (b) summer, and (c) winter temperature, and (d) annual, (e) summer, and (f) winter precipitation, 35,000 years ago compared to present resulting from orbital forcing. Gray shaded areas as in Fig. 4.2.

to reflect a change in atmospheric circulation patterns and the weakening of the Siberia anticyclone.

The imposition of moderately large ice sheets in the 35,000 year atmospheric response to orbital and ice sheet forcing experiment has a significant effect on high-latitude climates in North America and western Eurasia, producing a strong decrease in mean annual temperature and a southward shift in the precipitation belt generated by the westerlies. However, the ice sheets have comparatively little impact on the climate of South East Asia. This is consistent with the results of simulations for the early Holocene, which show that the impact of the small ice sheets characteristic of this interval only have an impact on the climate of immediately adjacent regions (Kutzbach et al. 1996).

The vegetation feedback had significant effects on the simulated climate 35,000 years ago. The expansion of forest vegetation into areas of mid-latitude China that are occupied by desert and/or steppe vegetation today resulted in a significant increase in land-surface albedo.

The decrease in albedo resulted in the simulation of decreased winter temperatures over most of the continental land mass. Although the mid-latitudes of China were up to 3°C warmer than today in the orbitally forced experiment, vegetation feedback resulted in a considerable expansion of the area affected. Vegetation feedback has comparatively little effect on summer temperatures, and as a result the additional impact of vegetation feedback on mean annual temperature is small ($\sim 1°C$).

4. Implications

There is a considerable uncertainty about the direction of future regional climate changes over East Asia as a consequence of anthropogenic changes in greenhouse gas concentrations in the atmosphere and land-use changes — in particular, will the climate become warmer-more humid or warmer-more arid. The accuracy of model-based simulations of climate projections cannot be assessed, but the models used to generate such projections can be tested by simulating past climates and comparing these simulations with paleoenvironmental reconstructions. Past time periods do not provide analogous for the future. Even intervals which were generally warmer than today were the result of very different changes in forcing from the values occurring today and expected in the future. Nevertheless, they provide opportunities to examine periods when regional climates were radically different from today.

There is a wealth of paleoenvironmental data from China that can be used to provide benchmarks for model evaluation and testing. The values of such data are considerably enhanced when they are archived in public-access databases and so can be used for mapping of regional patterns over large areas (e.g. Harrison 2003). Pollen-based vegetation maps, lake status data, and mapped patterns of loess accumulation rates all provide useful ways to examine climate model simulations. Continued improvement of these three data sets, and the creation of new syntheses for other kinds of paleoenvironmental records, will be necessary in order to continue the process of evaluating the models that will be used to simulate potential future climate changes and their impacts.

Literature Cited

An, Z. S., X. H. Wu, Y. C. Lu and S. M. Wang. 1990. A preliminary study on the paleoenvironmental change of China during the last 20,000 years. Pp 1–26 in *Loess, quaternary geology and global change (Part II)*, edited by D. S. Liu. Beijing: Science Press.

Chen, X., G. Yu and J. Liu. 2002. Climate simulation and mechanisms of temperature variations in Eastern Asia in the mid-Holocene. *Science in China* 32:335–345.

CLIMAP Members. 1981. Seasonal reconstructions of the Earth's surface at the last Glacial Maximum. *Geological Society of America Map Chart Series* 36.

Coe, M. T. and S. P. Harrison. 2002. Simulating the water balance of northern Africa during the mid-Holocene: An evaluation of the 6 ka BP PMIP experiments. *Climate Dynamics* 19:155–166.

Doherty, R., J. Kutzbach, J. Foley and D. Pollard. 2000. Fully coupled climate/dynamical vegetation model simulations over Northern Africa during the mid-Holocene. *Climate Dynamics* 16:561–573.

Farrera, I., S. P. Harrison, I. C. Prentice, G. Ramstein, J. Guiot, P. J. Bartlein, R. Bonnefille, M. Bush, W. Cramer, U. von Grafenstein, K. Holmgren, H. Hooghiemstra, G. Hope, D. Jolly, S.-E. Lauritzen, Y. Ono, S. Pinot, M. Stute and G. Yu. 1999. Tropical climates at the last glacial maximum: a new synthesis of terrestrial palaeoclimate data. I. Vegetation, lake-levels and geochemistry. *Climate Dynamics* 15:823–856.

Ganopolski, A., C. Kubatzki, M. Claussen, V. Brovkin and V. Petoukhov. 1998. The influence of vegetation–atmosphere–ocean interaction on climate during the mid-Holocene. *Science* 280:1916–1919.

Harrison, S. P. 2003. Contributing to global change science: The ethics, obligations and opportunities of working with paleoenvironmental databases. *Norsk Geografisk Tidsskrift* 57:1–8.

Harrison, S. P. and G. Digerfeldt. 1993. European lakes as paleohydrological and paleoclimatic indicators. *Quaternary Science Reviews* 12:233–248.

Harrison, S. P., G. Yu and P. Tarasov. 1996. Late Quaternary lake-level records from northern Eurasia. *Quaternary Research* 45:138–159.

Harrison, S. P., G. Yu, H. Takahara and I. C. Prentice. 2001. Paleovegetation — Diversity of temperate plants in east Asia. *Nature* 413:129–130.

Hewitt, C. D. and J. F. B. Mitchell. 1998. A fully coupled GCM simulation of the climate of the mid-Holocene. *Geophysical Research Letters* 25:361–364.

Joussaume, S. and K. E. Taylor. 1995. Status of the paleoclimate modeling intercomparison project (PMIP). Pp 425–430 in *Proceedings of the First International AMIP Scientific Conference*. WCRP Report No. 92. Monterey, California, USA.

Joussaume, S., K. E. Taylor, P. Braconnot and PMIP participants. 1999. Monsoon changes for 6,000 years ago: Results of 18 simulations from the Paleoclimate Modeling Intercomparison Project (PMIP). *Geophysical Research Letters* 26:859–862.

Kageyama, M., O. Peyron, S. Pinot, P. Tarasov, J. Guiot, S. Joussaume and G. Ramstein. 2001. The Last Glacial Maximum climate over Europe and western Siberia: A PMIP comparison between models and data. *Climate Dynamics* 17:23–43.

Kitoh, A. and M. Shigenori. 2001. A simulation of the last glacial maximum with a coupled atmosphere-ocean GCM. *Geophysical Research Letters* 28:2221–2224.

Kitoh, A. and M. Shigenori. 2002. Tropical Pacific climate at the mid-Holocene and the Last Glacial Maximum simulated by a coupled ocean-atmosphere general circulation model. *Paleooceanography* 17:1047–1052.

Kohfeld, K. E. and S. P. Harrison. 2003. Records of aeolian dust deposition on the Chinese Loess Plateau during the Late Quaternary. *Quaternary Science Reviews* 22:1859–1878.

Kutzbach, J. E., G. B. Bonan, J. A. Foley and S. P. Harrison. 1996. Vegetation and soils feedbacks on the response of the African monsoon response to orbital forcing in early to middle Holocene. *Nature* 384:623–626.

Levis, S., J. A. Foley and D. Pollard. 1999. CO_2, climate and vegetation feedbacks at the Last Glacial Maximum. *Journal of Geophysical Research* 104:31191–31198.

Liu, H. and G. X. Wu. 1997. Impacts of land surface on climate of July and onset of summer monsoon: A study with an AGCM plus SSiB. *Advances in Atmospheric Sciences* 14:290–308.

Peyron, O., J. Guiot, R. Cheddadi, P. Tarasov, M. Reille, J. de Beaulieu, S. Bottema and V. Andrieu. 1998. Climate reconstruction in Europe for 18,000 yr B.P. from pollen data. *Quaternary Research* 49:183–196.

Prentice, I. C. and T. Webb III. 1998. BIOME 6000: Reconstructing global mid-Holocene vegetation patterns from paleoecological records. *Journal of Biogeography* 25:997–1005.

Prentice, I. C., J. Guiot, B. Huntley, D. Jolly and R. Cheddadi. 1996. Reconstructing biomes from paleoecological data: A general method and its application to European pollen data at 0 and 6 ka. *Climate Dynamics* 12:185–194.

Prentice, I. C., D. Jolly and BIOME 6000 participants. 2000. Mid-Holocene and glacial-maximum vegetation geography of the northern continents and Africa. *Journal of Biogeography* 27:507–519.

Shi, Y. F. and G. Yu. 2003. Warm-humid climate and transgressions during 40–30 ka B.P. and the potential mechanisms. *Quaternary Sciences* 23:1–11.

Winkler, M. G. and P. K. Wang. 1993. The late-Quaternary vegetation and climate of China. Pp. 221–261 in *Global climates since the Last Glacial Maximum*, edited by H. E. Wright Jr., J. E. Kutzbach, T. Webb III, W. F. Ruddiman, F. A. Street-Perrott and P. J. Bartlein. Minneapolis: University of Minnesota Press.

Wu, G. X., H. Liu and Y. C. Zhao. 1996. A nine-layer atmospheric general circulation model and its performance. *Advances in Atmospheric Sciences* 13:1–17.

Wyputta, U. and B. J. McAvaney. 2001. Influence of vegetation changes during the last glacial maximum using the BMRC atmospheric general circulation model. *Climate Dynamics* 17:923–932.

Yu, G., S. P. Harrison and B. Xue. 2001a. *Lake Status Records from China: Data Base Documentation.* Technical Reports in Max-Planck-Institute for Biogeochemistry, No. 4. Jena: Max-Planck-Institute, Germany.

Yu, G., B. Xue, J. Liu, X. Chen and Y. Q. Zheng. 2001b. *Lake Records from China and the Paleoclimate Dynamics.* Beijing: Meteorology Press.

Yu, G., L. Y. Tang, X. D. Yang, X. K. Ke and S. P. Harrison. 2001c. Modern pollen samples from alpine vegetation on the Tibetan Plateau. *Global Ecology and Biogeography* 15:503–520.

Yu, G., B. Xue and S. M. Wang. 2002. Reconstruction of Asian paleomonsoon patterns in China over the last 40 kyrs: A synthesis. *The Quaternary Research* 41:23–33.

Chapter 5

EL NIÑO AND THE SOUTHERN OSCILLATION-MONSOON INTERACTION AND INTERANNUAL CLIMATE

CHONGYIN LI and RONGHUI HUANG

1. Introduction

The Asian monsoon is a key component of the Earth's climate system, affecting the livelihood of more than 60% of the world's population. The variability of the Asian monsoon, particularly its interannual variation, is closely related to the El Niño and the southern oscillation phenomenon (ENSO), which is regarded as one of the strongest signals of interannual climate variation. In this chapter, we summarize the relationship between ENSO and the Asian monsoon activity, including the impacts of ENSO on the Asian monsoon; and vice versa.

2. Impacts of ENSO on the Asian Monsoon

In the wintertime, the high-pressure system over Eurasia with surface pressure at the center often reaching over 1040 hPa together with the horizontal temperature gradient in the middle troposphere, forces cold air from Siberia eastward and southward. The movement of this cold air is generally referred to as the East Asian winter monsoon. The cold surges, that affect most parts of eastern and southern China, as well as the South China Sea, are a major facet of the East Asian winter monsoon. These surges not only dominate the weather over China and Southeast Asia, but also affect convection over the maritime continent (Chang and Lau 1980), the Southern Hemisphere monsoon (Davidson et al. 1983). Associated with the cold surges are the strong low-level north easterlies that extend all the way to the South China Sea (Wu and Chan 1995).

Based on previous observational studies, it was suggested that the pressure surge could represent a transient, gravity-wave-like motion due to a pressure-wind imbalance (Chang et al. 1983), which then allows a fast propagation of energy from the mid-latitudes to the tropics with a speed of $\sim 40\,\mathrm{m\,s^{-1}}$, which is faster than the advective speed (generally $\sim 10\,\mathrm{m\,s^{-1}}$). Theoretical analysis shows that the cold surges mainly result from topographic gravity waves with a phase speed of $\sim 30\,\mathrm{m\,s^{-1}}$ (Lu and Zhu 1990).

In contrast, the Asian summer monsoon, which is represented by the westerly or southwesterly over the South Asia and northwestern Pacific/East Asia regions, is composed of two related but slightly independent subsystems, the South Asian (Indian) monsoon system and the East Asian monsoon system (Tao and Chen 1987). The weather and climate in South Asia are dominated by the Indian monsoon system, while those in East Asia are dominated mainly by the East Asian monsoon system. Generally, the Asian summer monsoon breaks out first over the South China Sea in the middle of May. Then the northward movement of the summer monsoon leads to the onset of the East Asian summer monsoon, and its northwestward movement leads to the onset of the Indian summer monsoon.

The strong westerlies in the lower troposphere and strong easterlies in the upper troposphere predominate for the Indian summer monsoon, while the magnitude of the meridional wind is very close to that of the zonal wind in the East Asian summer monsoon region. The Asian summer monsoon system is illustrated in Fig. 5.1. The East Asian summer monsoon system includes the Australian High, the cross-equatorial flow, the south westerlies, the South China Sea monsoon trough, the south easterlies caused mainly by the subtropical high over the western Pacific, and the Mei-yu front.

2.1. East Asian summer monsoon and ENSO

2.1.1. Interannual variability of rainfall

The East Asian summer monsoon can bring large amounts of water vapor from the Pacific Ocean and the Indian Ocean to the East Asian monsoon region, so that continuous and heavy rainfall may occur in this region (Huang and Wu 1989). The variation and anomaly of the summer monsoon over East Asia has a substantial influence on industry, agriculture, and the daily lives of people in the region. The droughts and floods caused by

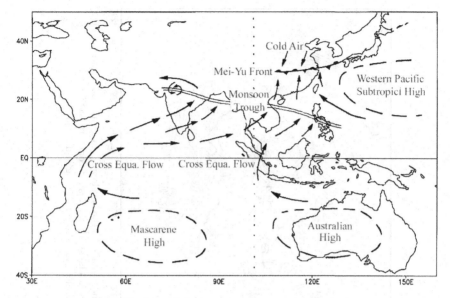

Fig. 5.1. Asian summer monsoon systems.

the monsoon anomaly have brought especially heavy economic loss to this region.

The interannual variation of summer precipitation is very evident in East Asia, particularly in eastern China (Huang and Fu 1996). Temporal variations of the summer precipitation anomaly in the Yangtze/Huaihe River valleys and North China show that interannual variations in these two regions are very clear and seem to have opposite phases. This is an important feature of the interannual variation in the East Asian summer monsoon.

2.1.2. Impact of ENSO on summer rainfall

The interannual variability of the East Asian summer monsoon is closely associated with tropical Pacific sea surface temperature. However, the impact of ENSO on summer monsoon rainfall anomaly in East Asia is different during the different stages of the ENSO (Huang and Zhou 2002). Figure 5.2 shows the composite distributions of summer precipitation anomalies (in percentages) in China for the period of 1951–2000. In summer for the developing stages of an El Niño event (Fig. 5.2(a)), drought can occur in northern China, while severe flooding can occur in the Yangtze/Huaihe

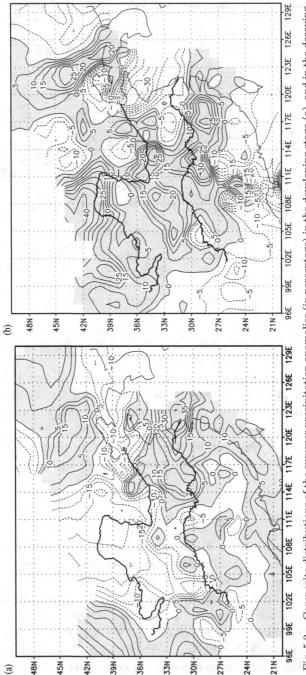

Fig. 5.2. Composite distributions of the summer precipitation anomalies (in percentage) in the developing stage (a), and in the decaying stage (b) of the El Niño events that occurred in the period from 1951 to 2000. The solid and dashed contours (contour interval: 5%) indicate the positive and negative anomalies, respectively; positive anomaly regions are shaded (Huang and Zhou 2002).

River valleys and along the lower reaches of the Yangtze River. However, during the decaying stages of an El Niño event (Fig. 5.2(b)), severe flooding can occur to the south of the Yangtze River, especially in the Dongting Lake and the Boyang Lake valleys, but drought may occur in the Yangtze/Huaihe River valleys.

Similarly, during the summer in the developing stage of a La Niña event, the positive monsoon rainfall anomalies occur to the south of the Yangtze River valley and along the upper reaches of the Yellow River, while negative rainfall anomalies occur in southern China and in the Yangtze/Huaihe River valleys. Moreover, during the summer in the decaying stage of a La Niña event, the distributions of rainfall anomalies are opposite.

The influence of ENSO cycle on the interannual variability of summer monsoon rainfall in East Asia is closely associated with anomalous water vapor transport during the different stages of an ENSO cycle (Zhang 2001).

After the mature phase of El Niño, strong rainfall can occur in the Yangtze River valley, South Korea and Japan because of anomalous convergence of water vapor transport into the region by the intensified southerly winds.

2.2. East Asian winter monsoon and ENSO

During the northern winter, the atmospheric flow over Asia is generally referred to as the East Asian winter monsoon, but the activity of the East Asian winter monsoon is affected by ENSO. The El Niño event can weaken the East Asian winter monsoon through a remote response and teleconnection process. A warm winter (weak East Asian winter monsoon) usually occurs in East Asia during the development to the mature phase of an El Niño, while the reverse occurs during the development to the mature phase of La Niña (Li 1989a, 1990).

Two processes may be responsible for the influence of ENSO on the East Asian winter monsoon. First, the Hadley and Ferrel circulations are both enhanced during an El Niño event. As a result, the westerlies in the mid-latitudes are strengthened, which then suppress the development of the 500 hPa trough over East Asia (Zhang et al. 1997). Second, an anticyclonic circulation usually exists over the western Pacific to the east of the Philippines during an El Niño winter, and the anomalous southerly wind therefore reduces the cold air outbursts southwards in East Asia and the western North Pacific region. That is, in the winter during the mature phase of an El Niño event, the East Asian winter monsoon tends to be weak. No coherent signal is detected in non-ENSO years (Mu 2001).

3. The Forced ENSO by an Anomalous Winter Monsoon

3.1. *Occurrence of ENSO and anomalous East Asian winter monsoon*

The studies have shown that an anomalous winter monsoon in East Asia plays an important role in the occurrence of El Niño by exciting a westerly wind anomaly and stronger intraseasonal oscillation over the equatorial western Pacific (Li 1989b; Li and Mu 1998a). Generally, a strong winter monsoon in East Asia can be represented by a 500 hPa trough over East Asia, a strong Siberian surface high, a low temperature anomaly in eastern China, and the northerly wind anomalies over the northwestern Pacific/East Asia region (see first four variables in Fig. 5.3). Following a strong winter monsoon in East Asia, there are westerly anomalies over the equatorial western Pacific prior to the occurrence of El Niño. These results indicate that a strong winter monsoon in East Asia is likely to excite an El Niño event. The East Asian winter monsoon is also found to be weak in the winter of an El Niño event, suggesting a reducing effect of the El Niño on the East Asian winter monsoon.

The composite pattern of La Niña shows the opposite result. Therefore, the linkages suggest that there is an interaction relationship between ENSO and the winter monsoon in East Asia. A continued strong winter monsoon in East Asia is likely to lead to the occurrence of an El Niño, and the winter monsoon is likely to be suppressed by El Niño during that winter of the mature phase. While a continued weak winter monsoon in East Asia is likely to lead to the occurrence of a La Niña, and the winter monsoon is likely to be enhanced by La Niña during that winter of the mature phase.

For the strongest El Niño event in 1997, the anomalies of mid-latitude atmospheric circulation in the East Asian region played an important role in the occurrence of the El Niño event (Yu and Rienecker 1998). During that event, the onset and developing processes of the westerly bursts over the equatorial western Pacific were largely triggered by strong winter monsoon. It also suggested that the East Asian winter monsoon plays an important role in the occurrence of an El Niño event.

It has been shown that the occurrence of ENSO is closely related to subsurface ocean temperature anomalies in the western Pacific warm

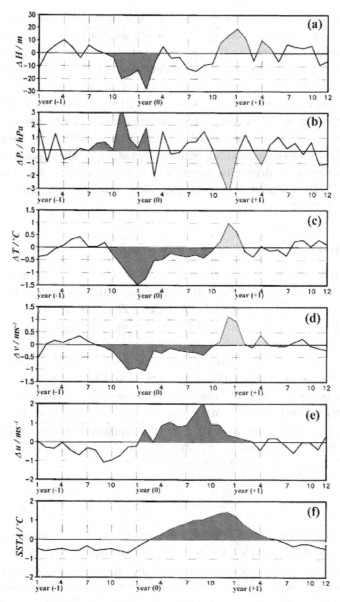

Fig. 5.3. For El Niño cases, the composite temporal variations of (a) 500 hPa height anomalies in the (30°–40°N, 100°–130°E) region, (b) sea level pressure anomalies in the (33°–50°N, 80°–110°E) region, (c) surface air temperature anomalies in the (30°–40°N, 120°–140°E) region, (d) surface meridional wind anomalies in the (25°–35°N, 120°–140°E) region, (e) zonal wind anomalies in the (5°S–5°N, 140°–160°E) region, and (f) the SSTA in Nino 3 region (Mu and Li 1999).

pool and its eastward propagation along the equator (Li and Mu 1999). The positive subsurface ocean temperature anomalies in the warm pool region always exist prior to the occurrence of an El Niño event. As positive subsurface ocean temperature anomalies propagate eastwards along the equator, there was westward propagation of negative subsurface ocean temperature anomalies along 10°N and 10°S latitudes, meanwhile the warm pool will be controlled by negative subsurface ocean temperature anomalies. It might be suggested that the ENSO is closely related to the interannual cycle of subsurface ocean temperature anomalies in the tropical pacific, while this cycle is driven by anomalous zonal wind over the equatorial western Pacific induced mainly by anomalous East Asian winter monsoon (Li and Mu 2002).

3.2. *Dynamical impact of anomalous winter monsoon*

It was indicated that anomalous anticyclone circulation forms over the tropical western Pacific (the Philippine Sea) after the occurrence of an El Niño event (Zhang et al. 1996; Wang and Zhang 2002), which might be regarded as an impact process of ENSO on the East Asian monsoon. But an anomalous cyclone circulation probably caused by strong (weak) winter monsoon anomaly in East Asia is found over the tropical western Pacific (the Philippine Sea) before the occurrence of an El Niño event.

To analyze dynamically the impact of anomalous East Asian winter monsoon on zonal wind anomaly over the equatorial western Pacific and atmospheric circulation anomaly over the tropical western Pacific, as a stationary state case near the equator ($y \to 0$), the disturbance equation of zonal wind (non-basic flow) will become

$$D_m u = -\frac{\partial p}{\partial x}.$$

It means that the variation of zonal wind depends on zonal pressure gradient near the equator. Note that when $\frac{\partial p}{\partial x} < 0$ over the equatorial western Pacific corresponding to strong winter monsoon, the anomalous westerly wind will be formed or enhanced; but corresponding to weak winter monsoon, when $\frac{\partial p}{\partial x} > 0$, the anomalous easterly wind is enhanced. Therefore, it can be suggested that a strong East Asian winter monsoon cannot only lead to the occurrence of an anomalous northerly over the northwestern Pacific/East Asia region but also an anomalous westerly over the equatorial western Pacific, which plays a key role in the occurrence of an El Niño event.

3.3. Numerical simulation of anomalous winter monsoon exciting ENSO

The impact of an anomalous winter monsoon in East Asia on the oceanic temperature in the tropical Pacific simulated in a coupled atmosphere-ocean model represents the response of the tropical Pacific to the forcing of an anomalous East Asian winter monsoon (Li and Mu 1998b). The results suggested that a strong winter monsoon in East Asia would excite positive sea surface temperature anomalies in the equatorial eastern-central Pacific. The simulation still shows that the weak winter monsoon in East Asia is conducive to excitation of the La Niña event.

By using a coupled atmosphere-ocean model, numerical simulation results also show that an anomalous strong East Asian winter monsoon leads to westerly anomalies over the equatorial western Pacific, which drives the eastward propagation of positive subsurface ocean temperature anomalies from warm pool to the equatorial eastern Pacific and hence the El Niño event occurs (Zhou and Li 1999).

4. Variability of Relationship between Asian Monsoon and ENSO

The negative correlation between the Indian monsoon rainfall and ENSO has been recognized for a long time (Rasmusson and Carpenter 1983). However, this negative correlation has been weakened rapidly since the late 1970s (Mehta and Lau 1997). This weakened relationship can be defined by the correlation between June–September all-India rainfall and the sea surface temperature anomaly in Nino 3 region. It is quite clear that the negative correlation shown in Fig. 5.4 has been consistently around the 1% significance level from 1856 until the 1970s, but this correlation was decreased and the significance level was about 10% in 1990s. The same situation was also documented for the East Asian monsoon and ENSO (Wu and Wang 2002).

Some possible causes of the monsoon–ENSO relationship change have also been studied in recent years. Webster and Palmer (1997) related this interruption of the ENSO–monsoon relationship to the chaotic nature of the monsoon, while Kumar et al. (1999) viewed the rapid weakening of the correlation as systematic and proposed that it may be due to changes in the Pacific Walker circulation and the warming of the Eurasian Continent. Some studies suggested that the weakened

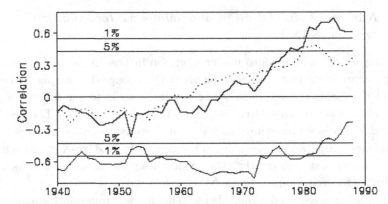

Fig. 5.4. Correlation (based on a 21-year sliding window) between the Indian summer monsoon rainfall and summer Nino 3 sea surface temperature (thin solid), winter western European surface air temperature (thick solid), and winter central Eurasian surface air temperature (dotted). The 1% and 5% significance levels are indicated as horizontal lines (Chang et al. 2001).

ENSO–monsoon relationship results from their interdecadal variation and the Pacific Decadal Oscillation (McCabe and Dettinger 1999). Recent studies (Chang et al. 2001) showed that the weakened relationship is more likely related to recent circulation changes over the North Atlantic (see Fig. 5.4).

5. Interannual Climate Variability

The discussion above has shown obvious interactions between the East Asian monsoon and ENSO, and that the climate variation in East Asia does naturally include the ENSO signal (see Li 1986; Weng et al. 1999). In this section, we will focus on the tropospheric biennial oscillation in the East Asian climate variability.

The quasi-biennial oscillation, a quasi-periodic variation (about 26 months) of the wind field in the equatorial lower stratosphere has been described (Read 1961). The phenomena of the quasi-biennial variation in the tropospheric circulation and climate were revealed in the 1990s (Ropelewski et al. 1992), and it was called the tropospheric biennial oscillation. A quasi-biennial variation feature of the precipitation in eastern China and other areas has been indicated in some studies (Huang 1988; Lau and Shen 1988).

Recent studies advance several mechanisms to explain the tropospheric biennial oscillation. We summarize only some aspects here, as the detailed information has already been published. Initially, the studies indicated that the tropospheric biennial oscillation might result from the influence of the quasi-biennial oscillation in the stratosphere because of its influence on tropical cyclones, summer rainfall, and subtropical high activity (Elsner et al. 1999). It has also been suggested that the tropospheric biennial oscillation is caused by the anomalies of winter snow cover over Eurasia and the Tibetan Plateau (Sankar et al. 1996). The ENSO has also been implicated as a possible cause of the tropospheric biennial oscillation, and it has been suggested that its origin is the interaction between the anomalous East Asian winter monsoon and the ENSO cycle (Li et al. 2001). The tropospheric biennial oscillation was studied by using a simple 5-box tropical atmosphere–ocean–land model (Chang and Li 2000) and the results show that it is an inherent result of the interactions between the northern summer and winter monsoons, and the equatorial Indian and Pacific Oceans. It further implies a potential role for the monsoon to interact with the ENSO.

6. Conclusions

The discussion above shows that there are interannual climate variations in East Asia, especially in the summer rainfall in eastern China. The ENSO-like mode and tropospheric biennial oscillation are major features of interannual climate variation in East Asia. The ENSO–Monsoon relationship not only points to the impact of ENSO on the East Asian monsoon, but also to the possible exciting of the East Asian monsoon, particularly the anomalous East Asian winter monsoon, on the occurrence of ENSO.

The Asian monsoon system exhibits prominent low frequency variations, particularly the intraseasonal oscillation with a time scale of 30–60 days. They may have strong impacts on the monsoon and the interaction with the ENSO (Li and Zhou 1994; Long and Li 2002). Further studies are needed to identify the physical processes responsible for the interdecadal changes of the ENSO–monsoon relationships, and they will improve our understanding of the ENSO–monsoon interaction.

As land surface processes such as increased snow cover and snow depth could also affect the ENSO–monsoon relationship, it is necessary to study their impact on the relationship.

Literature Cited

Chang, C. P. and K. M. Lau. 1980. Northeasterly cold surges and near-equatorial disturbances over the winter MONEX area during December 1974. Part II: Planetary-scale aspects. *Monthly Weather Review* 108:298–312.

Chang, C. P. and T. Li. 2000. A theory for the tropical tropospheric biennial oscillation. *Journal of Atmospheric Science* 57:2209–2224.

Chang, C. P., J. E. Millard and G. T. J. Chen. 1983. Gravitational character of cold surges during winter MONEX. *Monthly Weather Review* 111:293–307.

Chang, C. P., H. Patrick and J. Ju. 2001. Possible roles of Atlantic circulations on the weakening Indian monsoon rainfall — ENSO relationship. *Journal of Climate* 14:2376–2380.

Davidson, N. E., J. L. McBride and B. J. McAvaney. 1983. The onset of the Australian monsoon during winter MONEX: Synoptic aspects. *Monthly Weather Review* 111:495–516.

Elsner, J. B., A. B. Kara and M. A. Owens. 1999. Fluctuations in North Atlantic hurricane frequency. *Journal of Climate* 12:407–437.

Huang, J. Y. 1988. The spatial and temporal analysis of the representations of the quasi-biennial oscillation in precipitation over China. *Chinese Journal of Atmospheric Science* 12:323–330.

Huang, R. H. and Y. F. Fu. 1996. The interaction between the East Asian monsoon and ENSO cycle. *Climate and Environment Research* 1:38–54 (in Chinese with English abstract).

Huang, R. H. and Y. F. Wu. 1989. The influence of ENSO on the summer climate change in China and its mechanisms. *Advances in Atmospheric Science* 6:21–32.

Huang, R. H. and L. T. Zhou. 2002. Research on the characteristics, formation mechanism and prediction of severe climate disasters in China. *Journal of Natural Disasters* 11:1–9 (in Chinese with English abstract).

Kumar, K. K., B. Rajagopalan and M. A. Cane. 1999. On the weakening relationship between the Indian monsoon and ENSO. *Science* 284:2156–2159.

Lau, K. M. and P. J. Shen. 1988. Annual cycle, quasi-biennial oscillation, and southern oscillation in global precipitation. *Journal of Geophysical Research* 93:975–988.

Li, C. Y. 1986. El Nino and typhoon action over the Western Pacific. *Chinese Science Bulletin* 31:538–542.

Li, C. Y. 1989a. Warmer winter in Eastern China and El Nino. *Chinese Science Bulletin* 34:1801–1805.

Li, C. Y. 1989b. Frequent activities of stronger upper-level troughs in East Asia in wintertime and the occurrence of the El Nino event. *Science in China (B)* 32:976–985.

Li, C. Y. 1990. Interaction between anomalous winter monsoon in East Asia and El Niño events. *Advances in Atmospheric Science* 7:36–46.

Li, C. Y. and M. Q. Mu. 1998a. ENSO cycle and anomalies of winter monsoon in East Asia. Pp. 60–73 in *East Asia and Western Pacific Meteorology and Climate*, edited by C. P. Chang, J. C. L. Chan and J. T. Wang. Singapore: Word Scientific.

Li, C. Y. and M. Q. Mu. 1998b. Numerical simulations of anomalous winter monsoon in East Asia exciting ENSO. *Chinese Journal of Atmospheric Science* 22:393-403.

Li, C. Y. and M. Q. Mu. 1999. ENSO occurrence and sub-surface ocean temperature anomalies in the equatorial warm pool. *Chinese Journal of Atmospheric Science* 23:217-225.

Li, C. Y. and M. Q. Mu. 2002. A further study of the essence of ENSO. *Chinese Journal of Atmospheric Science* 26:309-328.

Li, C. Y. and Y. P. Zhou. 1994. Relationship between intraseasonal oscillation in the tropical atmosphere and ENSO. *Chinese Journal Geophysics* 37:213-223

Li, C. Y., S. Q. Sun and M. Q. Mu. 2001. Origin of the TBO — Interaction between anomalous East-Asian winter monsoon and ENSO cycle. *Advances in Atmospheric Science* 18:554-566.

Long, Z. X. and C. Y. Li. 2002. Interannual variation of tropical atmospheric 30-60 days low-frequency oscillation and ENSO cycle. *Chinese Journal of Atmospheric Science* 26:51-62.

Lu, W. S. and Q. G. Zhu. 1990. Theoretical study on the effect of Qinghai-Xizang Plateau on cold surges. *Acta Meteorologica Sinica* 4:620-628.

McCabe, G. Y. and M. D. Dettinger. 1999. Decadal variations in the strength of ENSO teleconnections with precipitation in the western United States. *International Journal of Climatology* 19:1399-1410.

Mehta, V. and K. M. Lau. 1997. Influence of solar irradiance on Indian monsoon-ENSO relationship at decadal-multidecadal time scales. *Geophysical Research Letters* 24:159-162.

Mu, M. Q. 2001. A further research on the cyclic relationship between anomalous East-Asian winter monsoon and ENSO. *Climate and Environment Research* 6:273-285 (in Chinese with English abstract).

Mu, M. Q. and C. Y. Li. 1999. ENSO signals in interannual variability of East Asian winter monsoon, Part I: Observed data analyses. *Chinese Journal of Atmospheric Science* 23:139-149.

Rasmusson, E. M. and T. H. Carpenter. 1983. The relationship between eastern equatorial Pacific sea surface temperatures and rainfall over India and Sri Lanka. *Monthly Weather Review* 111:517-528.

Reed, R. J. 1961. Evidence of the downward-propagating wind reversal in the equatorial stratosphere. *Journal of Geophysical Research* 66:813-818.

Ropelewski, C. F., M. S. Halpert and X. Wang. 1992. Observed tropospheric biennial variability and its relationship to the southern oscillation. *Journal of Climate* 5:594-614.

Sankar Rao, K. M. Lau and S. Yang. 1996. On the relationship between Eurasian snow cover and the Asian summer monsoon. *International Journal of Climatology* 15:605-616.

Tao, S. Y. and L. X. Chen. 1987. A review of recent research on the East Asian summer monsoon in China. Pp. 60~92 in *Monsoon Meteorology*, edited by C. P. Chang and T. N. Krishnamurti. Oxford: Oxford University Press.

Wang, B. and Q. Zhang. 2002. Pacific-East Asian teleconnection. Part II: How the Philippine Sea anomalous anticyclone is established during El Niño development. *Journal of Climate* 15:3252–3265.

Webster, P. J. and T. N. Palmer. 1997. The past and the future of El Nino. *Nature* 390:562–564.

Weng, H.-Y., K. M. Lau and Y.-K. Xue. 1999. Multi-scale summer rainfall variability over China and its long-term link to global sea surface temperature variability. *Journal of the Meteorological Society of Japan* 77:845–857.

Wu, M. C. and J. C. L. Chan. 1995. Surface features of winter monsoon surges over South China. *Monthly Weather Review* 123:662–680.

Wu, R. and B. Wang. 2002. A contrast of the East Asian summer monsoon–ENSO relationship between 1962–77 and 1978–93. *Journal of Climate* 15:3266–3279.

Yu, L. and M. Rienecker. 1998. Evidence of an extra-tropical atmospheric influence during the onset of the 1997–98 El Nino. *Geophysical Research Letters* 25:3537–3540.

Zhang, R. H. 2001. Relations of water vapor transports from Indian monsoon with those over East Asia and the summer rainfall in China. *Advances in Atmospheric Science* 18:1005–1017.

Zhang, R., A. Sumi and M. Kimoto. 1996. Impact of El Nino on East Asian monsoon: A diagnostic study of the 86/87 and 91/92 events. *Journal of the Meteorological Society of Japan* 74:49–62.

Zhang, Y., K. R. Sperber and J. S. Boyle. 1997. Climatology and interannual variation of the East Asian winter monsoon: Results from the 1979–95 NCEP/NCAR Reanalysis. *Monthly Weather Review* 125:2605–2619.

Zhou, G. Q. and C. Y. Li. 1999. Simulation on the relation between the subsurface temperature anomaly in western Pacific and ENSO by using CGCM. *Climate and Environmental Research* 4:352–363 (in Chinese with English abstract).

Part II

ATMOSPHERIC COMPOSITION

Chapter 6

RECENT TRENDS IN SUMMER PRECIPITATION IN CHINA AND MONGOLIA

TETSUZO YASUNARI, NOBUHIKO ENDO and BORJIGINTE AILIKUN

1. Introduction

China has experienced several massive floods since 1991. Large floods struck the Yangtze River basin in the summers of 1991, 1998, 1999, and 2002. Flood disasters occurred in the Chu (Pearl) River basin in the summers of 1994, 1996, and 1997, and floods damaged the Huaihe River basin in 1991 and again in 2003. The frequent appearance of major floods suggests that both total summer precipitation and the frequency of extreme precipitation events may be changing. It is also possible that recent global warming as well as anthropogenic land use changes are resulting in more frequent massive floods.

Meiyu is the rainy season in China and it starts in south China in May. Meiyu rains migrate northward before dissipating in late July. Except over south China (Yatagai and Yasunari 1995) summer precipitation produces more than 40% of the annual precipitation in China. Takahashi (1993) studied regional differences in the contribution of daily rainfall to total precipitation in eastern Asia during the Meiyu season using frequency distributions of daily precipitation and the dominant precipitation intensity class. Matsumoto and Takahashi (1999) showed that the contribution of heavy rain (daily totals exceeding 50 mm) to total summer (May–August) precipitation is large over the middle parts of the Yangtze and Huaihe River basins and along the south coast of China. They also examined seasonal changes in heavy precipitation events. The northward migration of heavy precipitation events was linked to the northward movement of the Meiyu rains.

Secular changes in precipitation have recently attracted the attention of many researchers in the context of climate change. IPCC (2001) noted that global precipitation trends were positive throughout the last century. However, trends may vary with regions and seasons. For example, Chen et al. (1991) found that, over the 20th century, precipitation decreased over most of China, especially over northern and northwestern China, but Zhai et al. (1999a) reported no significant trend in annual precipitation over China between 1951 and 1995. Yatagai and Yasunari (1994) found no increase in annual precipitation over most of China, but they showed that there was a clear increase in summer precipitation over the middle and lower reaches of the Yangtze River, and a decrease in summer precipitation over the middle reach of the Yellow River.

Temporal changes of precipitation totals can result from both changes in precipitation frequency and changes in precipitation intensity during each event. Iwashima and Yamamoto (1993) investigated long-term changes of extreme daily precipitation in Japan and found a sharp increase between the 1890s and 1980s. Similar changes have been found in other parts of the world. Groisman et al. (1999) applied a statistical model to summer daily rainfall data from 1951 to 1994 over the USA, former USSR, China, Canada, Norway, Mexico, Poland, and Australia, and found that mean summer precipitation and the frequency of summer precipitation events increased in the USA, Norway, and Australia, but not in China.

Zhai et al. (1999b) investigated trends in annual and extreme precipitation in China for 1951–1995 and found no significant trends in annual precipitation or 1-day or 3-day maximum rainfalls. They also showed that the number of rainy days per year decreased, so an increase in precipitation intensity (precipitation/number of rainy days) was apparent. They further showed that annual precipitation increased over northwestern China and decreased over northern China. Nevertheless, regional characteristics of the temporal changes of heavy precipitation and their relation to the temporal changes in total precipitation in China during summer have not been intensively studied. It is also necessary to examine changes in the atmospheric circulation that force changes in heavy precipitation.

Endo et al. (2005) have recently scrutinized regional characteristics in trends and interannual variability of various indices of summer precipitation using daily precipitation data from 1961 to 2000 for more than 500 stations in China. The indices include summer precipitation totals, precipitation intensity, number of rainy days, and the frequency and amount of heavy precipitation events. Their results are summarized here as the most-updated results of the recent trends and variability of summer precipitation characteristics, including their seasonal dependencies. The atmospheric circulation changes related to these precipitation changes are also briefly reviewed.

2. Analysis of Data

This study used daily observations of precipitation and temperature compiled by the China Meteorological Administration at 726 stations in China. Many stations are east of 100°E, but stations are also well distributed in northwestern China and on the eastern Tibetan Plateau. Only those stations with data in June, July, and August during 1961–2000 were selected. The Chinese monthly precipitation dataset (from the Carbon Dioxide Information Analysis Center of the U.S. Department of Energy), which includes monthly precipitation records at 267 stations, was used as a reference.

Summer precipitation totals were obtained by summing rainfall data on days when rainfall was equal to or greater than 0.1 mm. However, the number of trace precipitation (less than 0.1 mm) days per season was counted. In this study, precipitation intensity is the seasonal precipitation divided by the number of rainy days.

Daily rainfall data for each station were sorted into ascending order and grouped into 10 classes in order to investigate how the frequencies and amounts of precipitation varied. Each of the 10 classes had an interval width of 10%; the lightest precipitation class was C1, and the heaviest was C10.

To define regions for the examination of precipitation characteristics, we applied an Empirical Orthogonal Function (EOF) analysis to summer precipitation data at the 71 stations east of 105°E and south of 45°N. The different regions were, (i) north China (34°–42°N, 105°–120°E, including the Shandong Peninsula; 98 stations), (ii) the Yangtze River basin (27°–34°N, east of 105°E; 116 stations), (iii) south China (21°–27°N, east of 105°E; 73 stations), the area west of 105°E divided into (iv) northwestern China (north of 36°N; 81 stations) and (v) Tibet (south of 36°N; 80 stations), and (vi) northeastern China (north of 42°N; 98 stations). Regional precipitation averages were obtained simply from the arithmetic mean.

The geographic distributions of precipitation trends were estimated by a least square linear fit at each station. Regional time series of precipitation were also subjected to a simple linear regression in order to facilitate a comparison between regional precipitation characteristics and regional precipitation trends. The non-parametric Kendall–Tau test evaluated the significance of the trends. A significance level of 0.05 was used throughout this analysis.

3. Regional Trends

Trends are expressed as the percentage of the mean precipitation for 1961–2000. The largest increase, with values exceeding 40%, occurred over the middle and lower parts of the Yangtze River basin (Fig. 6.1). A decline in total precipitation occurred over northern China, especially over the Shandong peninsula. Total precipitation also decreased over Liaodong peninsula, but increased over most of northeastern China. Trends changed sign between coastal regions and inland regions. Over semi-arid and arid regions, such as Inner-Mongolia and northwestern China, the total precipitation increased. The total precipitation trend exceeded 30% in the

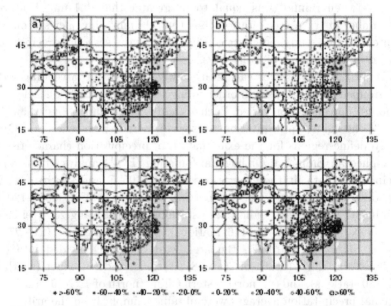

Fig. 6.1. Trends for 1961–2000 in (a) summer total precipitation, (b) rainy days, (c) precipitation intensity, and (d) class 10 precipitation. Trends expressed as percent of mean precipitation from 1961 to 2000. Solid circles and open circles are decreasing and increasing trends, respectively.

Taklimakan Basin. In contrast, small trends dominated over south China, the eastern Tibetan Plateau, and Yunnan province.

The number of rainy days declined in north and northeastern China (Fig. 6.1(b)). The decrease was large, with a value of about 40% near the Bohai Sea. On the other hand, the number of rainy days clearly increased over northwestern China and the Yangtze River basin. The largest trend, about 60%, was observed in the Taklimakan Basin. Decreases dominated over the Tibetan Plateau, Yunnan, and coastal southern China, but the trends were very small. The seasonal mean precipitation intensity increased at most stations (Fig. 6.1(c)), and was particularly large over the lower reaches of the Yangtze River and the Taklimakan Basin, regions where total precipitation and rainy days also increased. On the other hand, the intensity of rainfall also increased over northern China where the total precipitation and number of rainy days decreased.

The heaviest class interval, C10, which designates daily rainfall greater than the 90th percentile, increased significantly over the Yangtze River basin and in northeastern China (Fig. 6.1(d)). In contrast, C10 decreased over northern China, especially near the Bohai Sea. The largest trend, over northwestern China, exceeded 40% for the analysis period.

The total precipitation and number of rainy days increased at most stations in the middle and lower reaches of the Yangtze River basin; decreases occurred north of the Yangtze River basin. Figure 6.2 shows trends in the longitudinal band from 110°E to 120°E to highlight their latitudinal distribution. A statistically significant increase in the precipitation appeared in a narrow region from 27°N to 32°N (Fig. 6.2(a)). Rainy days also increased in the Yangtze River basin (Fig. 6.2(c)). Furthermore, increases in C10 occurred in a limited region from 28°N to 30°N (Fig. 6.2(d)). Increases in the number of precipitation events and the amount of rain caused by heavy precipitation were notable features in the Yangtze River basin. Total precipitation decreased north of 35°N, and a statistically significant decrease in rainy days occurred at many stations in northern China (Fig. 6.2(c)). Although total and Class 10 precipitation both decreased over northern China, statistically significant trends occurred at limited stations. Intensity increased at most stations (Fig. 6.2(b)). In the Yangtze River basin, the increase in precipitation was relatively larger than the increase in rainy days; thus, increased as defined. Over northern China, precipitation decreased by a smaller amount than rainy days, thus, the intensity of rainfall increased in that region.

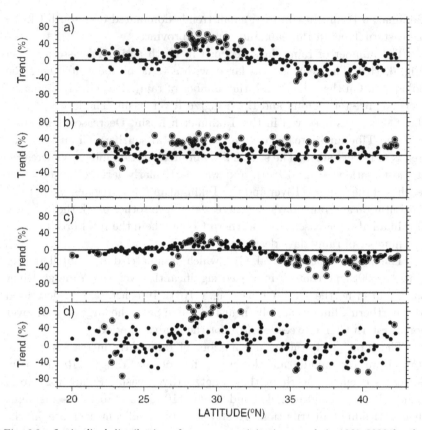

Fig. 6.2. Latitudinal distribution of summer precipitation trends in 1961–2000 for the longitudinal band from 110°E to 120°E. (a) Summer total precipitation, (b) precipitation intensity, (c) rainy days, and (d) class 10 precipitation. Trends expressed as percent of mean precipitation from 1961 to 2000. Double circle indicates statistical significance at the 0.05 confidence level.

Total precipitation increased over the Yangtze River basin, northwestern China, and southern China. Over the Yangtze River basin it increased by more than 25% of the climatological mean of 509.3 mm, that is, about 35 mm decade^{-1}, which was statistically significant at a 95% confidence level. The increase in rainfall over northwestern China (about 4.5 mm decade^{-1}) was smaller than that over the Yangtze River basin, but was still statistically significant. The intensity of rainfall over both the Yangtze River and northwestern China increased significantly, with a slight increase in rainy days. The increase in precipitation is therefore largely attributable to an increase in the number of heavy precipitation events.

Over northern and northeastern China and Tibet rainfall decreased. Over northeastern China, there were statistically significant increases in intensity and decreases in the number of rainy days suggesting a linkage between the two. The number of light precipitation events decreased as the number of heavy precipitation events increased in this region. The number of rainy days significantly decreased in Tibet and north China, while the intensity of rain slightly increased.

The regional trends for each of the 10 classes are shown in Fig. 6.3. The two upper-most classes showed significant increases for 1961–2000 over the Yangtze River basin (Fig. 6.3(a)), where the average daily rainfall was 248.0 mm for C10 and 108.5 mm for C9. In contrast, the value for class 1 decreased significantly to 0.5 mm. As described earlier, the total rainfall increased over northwestern China where significant increases occurred in classes 10 and 9 (Fig. 6.3(b)) which have precipitation values of 34.5 and 15.8 mm, respectively. Over southern China, the decreases in the lowest three precipitation classes were significant and heavy precipitation increased slightly. As noted above, precipitation decreased over Tibet and northern and northeastern China and most of the precipitation classes declined in these regions (Figs. 6.3(d), (e), and (f)). The lightest precipitation class decreased in all regions, and the lowest class average daily rainfall was less than 1 mm.

In Mongolia, increasing trends of summer rainfall are apparent in the relatively dry zones of eastern, southern, and western Mongolia, whereas decreasing trends are dominant in the relatively wet central and northern zones (Endo et al. 2006).

These overall tendencies in the dry zones in Mongolia are consistent with the trends in the dry zones (Taklimakan desert) of the inner Mongolian region. In particular heavier rainfall events (class 10) increased in eastern and western Mongolia, and a decreasing trend is apparent in north central Mongolia.

4. Interannual Variability of Heavy Rainfall

Figure 6.4(a) shows a time series of total precipitation and the contribution of heavy rainfall event (>50 mm) in northern China. It is apparent that there was year-to-year variability in the total precipitation data for northern China.

Nitta and Hu (1996) reported a decrease in precipitation around the middle stretch of the Yellow River, but their data did not show a long-term

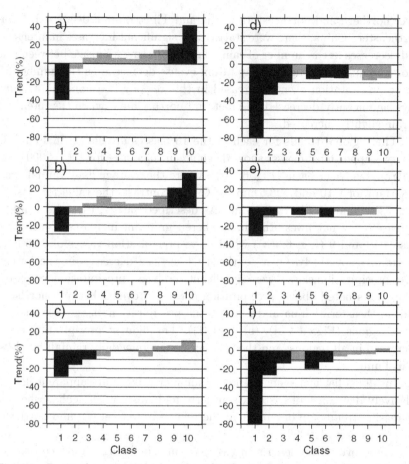

Fig. 6.3. Regional trends of summer precipitation for various classes of precipitation defined by 10 percentile class intervals for (a) Yangtze River basin, (b) northwestern China, (c) southern China, (d) northern China, (e) Tibet, and (f) northeastern China. Trends are expressed as percent of mean precipitation from 1961 to 2000. Solid bar indicates the trend is statistically significant at the 0.05 significance level (Endo et al. 2005).

trend after the mid-1960s. Decadal-scale variability signals were weak for their time series.

On an average there were about 35 wet days each year before 1980 in North China (Fig. 6.4(b)). Two stepwise decreases in numbers of wet days occurred, one around 1980 and the other in 1997. High rainfall events clearly occurred less frequently from 1980 to 1993 (84 events). The large decrease in wet days around 1980 coupled with the small change in total

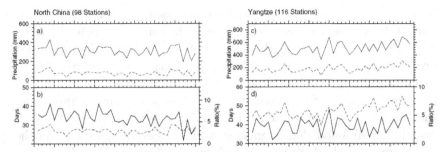

Fig. 6.4. (a) Time series of summer total rainfall (solid line) and amount from heavy rainfall (>50 mm day^{-1}; dashed line) for northern China (98 stations). (b) Time series of rainy days (solid line) and ratio between heavy rainfall and number of rainy days (dashed line) (c) and (d), as for (a) and (b), but for Yangtze River basin (116 stations).

precipitation suggests an increase in the seasonal average rainfall intensity, found in Fig. 6.1(c). In the late 1990s, the regional average number of wet days fell below 30 days, while the regional total frequency of heavy rainfalls (>50 mm) was 102 events after 1994.

Han and Gong (2003) analyzed 31 stations over the plains of northern China and found the period that experienced heavy rainfall events lengthened between 1980 and 1993, with earlier onset and later ending. Such a lengthening of the heavy rainfall period is unclear in the present results using 98 gauges distributed over a larger area of northern China.

The regional time series of total rainfall and the amount generated by heavy precipitation in the Yangtze River basin are shown in Figs. 6.4(c) and 6.4(d). The total yearly rainfall increased slightly until 1979. Fluctuations at longer time scales were small, but large-amplitude variations in rainfall occurred on shorter time scales from 1979 to the late 1980s. Total rainfall then continuously increased again in the 1990s. A linear increase in heavy precipitation occurred after 1979 (Fig. 6.4(c)). Decadal average percentages of the contribution of heavy precipitation to the total precipitation were 28.3%, 31.5%, and 34.2% during the 1970s, 1980s, and 1990s, respectively: the contribution of heavy precipitation to total precipitation increased after the late 1970s. This increase was coincident with the increase in the three upper precipitation classes (8, 9, and 10) in this region, as described above. The ratio of the number of days with rainfall >50 mm to wet days increased from 5% in the 1960s to 7% in the 1990s. Wet days alone show a large interannual variation with no clear trend (Fig. 6.4(d)).

The observations on 10 day rainfall amount and frequency of heavy rainfall events are consistent with an earlier onset and prolonged influence of Meiyu frontal activity in the Yangtze River basin in recent decades. More than 80 heavy rainfall events were observed in four summers in the 1990s but only once in 1969 and the number of heavy rainfall days increased during the past two decades. This change suggests more vigorous Meiyu frontal activity in the 1990s.

The time scale of the interannual variation of total precipitation and that of the heavy precipitation is short (2–3 years) after 1980. Nitta and Hu (1996) found that interannual variations in precipitation over the middle reach of the Yellow River and over the lower and middle reaches of the Yangtze River were out of phase in their first EOF component. The quasi-biennial signal became remarkable in the score time series of EOF-1 after the mid-1970s. In this context, the time-scale changes of interannual variation in total precipitation over the Yangtze River basin are consistent with the results of Nitta and Hu (1996).

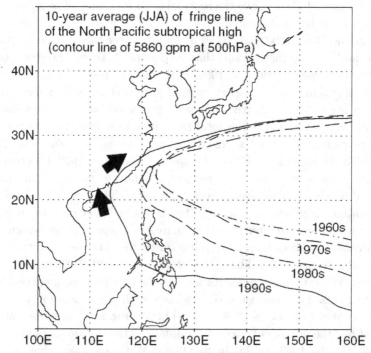

Fig. 6.5. Decadal change of fringe line of the north Pacific subtropical high, as indicated with contours of 10-year average of 5,860 geopotential meter height at 500 h Pa. Presumed moisture flux along the periphery of the high is also schematically shown with arrows (Endo 2004).

The relatively dry 1980s and the relatively wet 1990s may have resulted from changes in the large-scale circulation on decadal time scales. Figure 6.5 shows the decadal change of the fringe line of the north Pacific subtropical high, as indicated with 10-year average of 5,860 geopotential meter height of 500 h Pa. This diagram shows the westward extension of the subtropical high toward the China continent from 1960s to 1990s, which may have facilitated the intensification of low-level moisture flux and convergence along the periphery of the subtropical high over southern China and increase of precipitation and convection at the lower and middle reaches of the Yangtze River basin.

5. Discussion

Based on Endo et al. (2005, 2006), the recent trends total precipitation, the number of rainy days, and intensity of precipitation during summer were summarized from daily rainfall datasets for China and Mongolia from 1961 to 2000. Total precipitation significantly increased in summer in the Yangtze River basin and northwestern China and decreased in the other regions. The number of rainy days increased in the Yangtze River basin and over northwestern China, but decreased over Tibet, northern and northeastern China. When daily precipitation totals were grouped into 10 classes at 10-percentile intervals and simple linear regression analysis was applied, the upper 20-percentile totals showed a statistically significant increase over the Yangtze River basin, arid and semi-arid regions of northwestern China, and western and eastern Mongolia. Most of the 10 classes of precipitation decreased over Tibet and north and northeastern China.

The results suggest that the times of active Meiyu frontal activity have lengthened over the Yangtze River basin in the recent decade. Over northern China, heavy rainfall (exceeding 50 mm day^{-1} was less frequent in the 1980s than in the 1970s and 1990s. Summer precipitation totals increased as the frequency of heavy rain increased in the Yangtze River basin. Total summer precipitation and the contributions of heavy rainfall were obviously large in south China during the 1990s. Meiyu frontal activity in mid-June was especially active during the 1960s and 1990s in south China. Interannual variations in heavy rainfall and summer precipitation totals in August may relate to changes in tropical cyclone activity. The increase in precipitation and heavy rainfall events over south China and the Yangtze River basin during the 1990s may have, at least partly, been associated with the westward extension of the north Pacific subtropical high.

It has been suggested that the changes in precipitation, and the tendency toward increased floods and droughts in certain regions are caused by the large amounts, in the atmosphere, of black carbon particles which have been generated from industrial pollution, traffic, and burning of coal and biomass fuels (Menon et al. 2002; Menon 2004; Qian et al., Chapter 8, this volume). Menon et al. (2002) conducted simulations to monitor the effects of black carbon on the hydrologic cycle over China using a climate computer model and aerosol data from 46 stations in China. The results indicated that black carbon affects regional climate by absorbing sunlight and heating the air, which rises, forming clouds and causes rain to fall in heavily polluted regions. The rising air in southern China is balanced by sinking air and drying in northern China. The simulations indicated that increased black carbon over southern China would create a clear tendency toward the flooding in southern China and increasing drought over northern China, similar to the recent occurrence.

Literature Cited

Chen, L. X., Y. N. Shao, M. Dong, Z. H. Ren and G. S. Tian. 1991. Preliminary analysis of climate variation during the last 39 years in China. *Advances in Atmospheric Sciences* 8:279–288.

Endo, N. 2004. Long-term trend of low level cloud over China during summer season. *Kaiyo Monthly* 36:272–278 (in Japanese).

Endo, N., B. Ailikun and T. Yasunari. 2005. Changes in precipitation amounts and number of rainy days and heavy rainfall events during summer in China. *Journal of the Meteorological Society of Japan* 83:621–631.

Endo, N., T. Kadota, J. Matsumoto, B. Ailikun and T. Yasunari. 2006. Summer rainfall characteristics in Mongolia: Its climatology and trends. *Journal of the Meteorological Society of Japan* 84:543–551.

Groisman, P. Y., T. Karl, D. R. Easterling, R. W. Knight, P. F. Jamson, K. J. Hennessy, R. Suppiah, C. M. Page, J. Wibig, K. Fortuniak, V. N. Razuvaev, A. Douglas, E. Forland and P. M. Zhai. 1999. Changes in the probability of heavy precipitation: Important indicators of climatic change. *Climate Change* 42:243–283.

Han, H. and D. Gong. 2003. Extreme climate events over northern China during the last 50 years. *Journal of Geographical Sciences* 13:469–479.

IPCC (Intergovernmental Panel on Climate Change). 2001. *Climate Change 2001: The Scientific Basis*, edited by J. T. Houghton, Y. Ding, D. J. Griggs, M. Noguer, P. J. van der Linden, X. Dai, K. Maskell and C. A. Johnson. Cambridge, U.K.: Cambridge University Press.

Iwashima, T. and R. Yamamoto. 1993. A statistical analysis of the extreme events: Long-term trend of heavy daily precipitation. *Journal of the Meteorological Society of Japan* 71:637–640.

Matsumoto, J. and K. Takahashi. 1999. Regional difference of daily rainfall characteristics in East Asian summer monsoon season. *Geographical Review of Japan* 72B:193–201.

Menon, S. 2004. Current uncertainties in assessing aerosol effects on climate. *Annual Review of Environmental Resource* 29.

Menon, S., J. Hansen, L. Nazarenko and Y. F. Luo. 2002. Climate effects of black carbon aerosols in China and India. *Science* 297:2250–2253.

Nitta, T. and Z. Hu. 1996. Summer climate variability in China and its association with 500 h Pa height and tropical convection. *Journal of the Meteorological Society of Japan* 74:425–445.

Takahashi, H. 1993. Regional difference and variability of the contribution of daily precipitation to total precipitation amount during the Baiu season in East Asia. *Geographical Science* 48:20–32.

Yatagai, A. and T. Yasunari. 1994. Trends and decadal-scale fluctuations of surface air temperature and precipitation over China and Mongolia during the recent 40 year period (1951–1990). *Journal of the Meteorological Society of Japan* 72:937–957.

Yatagai, A. and T. Yasunari. 1995. Interannual variations of summer precipitation in the arid/semi-arid regions in China and Mongolia. Their regionality and relation to the Asian summer monsoon. *Journal of the Meteorological Society of Japan* 73:909–923.

Zhai, P. M., A. Sun, F. Ren, X. Liu, B. Gao and Q. Zhang. 1999a. Changes of climate extremes in China. *Climate Change* 42:203–218.

Zhai, P. M., F. M. Ren and Q. Zhang. 1999b. Detection of trends in China's precipitation extremes. *Acta Meteorologica Sinica* 57:208–216.

Chapter 7

WARMING IN EAST ASIA AS A CONSEQUENCE OF INCREASING GREENHOUSE GASES

ZONGCI ZHAO, AKIO KITOH and DONG-KYOU LEE

1. Introduction

Houghton et al. (2001) reported that the annual mean surface air temperature averaged over the globe had increased by about 0.6°C over the 20th century. There is a new and strong evidence that most of the warming observed over the last 50 years is attributable to human activities. The atmospheric concentrations of carbon dioxide, methane, and nitrous oxide have increased by 31%, 151%, and 17% since 1750, respectively, and it is expected that the concentrations of atmospheric greenhouse gases and aerosols, as well as their radiative forcing, will continue to increase as a result of human activities throughout the 21st century. It also pointed out that the confidence in the ability of climate models to project future climate is increasing, and that as a result of human influences the annual mean surface air temperature averaged globally is expected to increase by 1.4°–5.8°C over the period 1990–2100. These projections were obtained from the full range of scenarios based on a number of climate models (Houghton et al. 2001).

The anthropogenic emissions over East Asia have increased, during the last several decades, in the same way as the global emissions. Since 2000 many scientists have focused on the impacts of human activities and global warming on climate change in East Asia using general circulation and regional climate models. Their research concentrated on temperature, precipitation, extreme events, and the East Asian monsoon (Kitoh et al. 1997; JMA 2000; Guo et al. 2001; Wang 2001; Gao et al. 2002; Ding and Xu 2003; Zhao et al. 2004). A number of researches pointed out a certain capability of the simulations by the climate models over East Asia (JMA 2000; Lee and Suh 2000; Guo et al. 2001; Kato et al. 2001).

In this chapter, we summarize and review observed and projected results of the increase of greenhouse gases and sulfate aerosols, in recent years in East Asia, on regional aspects of global warming. East Asia is defined here as the region about 15°–60°N and 70°–150°E. A number of general circulation models provided by scientists in Australia, Canada, China, Germany, Japan, Korea, United Kingdom, and USA with the Intergovernmental Panel on Climate Change scenarios have been employed by the East Asian researchers to simulate the effects of changes in greenhouse gas concentration on air temperature etc. (e.g. JMA 2000; Zhao and Xu 2002; Ding and Xu 2003; Min et al. 2004). The scenarios studied included doubled CO_2 concentration, business as usual, six developed by the Intergovernmental Panel on Climate Change (designated IS92; Leggett et al. 1992), one with greenhouse gas forcing only (GG), one with greenhouse gas and sulfate (aerosol) forcing (GS), and 40 from the Intergovernmental Panel on Climate Change Special Report on Emission Scenarios (SRES; Nakicenovic et al. 2000) with various times of reaching doubled CO_2 concentration.

2. Warming in East Asia over the 21st Century

Based on the observations for the 20th century, it is found that the surface air temperature over East Asia had increased by about 0.84°C per 100 years for 1900–1999 which was warmer than the increase in average globe temperature (Zhao and Xu 2002). Studies by Oh et al. (1998), JMA (2000), Yu et al. (2001), Zhai and Pan (2003), Chung et al. (2003), Chae (2003) and Ha (2003) provided details of the temperature increases for different parts of East Asia. The control runs by certain general circulation models did not give the linear trends expected and this was interpreted to mean that human activities were most likely responsible for the warming in East Asia for the 20th century, especially for the last 50 years (Zhao and Xu 2002).

The simulations and projections by the general circulation models suggest that we need to be concerned about climate change in East Asia during the 21st century. Simulations and projections by more than 40 climate models using different scenarios suggest that the temperature over East Asia might increase by 3°–5°C per 100 years (Table 7.1; Zhao et al. 2003; Ding and Xu 2003; Min et al. 2004). The projections for regional

Table 7.1. Annual average surface temperature anomalies in East Asia for the 21st century as projected by the GCM7 (CCC, CCSR, CSIRO, DKRZ, GFDL, HADL, and NCAR) with GG, GS, and SRES A2 B2 scenarios (relative to the mean of 1961–1990) (unit: °C).

	2010	2020	2030	2040	2050	2060	2070	2080	2090	2099
GG	1.39	1.85	2.58	2.60	3.08	3.57	4.05	4.86	5.33	5.54
GS	1.22	1.39	1.77	1.67	2.28	2.55	3.13	3.61	4.27	4.67
SREA2	0.74	1.10	1.37	1.76	1.95	2.63	3.47	4.08	4.38	5.42
SREB2	0.73	1.22	1.49	1.66	2.22	2.53	2.63	3.14	3.33	3.43
S7A2		1.2			2.5			4.1		
S7B2		1.4			2.4			3.2		

*GG and GS (Zhao and Xu, 2002); SREA2 and SREB2 (Ding and Xu, 2003); S7A2 and S7B2 (Min et al., 2004, Lee personal communications).

distribution indicate a warming of 4°–7°C over the north of East Asia, and 2°–5.5°C over the south of East Asia in 2071–2100 relative to the mean surface temperature of 1961–1990.

In addition, over East Asia, significant warming in the winters of the late 21st century might cause shorter cold spells, and warm summers might cause longer hot spells (Ding and Xu 2003; Min et al. 2004). In some studies it was pointed out that the diurnal range of temperatures in China and Japan might decrease, because the minimum temperature is likely to increase due to human induced emissions (JMA 2000; Zhao et al. 2003).

3. Precipitation Change

Precipitation change is an important issue in the East Asian monsoon regions. Floods have caused huge life and economic losses in the past, and continuing droughts have damaged agriculture and water resources.

Some studies have concentrated on the projections of precipitation over East Asia for the 21st century, while others have tried to determine the changes in the geographical distribution of the precipitation. Most of the projections, with over 40 climate models and scenarios, suggest that the annual precipitation over East Asia could increase by about 23–151 mm for the period 2071–2100 relative to 1961–1990. A few models and scenarios projected decreasing precipitation for the 21st century (Table 7.2). The modeling studies indicated that precipitation over most of the regions in East Asia would increase. The studies indicate a 20–50% increase in precipitation for the northern part of East Asia and a slight decrease of

Table 7.2. Annual average precipitation anomalies in East Asia for the 21st century as projected by the GCM7 (CCC, CCSR, CSIRO, DKRZ, GFDL, HADL, and NCAR) with GG, GS, and SRES A2 B2 scenarios (relative to the mean of 1961–1990) (unit: mm).

	2010	2020	2030	2040	2050	2060	2070	2080	2090	2099
GG	38	30	46	50	53	74	55	65	82	74
GS	22	0	−7	−11	10	−11	−8	34	31	34
SREA2	21	−11	−37	−22	21	24	29	49	89	151
SREB2	10	32	47	36	37	65	27	59	99	23
S7A2		0.4%			2.2%			5.0%		
S7B2		1.4%			2.6%			4.0%		

*GG and GS (Zhao and Xu, 2002); SREA2 and SREB2 (Ding and Xu, 2003); S7A2 and S7B2 (Min et al., 2004, Lee personal communications).

about 1–5% in some small areas along the lower branch of the Yangtze River, and the central and southern parts of Japan in the 21st century.

In addition, a few studies with the global or regional climate models coupled with human related emissions indicated that the frequency of floods over East Asia might increase in the 21st century (Gao et al. 2002; Zhao et al. 2004).

4. Other Climate Phenomena

In addition to the studies on changes in temperature and precipitation in East Asia for the 21st century, several climate models have been used to estimate the effect of human activities on maximum and minimum temperature, the number of heavy rainfall days, annual typhoon numbers, East Asian monsoon, and the sea surface temperature over the Kuroshio, warm pool, and El Nino regions of the tropical Pacific Ocean.

Studies of observed data by Guo et al. (2004) showed that both winter and summer monsoons weakened during the last 30 years and evaluation of the evidence by the climate models suggested that this might be due to both natural and human related emissions. Projections using the climate models coupled with human activities suggested that the winter monsoon would weaken and the summer monsoon would get stronger in the 21st century (Bueh et al. 2003; Ding and Xu 2003; Zhao et al. 2004). With the aid of the general circulation models, studies have been made on the effect of human activities on annual typhoon numbers over the northwestern Pacific Ocean during the last 50 years (e.g., Wang 2001; Chan and Liu 2003; Zhao et al. 2004). The results of the projections were very different depending on models. Some models predicted that the annual typhoon numbers over the northwestern Pacific Ocean might decrease (e.g., Tsutsui et al. 1999; Sugi et al. 2002), whereas a regional climate model with a doubling of greenhouse gas emission indicated that annual typhoon numbers over the South China Sea and western parts of the northwestern Pacific Ocean would increase slightly (Gao et al. 2002).

Studies on the changes of sea surface temperature over the tropical Pacific Ocean suggested that anthropogenic greenhouse gas emissions would cause warming over the Kuroshio, warm pools, and El Nino regions in the 21st century. The warming in the EL Nino regions might be larger than in the Kuroshio region and the warm pool (Choi et al. 2002; Zhao et al. 2004). These results suggest that the gradients in sea surface temperature between the western and eastern tropical Pacific oceans might decrease due to human related emissions.

Climate changes over East Asia for the 20th and 21st centuries, as simulated and projected by the climate models with different human related emissions are summarized in Table 7.3. The observed results are also provided in Table 7.3 for comparison with the simulated results. The results suggest warming and slightly increased precipitation over East Asia for the 21st century due to the anthropogenic emissions.

It has been suggested that the levels of warming over East Asia in the 21st century due to human activities will compare with the paleoclimate and historical climate changes. The warming by the end of the 21st century might be larger than the Medieval Warm Period and similar to Holocene Maximum Period (Table 7.4).

5. Conclusions

About 40 climate models and scenarios were used in the projections of temperature change over East Asia. The various models and scenarios projected a warming range of 2°–5°C over East Asia by the end of the 21st century. The warming effect might be larger than Medieval Warm Period and similar to Holocene Maximum Period. Corresponding to the warming due to human influence, the East Asian winter monsoon might weaken and summer monsoon might strengthen, and more frequent floods and droughts might appear in some parts of East Asia.

There are many uncertainties in the studies of the human induced emissions, and the use of climate models to simulate climate change and project what might happen in the future, especially for regional scales. There needs to be a coupling of modern understanding, from observation

Table 7.3. Simulated and projected climate change in East Asia for the 20th and 21st centuries (base line 1961–1990).

Phenomena	Observations in the 20th century	Simulations in the 20th century	Projections in the 21st century
Surface air temperature	0.84°C 100 yr^{-1} (Jones et al. and Hulme et al. personal communication)	0.44°–1.43°C 100 yr^{-1} as simulated by more than 40 models and scenarios	2.95°–4.99°C 100 yr^{-1} as projected by more than 40 models and scenarios
T_{max} and T_{min}	T_{max} 0.3°–0.5°C 50 yr^{-1}, T_{min} 1.3°–1.5°C 50 yr^{-1} (JMA* 2000; Wang 2001; Yu et al. 2001; Zhai and Pan 2003)	T_{max} 0.4°–1.1°C 50 yr^{-1}, T_{min} 0.5°–1.3°C 50 yr^{-1} as simulated by about 16 models and scenarios	3°–5°C 100 yr^{-1} as projected by about 16 models and scenarios
Precipitation	No obvious linear trends (Hulme et al. 1994; Wang 2001)	No obvious linear trends as simulated by more than 40 models and scenarios	Increase of 23–151 mm by the end of 21st century as projected by about 30–35 models and scenarios
Frequency of floods and droughts			Increasing as projected by a few models
East Asian monsoon	Winter: weaker Summer: weaker (Guo et al. 2004)	A few models and scenarios predict winter slightly weaker, summer weaker	18 models and scenarios project winter weaker, summer: stronger
Typhoon numbers over the northwestern Pacific Ocean	Reduced 3–4 times 53 yr^{-1} (Wang 2001)	Reduced 2–3 times 53 yr^{-1} as simulated by a few models	A few models project reduction by ~10–30% by the end of 21st century; A regional model with doubled CO_2 projects increase 1–2 times yr^{-1} near China

*Japan Meteorological Agency.

Table 7.4. Status of climate change over East Asia and China for the 21st century to compare with the key paleoclimatic change periods (possible reasons based on Zhao et al. 2003).

Time scales	Temperature change (°C)	Precipitation change (mm)	Possible reasons for change
Rising periods of Tibetan Plateau	Amplitude > 10	Amplitude > 200	Rising of mountain, continents drift, solar activity, and others
Last Glacial Maximum	−5 to −10	−50 to −100	Changes of Earth's orbital parameters
Holocene Maximum	+1 to +4	+20 to +100	Changes of Earth's orbital parameters
Little Ice Age	−0.4 to −0.9	±50–100	Solar activity, volcanic activity, interactions, and feedbacks in global climate system
Medieval Warm Period	+0.3 to +0.5	±50–100	Solar activity, volcanic activity, interactions, and feedbacks in global climate system
20th century (1900–1999)	+0.6 to +0.9 100 yr^{-1}	+5 to +10 100 yr^{-1}	Solar activity, volcanic activity, human activity, interactions, and feedbacks in global climate system
21st century (2000–2099)	2099: mean 4.7–5.5 Range: 3.3–7.1	2099: mean: 34–74 Range: −70 to 211	Greenhouse gases and sulfate aerosols

and model simulation, with the patterns of past changes identified in the paleoclimatic records. This would allow a better understanding of the paleoclimatic records and how the climate system varies, and enable us to more confidently predict future climate change.

Literature Cited

Bueh, C., U. Cubasch and S. Hagemann. 2003. Impacts of global warming on changes in the East Asian monsoon and the related river discharge in a global time-scale experiment. *Climate Research* 24:47–57.

Chae, S. S. 2003. On the climate change in DPR of Korea during the 20th century. Pp. 241–244 in *Proceedings of International Symposium on Climate Change*. Beijing: China Meteorological Press.

Chan, J. C. L. and K. S. Liu. 2003. Global warming does not lead to more intense tropical cyclones. Pp. 72–75 in *Proceedings of International Symposium on Climate Change*. Beijing: China Meteorological Press.

Choi, B.-H., D.-H. Kim and J.-W. Kim. 2002. Regional responses of climate in the north western Pacific Ocean to gradual global warming for a CO_2 quadrupling. *Journal of Meteorological Society, Japan* 80:1427–1442.

Chung, Y. S., M. B. Yoon and H. S. Kim. 2003. On climate variations and changes observed in South Korea. Pp. 226–231 in *Proceedings of International Symposium on Climate Change*. Beijing: China Meteorological Press.

Ding, Y. and Y. Xu. 2003. Projections of climate change over China for the 21st century by the GCMs with SRES scenarios. Report at Key Project on Climate Change in China, December, Beijing, China.

Gao X., Z. C. Zhao and F. Giorgi. 2002. Changes of extreme events in regional climate simulations over East Asia. *Advances in Atmospheric Sciences* 19:927–942.

Guo, Q., J. Cai, X. Shao and W. Sha. 2004. Changes of summer monsoon over East Asia for 1873–2000. *Advances in Atmospheric Sciences* 28:206–215 (in Chinese).

Guo Y., Y. Yu, X. Liu and X. Zhang. 2001. Simulation of climate change induced by CO_2 increasing for East Asia with IAP/LASG GOALS model. *Advances in Atmospheric Sciences* 18:53–66.

Ha, J. S. 2003. Long-term variation of droughts in spring and summer seasons and of the anomalous temperature in DPR of Korea over the last 1000 years. Pp. 245–247 in *Proceedings of International Symposium on Climate Change*. Beijing: China Meteorological Press.

Houghton, J. T., Y. Ding, D. J. Griggs, M. Noguer, P. J. van der Linden and D. Xiaosu. 2001. *Climate Change 2001: The Scientific Basis*, Contribution of Working Group I to the Third Assessment Report of the Intergovernmental Panel on Climate Change. Cambridge: Cambridge University Press.

JMA (Japan Meteorological Agency). 2000. *Information of Global Warming, Vol. 3 — Climate Change Due to Increase of CO_2 and Sulfate Aerosol Projected with a Coupled Atmosphere Ocean Model* (in Japanese).

Kato, H., K. Nishizawa, H. Hirakuchi, S. Kadokura, N. Oshima and F. Giorgi. 2001. Performance of the RegCM2.5/NCAR-CSM nested system for the simulation of climate change in East Asia caused by global warming. *Journal of Meteorological Society, Japan* 79:99–121.

Kitoh, A., S. Yukimoto, A. Noda and T. Motoi. 1997. Simulated changes in the Asian summer monsoon at times of increased atmospheric CO_2. *Journal of Meteorological Society, Japan* 75:1019–1031.

Lee, D.-K. and M. S. Suh. 2000. Ten-year east Asian summer monsoon simulation using a regional climate model (RegCM). *Journal Geophysical Research* 105(D24):29565–29577.

Leggett, J., W. J. Pepper and R. J. Swart. 1992. Emissions scenarios for IPCC: An update. Pp. 69–95 in *Climate change 1992. Supplementary report to the IPCC scientific assessment*, edited by J. T. Houghton, B. A. Callander and S. K. Varney. Cambridge: Cambridge University Press.

Min, S. K., E. H. Park and W. T. Kwon. 2004. Future projections of East Asian climate change from multi-AOGCM ensemble of IPCC SRES scenario simulation. *Journal of Meteorological Society, Japan* 82:1187–1211.

Nakicenovic, N., J. Alcamo, G. Davis, B. de Vries, J. Fenhann, S. Gaffin, K. Gregory, A. Grübler, T. Y. Jung, T. Kram, E. L. La Rovere, L. Michaelis, S. Mori, T. Morita, W. Pepper, H. Pitcher, L. Price, K. Riahi, A. Roehrl, H.-H. Rogner, A. Sankovski, M. Schlesinger, P. Shukla, S. Smith, R. Swart, S. van Rooijen, N. Victor and D. D. Zhou. 2000. *Special report on emissions scenarios for the Intergovernmental Panel on Climate Change*. Cambridge: Cambridge University Press.

Oh, S. N., K. J. Ha, K. Y. Kim and J. W. Kim. 1998. Effects of land hydrology in northeastern Asia in a doubling CO_2 climate experiment. *Journal of Korean Meteorological Society* 34:293–305 (in Korean).

Sugi, M., A. Noda and N. Sato. 2002. Influence of the global warming on tropical cyclone climatology: An experiment with the JMA global model. *Journal of Meteorological Society, Japan* 80:249–272.

Tsutsui, J., A. Kasahara and H. Hirakuchi. 1999. The impacts of global warming on tropical cyclones — a numerical experiment with the T42 version of NCAR CCM2. in *Proceedings of the 23rd Conference on Hurricanes and Tropical Meteorology*. Dallas: American Meteorological Society.

Wang S. W. 2001. *Advances on Climatology in the Modern Times*. China: Meteorological Press (in Chinese).

Yu, S., V. K. Saxena and Z. C. Zhao. 2001. A comparison of signals of regional aerosol-induced forcing in eastern China and the southeastern United States, *Geophysical Research Letters* 28:713–716.

Zhai, P. and X. Pan. 2003. Trends in temperature extremes during 1951–1999 in China. *Geophysical Research Letters* 30:1913.

Zhao Z. C. and Y. Xu. 2002. Detection and scenarios of temperature change in East Asia. *World Resource Review* 14:321–333.

Zhao, Z. C., A. Sumi and J. W. Kim. 2003. Characteristics of extreme climate change over East Asia and China for the several key periods. *World Resource Review* 15:289–304.

Zhao, Z. C., A. Sumi, C. Harada and T. Nozawa. 2004. Detection and projections of floods/droughts and typhoons over East Asia and China for the 20th and 21st centuries. Pp. 135–140 in *Proceedings of National Committee on Population and Resource*. Beijing: China Meteorological Press (in Chinese).

Chapter 8

CLIMATE IMPACTS OF ATMOSPHERIC SULFATE AND BLACK CARBON AEROSOLS

YUN QIAN, QINGYUAN SONG, SURABI MENON, SHAOCAI YU, SHAW LIU,
GUANGYU SHI, RUBY LEUNG and YUNFENG LUO

1. Introduction

Although the global average surface temperature has increased by about 0.6°C during the last century (IPCC 2001), some regions such as East Asia, Eastern North America, and Western Europe have cooled rather than warmed during the past decades (Jones 1988; Qian and Giorgi 2000). Coherent changes at the regional scale may reflect responses to different climate forcings that need to be understood in order to predict the future net climate response at the global and regional scales under different emission scenarios.

Atmospheric aerosols play an important role in global climate change (IPCC 2001). They perturb the earth's radiative budget directly by scattering and absorbing solar and long wave radiation, and indirectly by changing cloud reflectivity, lifetime, and precipitation efficiency via their role as cloud condensation nuclei. Because aerosols have much shorter lifetime (days to weeks) compared to most greenhouse gases, they tend to concentrate near their emission sources and distribute very unevenly both in time and space. This non-uniform distribution of aerosols, in conjunction with the greenhouse effect, may lead to differential net heating in some areas and net cooling in others (Penner et al. 1994).

Sulfate aerosols come mainly from the oxidation of sulfur dioxide (SO_2) emitted from fossil fuel burning. Black carbon aerosols are directly emitted during incomplete combustion of biomass, coal, and diesel derived sources. Due to the different optical properties, sulfate and black carbon affect climate in different ways.

Because of the massive emissions of sulfur and black carbon that accompany the rapid economic expansion in East Asia, understanding the effects of aerosols on climate is particularly important scientifically and politically in order to develop adaptation and mitigation strategies.

2. Aerosol-Induced Radiative Forcing and Climate Change Signals in Long-Term Observational Records

The competing radiative effects on climate include the greenhouse effect (due to infrared absorbers) and the warming or cooling effects of aerosols (Schwartz 1996). Increasing infrared absorbers such as greenhouse gases and absorbing aerosols (black carbon and dust) have led to an increase in daytime maximum and nighttime minimum temperatures, resulting in increased daily average temperature. Hansen et al. (1997) and Ackerman et al. (2000) reported that airborne absorbing aerosols have the capacity to raise temperature more effectively than CO_2 by transferring the absorbed energy to the surface and reducing large scale cloud cover (cloud-burning effect). IPCC (2001) showed that increasing global surface temperatures would likely to induce a more active hydrological cycle and increase the water holding capacity in the atmosphere. In contrast, an increase in sulfate aerosols and low cloud amount would result in a decrease of daytime maximum temperature with a cooling effect (Yu et al. 2001).

The rates of change of the annual mean daily average, maximum, minimum temperatures and diurnal temperature range in eastern China are 0.080, −0.020, 0.18, and −0.20°C $10\,\text{yrs}^{-1}$, respectively (Table 8.1), on the basis of climate data at 72 stations from 1951 to 1994. Based on the annual mean daily temperature, there was a mild warming trend in eastern China. Seasonally, the winter mean daily temperature has increased by 0.25°C $10\,\text{yrs}^{-1}$ while the summer mean daily temperature has decreased by 0.05°C $10\,\text{yrs}^{-1}$. The diurnal temperature range for all seasons showed

Table 8.1. The mean temperature change rates (°C $10\,\text{yrs}^{-1}$) at 72 stations over eastern China (Yu et al. 2001).

T	Annual	Winter Dec.–Feb.	Spring Mar.–May	Summer June–Aug.	Fall Sept.–Nov.
Daily	0.08 ± 0.1	0.25 ± 0.1	0.07 ± 0.2	−0.05 ± 0.1	0.03 ± 0.1
Max.	−0.02 ± 0.1	0.12 ± 0.1	−0.03 ± 0.2	−0.13 ± 0.2	−0.04 ± 0.1
Min.	0.18 ± 0.2	0.38 ± 0.2	0.17 ± 0.2	0.04 ± 0.2	0.11 ± 0.2
DTR	−0.20 ± 0.2	−0.26 ± 0.2	−0.21 ± 0.2	−0.17 ± 0.2	−0.15 ± 0.2

decreasing trends at most stations in eastern China, which is consistent with the results of Zhai et al. (1999). Liu et al. (2002) also found a gradual decrease of diurnal temperature range of about 1.2°C from 1930 to 1990 over Taiwan Island because of a persistent increase of nighttime minimum temperature and a nearly constant daytime maximum temperature.

Qian et al. (1996) and Qian and Giorgi (2000) investigated the possible connection between observed temperature trends and anthropogenic aerosol effects in China. Figure 8.1 shows the spatial pattern of observed aerosol

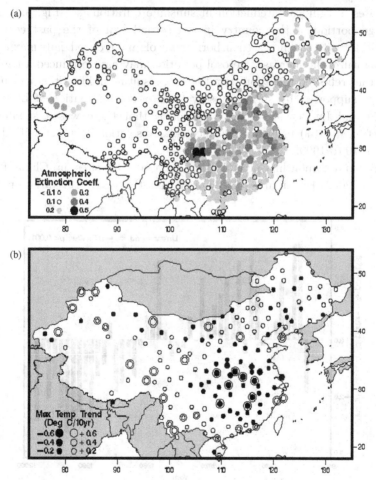

Fig. 8.1. (a) Aerosol extinction coefficient (km^{-1}) averaged for the period 1981–1998 across China. (b) Trends in summer mean daily maximum temperature for 1954–1998 (from Kaiser and Qian 2002).

extinction coefficient and the average daily maximum temperature trend from 1954 to 1998. It is obvious that both heavy aerosol loading and large cooling are mainly located over eastern China and Sichuan Basin. The average temperature and daily maximum temperature over southwest China showed statistically significant cooling trends of −0.1 and −0.14°C 10 yrs^{-1}, respectively, indicating a consistency with the expected cooling effects of reflective aerosols associated with the observed upward trend in the aerosol extinction coefficient.

Analysis of newly available long-term data (Kaiser and Qian 2002) indicates a significant reduction in sunshine duration (see Fig. 8.2) over a large portion of the country in the second half of the past century. Consistent with the increased airborne aerosols and reduced daily maximum temperature, the increased aerosol pollution may have produced a fog-like haze that reflected radiation from the sun, resulting in lower temperatures. This is supported by surface observations of solar radiation of Li et al. (1995) and Luo et al. (2000) who found that there was a significant reduction (~10%) in average surface solar insolation over most of China from 1970 to 1990.

Based on monthly precipitation data at 160 sites in China from 1951 to 2000, Hu et al. (2003) found that the long-term variation was

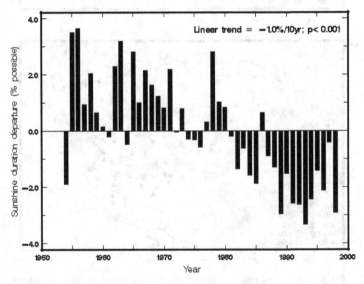

Fig. 8.2. Change in the sunshine duration in China relative to the average over 45 years (from Kaiser and Qian 2002).

highly seasonally and regionally dependent. Summer precipitation had a decreasing trend in the north and an increasing trend in the central part (Gemmer et al. 2004). Gong et al. (2004) found that compared to the 1950s, rainy days have been reduced by about 8 days in the 1990s. Liu et al. (2002) found that there was an increase in precipitation intensity in Taiwan as a result of a decreasing rainy hours and a relatively constant amount of precipitation. They suggested that the decrease of rainy hours was probably related to the increase of cloud droplets due to an increase in anthropogenic cloud condensation nuclei. Xu (2001) attributed the abnormal summer climate pattern of "north drought with south flooding" to the possible cooling effect of sulfate aerosol in the central eastern China. However, a global climate simulation (Menon et al. 2002) revealed that their model can reproduce the pattern of observed precipitation anomaly in China only when black carbon is included. Anthropogenic aerosols may play an important role in the hydrological cycle, but consistent signals are difficult to detect in the observed precipitation records.

Generally, it is expected that increased anthropogenic aerosol particles will increase the number of cloud droplets, decrease their size, and increase the amount and lifetime of clouds. However, Ackerman et al. (2000) proposed a different mechanism by which aerosols can reduce cloud cover. The detection of long-term change in the cloud cover is difficult because of large uncertainty in the observations. If diurnal temperature range can be used as a proxy measure of cloud cover change, Sun et al. (2000) showed that cloud cover has increased over China, as suggested by a decrease in diurnal temperature range. It is clear that more modeling and observational studies on the effect of aerosols on cloud cover are needed.

3. Sources of Sulfate and Black Carbon Aerosols

Sulfate and black carbon aerosols have both natural and anthropogenic sources. In this chapter, we focus on the effect of aerosols emitted by human activities.

3.1. *Generation of sulfate and black carbon in the atmosphere*

Asia is characterized by a large increase in sulfur emissions in the recent decades due to its rapid industrialization and population growth.

Most anthropogenic sulfur is discharged into the atmosphere in the form of sulfur dioxide (SO_2). About 50–80% of SO_2 is oxidized to sulfate, while the rest is dry-deposited at the Earth's surface (Roelofs et al. 1998). There are two pathways for the oxidation of SO_2 to sulfate in the atmosphere: photochemical reactions in clear air and heterogeneous oxidation reactions in clouds and rain drops. Qian et al. (2001) found that aqueous phase conversion of SO_2 to sulfate and wet removal are the primary factors that regulate the amounts of sulfate in the atmosphere.

On the contrary, the number and size characteristics of black carbon aerosol are more dependent on features related to its emission sources, such as fuel types and combustion technologies. It is critical for climate and long-range pollution transport studies that emission of sub-micron black carbon particles to be separated from the total emission since black carbon particles or fly ash emitted from coal burning, with diameter as large as several hundred microns, deposit quickly.

In some earlier studies, fossil fuel black carbon was simply assumed to be proportional to the mass of sulfate. This practice can generate bias in estimating black carbon emission since sulfate and black carbon have different production mechanisms and atmospheric characteristics. The approach of using emission factors based on fuel types and combustion and control technologies to calculate black carbon emission from fossil fuel usage is more reasonable (Bond et al. 2004).

3.2. Sulfur dioxide and black carbon emission

The annual emission of sulfur from East Asia in 1997 was 27.8 Tg. Emission from China alone was $\sim 25\,\text{Tg yr}^{-1}$, followed by the Republic of Korea (~ 1.28 Tg) and Japan (0.75 Tg) (Streets et al. 2000). While the burning of coal and oil is the major emission source, the industrial and power sectors contribute the most SO_2 in China (about 53% and 27%, respectively). Transportation contributes less than 2% of the total SO_2 emissions (Streets and Waldhoff 2000). Regions with the highest SO_2 emissions in China are mostly centralized in the eastern and southwestern parts of China (Akimoto and Hirohito 1994; Streets and Waldhoff 2000). Indeed, these two areas have particularly strong forcing effects due to sulfate and black carbon (Qian and Giorgi 2000).

Streets et al. (2003) developed an inventory of air pollutants in 2000 and found that black carbon emissions in China from fossil fuel and biofuel burning were 1.05 and $1.17\,\text{Tg yr}^{-1}$, respectively. Japan, the second largest emitter of black carbon in East Asia, emitted only about $0.053\,\text{Tg yr}^{-1}$. By sector, more than 80% of black carbon emission in China is from the residential sector; industry, field combustion, and transport are 7.2%, 5.6%, and 3.2%, respectively (Streets et al. 2001). The sources of black carbon are very different from sulfate by both fuel type and sector. Sulfate is largely emitted in the populated and industrial areas like the eastern parts of China, while black carbon emissions are mostly concentrated in agricultural regions such as Sichuan, Hebei, and Hubei provinces.

Streets et al. (2003) investigated the trends in sulfate and black carbon emission in China between 1995 and 2000 and found that SO_2 decreased from 25.2 to 20.4 Tg yr^{-1} and black carbon decreased from 1.34 to 1.05 Tg yr^{-1}. The reduction, especially for black carbon, is attributed to the economic downturn in the late 1990s and improved energy efficiency. Streets et al. (2001) predict that black carbon emission from China in the next 20 years will keep decreasing at an average rate of 4.7 Gg yr^{-1} or 0.36% yr^{-1}, while total energy consumption of China will increase by about 50% from 1995 to 2020 (Table 8.2).

Table 8.2. Summary of energy and emission estimates by sector and fuel type in China (from Streets et al. 2001).

Sector	Fuel	Energy use (PJ)		BC emissions (Gg)	
		1995	2020	1995	2020
Residential	Coal	3,872	4,848	605.4	534.8
	Oil	432	2,088	1.0	5.5
	Biofuel	7,939	6,016	512.0	386.8
	Subtotal	12,243	12,952	1,118.4	927.1
Industry	Coal	13,171	18,257	82.5	80.6
	Oil	2,040	2,513	11.1	14.5
	Biofuel	600	482	3.6	1.4
	Subtotal	15,811	21,252	97.2	96.5
Power generation	Coal	10,080	18,054	1.5	0.1
	Oil	731	607	6.1	4.8
	Biofuel	89	226	0.7	0.5
	Subtotal	10,900	18,887	8.3	5.4
Transport: road	Gasoline	1,208	4,047	2.3	7.6
	Diesel	508	2,798	13.3	73.3
Transport: other vehicles	Gasoline	100	139	0.2	0.3
	Diesel	764	1,644	20.0	43.1
	Coal	234	277	3.2	3.7
Transport: ships	Diesel	138	328	3.6	8.6
	Heavy fuel oil	87	308	0.8	2.7
	subtotal	3,039	9,541	43.4	139.3
Field combustion	Crop reside	N/A	N/A	74.7	56.1
Total		41,993	62,632	1,342.0	1,224.4

3.3. *Spatial distribution and lifetime*

In simulations using a global tropopheric chemistry model, Barth and Church (1999) found that sulfate from southeast China travels eastward and encircles the Earth during summer and autumn. They also qualitatively analyzed the black carbon aerosol from southeast China and determined that it has a lifetime similar to sulfates when the emitted black carbon is assumed hydrophilic. When the emitted black carbon is hydrophobic the lifetime is 3.5 days longer than sulfate. The difference between the sulfur and carbon aerosols is attributed to the aqueous production pathway of sulfate and its availability to rain out. Long-range transport of aerosols can thus affect the atmospheric burdens in distant regions.

4. Radiative Forcing

Radiative forcing is widely used to determine the relative importance of perturbations of atmospheric species due to anthropogenic activity (Ramaswamy 2001). The direct effect of aerosols refers to scattering or absorption of radiation by aerosol particles without the modification of cloud properties. Indirect forcing is broadly defined as the overall process by which aerosols perturb the Earth atmosphere radiation balance by modulation of cloud properties and cloud-precipitation processes.

Based on a series of regional climate simulations, Giorgi et al. (2002) assessed the direct radiative forcing and surface climatic effects of anthropogenic sulfate and fossil fuel soot over eastern Asia. The simulations showed that anthropogenic sulfate induces a negative radiative forcing that varies spatially from -1 to $-8\,\mathrm{W\,m^{-2}}$ in the winter and -1 to $-15\,\mathrm{W\,m^{-2}}$ in the summer.

The highest values occur over the Sichuan Basin of southwest China and some areas of eastern and northeastern China. Fossil-fuel soot exerts a positive atmospheric radiative forcing of 0.5–$2\,\mathrm{W\,m^{-2}}$. Zhou et al. (1998) and Luo et al. (2000) estimated the radiative forcing of aerosol in China between -5 and $-13\,\mathrm{W\,m^{-2}}$ with a maximum forcing over the Sichuan Basin and a minimum forcing over the Tibetan Plateau. Their results were consistent with the distribution of aerosol optical depth. Over mainland China, the direct radiative forcing of sulfate aerosols was 0 to $-7\,\mathrm{W\,m^{-2}}$ (Wang et al. 2003), with larger negative forcings over central and eastern China. Black carbon induced a positive "radiative forcing" at the top of atmosphere and negative net radiation flux at the surface with a distribution pattern that is consistent with the distribution of aerosol optical depth (Wu et al. 2004).

Using aerosol optical depths calculated from a coupled regional climate/air quality model and inferred solar radiation measurements over a 12-year period at several meteorological stations in China, Chameides et al. (1999) assessed the effect of atmospheric aerosols and regional haze from air pollution on the yields of rice and winter wheat grown in China. They found that the direct effect of regional haze results in a 5–30% reduction in solar irradiance reaching some of China's most productive agricultural regions and reduces optimal yields of around 70% of the crops grown in China by at least 5–30%.

Giorgi et al. (2003) assessed the relative importance of direct and indirect effects of anthropogenic sulfate on the climate of East Asia. Under present day sulfur emissions, the direct aerosol effects prevail during the cold season, while the indirect effects dominate in the warm season (when cloudiness is maximum over the region). In general, the surface impact of the indirect aerosol effects is in the same direction, and thus tends to reinforce the impact of the direct effects. The radiative forcing of both direct and indirect aerosol effects is negative, which results in a cooling of the surface (Giorgi et al. 2003). The indirect effects show little sensitivity to doubling of the aerosol amounts. This has the important implication that while the direct effects can be expected to become increasingly important as the pollution emissions increase with future economic growth, the indirect effects may show an asymptotic behavior.

5. Simulation of Effects on the Regional Climate and Hydrological Cycle

In general, simulations by interactively coupled regional climate and atmospheric chemistry/aerosol models showed the following (Giorgi et al. 2002, 2003; Qian et al. 2003): Direct and indirect aerosol effects result in a cooling of the surface in the range of $-0.1°C$ to $-1°C$. Overall, the indirect effects appear necessary to explain the observed temperature record over some regions of China, at least in the warm season. The aerosol forcing and surface cooling tend to inhibit precipitation over the region, although this effect is relatively small in the simulations. The aerosol induced cooling is mainly due to a decrease in daytime maximum temperature. The maximum cooling occurred over the Sichuan Basin, and is statistically significant.

In addition, the feedbacks and interactions between aerosols and climate show a general positive feedback in the winter (dry season) due to a sulfate aerosol induced decrease in precipitation and removal, and the intensity of feedback increases with aerosol loading (Qian and Giorgi 1999). Menon et al. (2002) performed several global climate simulations to differentiate between the climatic effects of sulfate and black carbon aerosols using realistic aerosol distributions. Their results suggested that black carbon aerosols can have a significant influence on regional climate through changes in the hydrological cycle and large-scale circulation. Results of an extension of this work suggest that the climate effects of aerosols (especially absorbing aerosols) are stronger when they are located at higher levels in the atmosphere (Menon 2004).

Reduction in local planetary albedo due to absorbing aerosols and seasonal variations in cloud droplet sizes have also been linked with precipitation changes over East Asia (Kawamoto and Nakajima 2003; Kruger and Graβl 2004). While some observed trends may be related to large-scale dynamical changes, observational evidence, supported by modeling results, does indicate that black carbon aerosols play a role in these trends.

6. Conclusions

Atmospheric aerosols that originate from a large variety of sources are distributed across a wide spectrum of particle size. They have much shorter atmospheric lifetime than most greenhouse gases, and their concentrations and composition have large spatial and temporal variability. It is a great challenge to adequately characterize the nature and distribution of atmospheric aerosols and to include their effects in climate models for accurate climate simulations and predictions. To reduce the uncertainties of assessing the impacts of atmospheric sulfate and black carbon aerosols on the regional climate of East Asia, significant progress is needed in both measurement and modeling of aerosols and improving the links between measurements and modeling.

6.1. *Measurements*

There is a need for countries in East Asia to develop and support a network of systematic ground-based observations of aerosol properties in the atmosphere that include a variety of physical and chemical measurements ranging from local *in situ* to remote sensing of total column or vertical profile of aerosol properties. As recommended by Penner (2001), a common strategy for aerosol measurements needs to be developed for a selected set of regionally representative sites. A comprehensive suite of aerosol and gas measurements should be taken that are long-term in scope, and a less comprehensive set of observations would be taken to provide background information for intensive shorter-term process-oriented studies.

Although not covered in this chapter, organic carbon, a twin product of black carbon in the combustion processes, is a negative forcing agent similar to sulfate. Insufficient consideration of organic carbon's negative effect in black carbon climate forcing studies could lead to controversial conclusions (Jacobson 2001, 2002; Feichter et al. 2003; Chock et al. 2003). Measurements need to resolve the relationships between organic and black carbon and their thermal and optical properties as a function of source to allow quantitative evaluation of their effects. Also, the hygroscopicity of freshly emitted aerosols needs to be measured to characterize their impacts on clouds. Intercomparison of current measurements would be useful.

There are very few systematic vertical profile measurements of size-segregated or even total atmospheric aerosol physical, chemical, and optical properties. There is no climatological database and few simultaneous measurements of these parameters together with aerosol amounts that can be used to evaluate the performance of process models and climate models that include aerosols as active constituents.

To understand how aerosols influence cloud-precipitation processes, there is a need for more information on cloud microphysical properties (e.g., size distributed droplet number concentration and chemical composition, hydrometer type) and macrophysical properties (e.g., cloud thickness, cloud liquid water content, precipitation rate, total column cloud fraction, cloud albedo). Emphasis should be placed on reducing uncertainties related to scaling-up of the processes of aerosol–cloud interactions from individual clouds to the typical resolution of a climate model. An integrated strategy for reducing uncertainties should also include high quality measurement of aerosols from space.

6.2. *Modeling*

Internal or external mixing of black carbon may have a large impact on the optical properties of the particles (Haywood et al. 1997; Myhre et al. 1998; Chung and Seinfeld 2002; Liu et al. 2002). More realistic treatments of aerosol mixing state, aging, and size distribution are needed to accurately describe the direct and indirect effects and the removal rate for black carbon. A better characterization of optical properties of internally mixed aerosol particles and their wavelength dependence as a function of source strength, as well as their vertical distribution would narrow the uncertainty in quantifying the climate forcing of black carbon.

Much more research is needed to improve our understanding of aerosol–cloud interactions. Large uncertainty in estimating aerosol and greenhouse gas forcing in climate models arises from different treatments of clouds, which affect not only cloud radiative forcing, but also aerosol distribution through wet removal that depends, for example, on the distribution of precipitation intensity, ice content in clouds, vertical velocities in clouds, and stratiform versus convective precipitation. In global climate models, this problem is further confounded by the need to parameterize subgrid clouds and precipitation because of the coarse spatial resolution.

Almost all previous attempts to represent aerosol–cloud interactions and their indirect climate effects in climate models have been limited to the effects of aerosol on stratiform cloud. These studies need to be extended to convective clouds as they are quite common in East Asia. Modeling studies of the interactions between convective clouds and aerosols are rare because climate models do not explicitly resolve convective cloud systems. Song and Leighton (1998) developed a parameterization scheme for convective clouds' impact on sulfate production. This scheme, factoring in the dynamic features of convective storms, has been applied in a Canadian global climate model, and it significantly improved the sulfate mass prediction in the accumulation mode (Song 1998). More work is needed to develop parameterization schemes for representing aerosol impacts on convective clouds in climate models.

Many recent studies have shown that aerosols perturb not only the radiation balance of the climate system, but also affect the regional and global hydrological cycles that could have serious global and regional consequences (Ramanathan et al. 2001). This again highlights the importance of improving our understanding and modeling of cloud and precipitation processes. Improved linkage between cloud microphysics and aerosols in climate models are obviously needed to test different hypotheses of how aerosols affect precipitation through changes in microphysical processes such as coalescence, ice nucleation, and riming in mixed phase clouds.

Modeling studies using global and regional climate models have produced a wide range of estimates of aerosol direct and indirect effects.

Vigorous model intercomparison should be performed to determine the main sources of discrepancy and compare model discrepancy to the interannual and decadal variability of sulfate and black carbon aerosols.

6.3. Impact of socio-economic development

Besides climate effects, aerosols could have socio-economic impacts through changes in agricultural yields (from reduced surface radiation), health effects from air-borne aerosols, and reduced visibility due to atmospheric haze (Zhou et al., Chapter 21, this volume).

The impacts of sulfate and black carbon aerosols on East Asia's environment and climate are closely intertwined with this region's socio-economic development. The burning of fossil fuels is the major anthropogenic emission source of the two aerosols. Issues such as energy efficiency and energy independence will have significant impacts on emission. Advanced technologies of bio-fuel, coal gasification, desulfurized coal, advanced diesel engines, and hybrid or fuel cell vehicles will partially alleviate the energy pressure of East Asia and reduce sulfur and black carbon emission, especially in China (Jiang and Wen, Chapter 22, this volume).

Governments can play an important role in regulating and reducing pollutant emission. Japan and Korea have already promulgated stringent emission regulations. China has been making rapid progress in improving fuel economy and setting emission standards that could be more stringent than the current standards of North America. With all these factors working together, the emission of sulfate and black carbon from East Asia could be reduced significantly despite rapid economic development.

Literature Cited

Ackerman A. S., O. B. Toon, D. E. Stevens, A. J. Heymsfield, V. Ramanathan and E. J. Welton. 2000. Reduction of tropical cloudiness by soot. *Science* 288:1042–1047.

Akimoto, H. and N. Hirohito. 1994. Distribution of SO_2, NO_x and CO_2 emissions from fuel combustion and industrial activities in Asia with 1 × 1 resolution. *Atmospheric Environment* 28:213–225.

Barth, M. C. and A. C. Church. 1999. Regional and global distributions and lifetimes of sulfate aerosols from Mexico City and southeast China. *Journal of Geophysical Research-Atmosphere* 104:30231–30239.

Bond, T., D. G. Streets, K. F. Yarber, S. M. Nelson, J.-H. Woo and Z. Klimont. 2004. A technology-based global inventory of black and organic carbon emissions from combustion. *Journal of Geophysical Research-Atmosphere* 109:doi:10.1029/2003JD003697.

Chamedies, W. L., H. B. Yu, S. C. Liu and coauthors. 1999. A case study of the effects of atmospheric aerosols and regional haze on agriculture: An opportunity to enhance crop yields in China through emission controls? *Proceedings of the National Academy of Sciences* 96:13626–13633.

Chock, D., Q. Song, H. Hass, B. Schell and I. Ackermann. 2003. Comment on "Control of fossil-fuel particulate black carbon and organic matter, possibly the most effective method of slowing global warming" by M. Z. Jacobson. *Journal of Geophysical Research-Atmosphere* 108:4769, doi:10.1029/2003JD003629.

Chung, S. H. and J. H. Seinfeld. 2002. Global distribution and climate forcing of carbonaceous aerosols. *Journal of Geophysical Research-Atmosphere* 107:doi:10.1029/2001JD001397.

Feichter, J., R. Sausen, H. Graßl and M. Fiebig. 2003. Comment on "Control of fossil-fuel particulate black carbon and organic matter, possibly the most effective method of slowing global warming" by M. Z. Jacobson. *Journal of Geophysical Research-Atmosphere* 108:doi:10.1029/2002JD003223.

Gemmer, M., S. Becker and T. Jiang. 2004. Observed monthly precipitation trends in China 1951–2002. *Theory and Applied Climatology* 77:39–45.

Giorgi, F., X. Q. Bi and Y. Qian. 2002. Direct radiative forcing and regional climate effects of anthropogenic aerosols over East Asia: A regional coupled climate-chemistry/aerosol model study. *Journal of Geophysical Research-Atmosphere* 107:4439, 10.1029/2001JD001066.

Giorgi, F., X. Q. Bi and Y. Qian. 2003. Indirect vs. direct effects of anthropogenic sulfate on the climate of East Asia as simulated with a regional coupled climate-chemistry/aerosol model. *Climatic Change* 58:45–376.

Gong, D. Y., P. J. Shi and J. A. Wang. 2004. Daily precipitation changes in the semi-arid region over northern China. *Journal of Arid Environments* doi:10.1016/j.jaridenv.2004.02.006.

Hansen, J. E., M. Sato and R. Ruedy. 1997. Radiative forcing and climate response. *Journal of Geophysical Research-Atmosphere* 102:6831–6864.

Haywood, J. M., D. L. Roberts, A. Slingo, J. M. Edwards and K. P. Shine. 1997. General circulation model calculations of the direct radiative forcing by anthropogenic sulfate and fossil-fuel soot aerosol. *Journal of Climate* 10:1562–1577.

Hu, Z.-Z., S. Yang and R. Wu. 2003. Long-term climate variations in China and global warming signals. *Journal of Geophysical Research-Atmosphere* 108:4614, doi:10.1029/2003JD003651.

IPCC (Intergovernmental Panel on Climate Change) 2001. *Climate Change 2001. The Scientific Basis*. Contribution of Working Group I to the Third Assessment Report of the Intergovernmental Panel on Climate Change, edited by J. T. Houghton, Y. Ding, D. J. Griggs, M. Nogure, P. J. van der Linden and X. Dai, Cambridge, U.K.: Cambridge University Press.

Jacobson, M. Z. 2001. Strong radiative heating due to the mixing state of black carbon in atmospheric aerosols. *Nature* 409:695–697.

Jacobson, M. Z. 2002. Control of fossil-fuel particulate black carbon and organic matter, possibly the most effective method of slowing global warming. *Journal of Geophysical Research-Atmosphere* 107:4410, doi:10.1029/2001JD001376.

Jones, P. D. 1988. Hemispheric surface air temperature variations: Recent trends and an update to 1987. *Journal of Climate* 1:654–660.

Kaiser, D. P. and Y. Qian. 2002. Decreasing trends in sunshine duration over China for 1954–1998: Indication of increased haze pollution? *Geophysical Research Letters* 29:2042, doi:10.1029/2002GL016057.

Kawamoto, K. and T. Nakajima. 2003. Seasonal variation of cloud particle size as derived from AVHRR remote sensing. *Geophysical Research Letters* 30:1810, doi:10.1029/2003GL017437.

Kruger, O. and H. Graßl. 2004. Albedo reduction by absorbing aerosols over China. *Geophysical Research Letters* 31: L02108, doi10.1029/2003GL019111.

Liu, S. C., C. Wang, C. Shiu, H. Chang, C. Hsiao and S. Liaw. 2002. Reduction in sunshine duration over Taiwan: Causes and implications. *Terrestrial Atmospheric and Oceanic Sciences* 13:523–545.

Li, X. W., X. J. Zhou and W. L. Li. 1995. The cooling of Sichuan province in recent 40 years and its probable mechanism. *Acta Meteorologica Sinica* 9:57–68.

Luo, Y. F., D. R. Lu and Q. He. 2000. Characteristics of atmospheric aerosol optical depth variation over China in recent 30 years. *Chinese Science Bulletin* 45:1328–1334.

Menon, S. 2004. Current uncertainties in assessing aerosol effects on climate. *Annual Review of Environmental Resource* 29.

Menon, S., J. Hansen, L. Nazarenko and Y. F. Luo. 2002. Climate effects of black carbon aerosols in China and India. *Science* 297:2250–2253.

Myhre, G., F. Stordal, K. Restad and I. S. A. Isaksen. 1998. Estimation of the direct radiative forcing due to sulfate and soot aerosols. *Tellus* 50B:463–477.

Penner, J. E., R. Charlson, J. Hales, N. Laulainen, R. Leifer, T. Novakov, J. Ogren, L. F. Radke, S. Schwartz and L. Travis. 1994. Quantifying and minimizing uncertainty of climate forcing by anthropogenic aerosols. *Bulletin of American Meteorological Society* 75:375–400.

Penner, J. E. 2001. Aerosols, their direct and indirect effects. In *Climate Change 2001, The Scientific Basis*, Contribution of Working Group I to the Third Assessment Report of the Intergovernmental Panel on Climate Change, edited by J. T. Houghton, Y. Ding, D. J. Griggs, M. Noguer, P. J. van der Linden and X. Dai. Cambridge, U.K.: Cambridge University press.

Qian, Y. and F. Giorgi. 1999. Interactive coupling of regional climate and sulfate aerosol models over eastern Asia. *Journal of Geophysical Research-Atmosphere* 104:6477–6499.

Qian, Y. and F. Giorgi. 2000. Regional climatic effects of anthropogenic aerosols? The case of Southwestern China. *Geophysics Research Letters* 27:3521–3524.

Qian, Y., C. B. Fu, R. M. Hu and Z. F. Wang. 1996. Effects of industrial SO_2 emission on temperature variation in China and East Asia. *Climatic and Environmental Research* 2:143–149.

Qian, Y., L. R. Leung, S. J. Ghan and F. Giorgi. 2003. Effects of increasing aerosol on regional climate change in China: Observation and modeling. *Tellus* 55B:914–934.

Qian, Y., F. Giorgi, Y. Huang, W. L. Chameides and C. Luo. 2001. Regional simulation of anthropogenic sulfur over East Asia and its sensitivity to model parameters. *Tellus* 53B:171–191.

Ramanathan, V., P. J. Crutzen, J. T. Kiehl and D. Rosenfeld. 2001. Aerosols, climate and the hydrological cycle. *Science* 292:2119–2124.

Ramaswamy, V. 2001. Radiative forcing of climate change. In *Climate Change 2001, The Scientific Basis*, Contribution of Working Group I to the Third Assessment Report of the Intergovernmental Panel on Climate. Change, edited by J. T. Houghton, Y. Ding, D. J. Griggs, M. Noguer, P. J. van der Linden and X. Dai. Cambridge, U.K.: Cambridge University Press.

Roelofs. G. J., J. Lelieveld and L. Ganzeveld. 1998. Simulation of global sulfate distribution and the influence on effective cloud drop radii with a coupled photochemistry-sulfur cycle model. *Tellus* 50B:224–242.

Schwartz, S. E. 1996. The whitehouse effect-shortwave radiative forcing of climate by anthropogenic aerosols: An overview. *Journal of Aerosol Sciences* 27:359–382.

Song, Q. 1998. A parameterization of in-cloud sulphate production. Ph.D Thesis, McGill University.

Song, Q. and H. G. Leighton. 1998. A parameterization of heterogeneous sulphate production for use in GCMs and regional climate models. *Journal Atmospheric Research-Atmosphere* 108, 8809, doi:10.1029/2002JD003093.

Streets, D. G., T. Bond, G. Carmichael, S. Fernandes, Q. Fu, D. He, Z. Klimont, S. Nelson, N. Tsai, M. Wang, J.-H. Wang and K. Yarber. 2003. An inventory of gaseous and primary aerosol emissions in Asia in the year 2000. *Journal of Geophysical Research-Atmosphere* 108:8809, doi:10.1029/2002JD003093.

Streets, D. G., and S. T. Waldhoff. 2000. Present and future emissions of air pollutants in China: SO_2, NO_x and CO. *Atmospheric Environment* 34:363–374.

Streets, D. G., S. Gupta and S. T. Waldhoff. 2001. Black carbon emissions in China. *Atmospheric Environment* 35:4281–4296.

Streets, D. G., N. Tsai, H. Akimoto and K. Oka. 2000. Sulfur dioxide emissions in Asia in the period 1985–1997. *Atmospheric Environment* 34:4413–4424.

Sun, B., P. Y. Groisman and R. S. Bradley. 2000. Temporal changes in the observed relationship between cloud cover and surface air temperature. *Journal of Climate* 13:4341–4357.

Wang, T. J., J. Z. Min, Y. F. Xu and K. S. Lam. 2003. Seasonal variations of anthropogenic sulfate aerosol and direct radiative forcing over China. *Meteorology and Atmospheric Physics* 84:185–198.

Wu, J., W. M. Jiang, C. B. Fu, B. K. Su, H. N. Liu and J. P., Tang. 2004. Simulation studies of the radiative effect of black carbon aerosol and regional climate responses over China. *Advances in Atmospheric Sciences* 21:637–649.

Xu, Q. 2001. Abrupt change of the mid-summer climate in central east China by the influence of atmospheric pollution. *Atmospheric Environment* 35:5029–5040.

Yu, S. C., V. K. Saxena and Z. C. Zhao. 2001. A comparison of signals of regional aerosol-induced forcing in eastern China and the southeastern United States. *Geophysical Research Letters* 28:713–716.

Zhai, P. M., F. Ren and Q. Zhang. 1999. Detection of trends in China's precipitation extremes. *Acta Meteorologica Sinia* 57:208–216.

Zhou, X. J., W. L. Li and Y. F. Luo. 1998. Numerical simulation of the aerosol radiative forcing and regional climate effect over China. *Scientia Atmospherica Sinica* 22:418–427.

Chapter 9

LONG-RANGE TRANSPORT AND DEPOSITION OF DUST AEROSOLS OVER THE OCEAN, AND THEIR IMPACTS ON CLIMATE

ZIFA WANG, GUANGYU SHI and YOUNG-JOON KIM

1. Dust Aerosols over East Asia

In recent years, increasing attention has been paid to dust storms in northern China. Movement of air can lift soil and sand particles and transport them to long distances. In the long term, this leads to redistribution of surface soils, a modification of the atmosphere, and brings great global change. As an example, the Loess Plateau in China was the result of ancient dust storms. Ancient Chinese made lots of records on this important natural phenomenon in historic books, which were used by Zhang (1982) to rebuild the three-century time-series of dust deposition over China from the Tang Dynasty to the present day. Statistical analysis and climatic explanations of dust storms were investigated in 1970s, followed by a study of the chemical and physical characteristics of dust storms and their long-range transport (Zhang et al. 2003). Automated Mie scattering Lidar was used to determine the intensities and vertical profiles of Asian dust with Lidar images of the normalized aerosol backscatter extinction coefficient, and depolarization ratio obtained at Beijing (Sugimoto et al. 2003). Satellite measurements using the Total Ozone Mapping Spectrophotometer, Sea viewing Wide Field of view Sensor, and others are also very useful to obtain real-time and high-resolution data for dust production and transport.

The Asian Pacific Regional Aerosol Characterization Experiment was the first systematic international project focusing on comprehensive surface, airborne, shipboard, and satellite measurements of Asian aerosol chemical composition, size, optical properties, and radiative impacts

(Huebert et al. 2003). The Aeolian Dust and its Impacts on Climate program initiated by the Chinese Academy of Sciences and the Ministry of Science of Japan in 2001 examined the processes involved in dust production and transportation and its impacts on regional and global change (Mikami et al. 2002).

2. Long-Range Transport of Asian Dust Particles

Mineral particles are emitted into the atmosphere as a result of high surface winds and can be transported aloft through boundary layer convection and vertical motion associated with frontal boundaries. There are several ways to define potential natural dust emission areas. For example, Shi et al. (2002) defined the potential source regions as the following five types of land: (1) deserts over northern China (Fig. 9.1); (2) downstream of dry rivers or lakes covering northern China, Mongolia and Kazakstan; (3) overused farmland in oases; (4) forests of poplar mainly located in southern Xinjiang and Ejinaqi areas; and (5) bare lands affected by human activities. Another way used by Uno et al. (2003) to define desert and semi-desert areas as source regions is by combining the United States Geological Survey vegetation database with the Total Ozone Mapping Spectrophotometer aerosol index of climatology to gain the potential dust emission database.

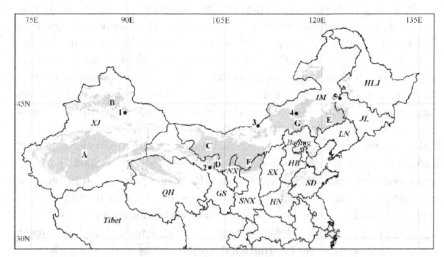

Fig. 9.1. Distribution of the main deserts in northern China. (A) Taklimakan Desert; (B) Gurbantunggut Desert; (C) Badain Jaran Desert; (D) Tengger Desert; (E) Horqin Desert; (F) Mu Us Desert; (G) Hunshandake Desert (Zou and Zhai 2004).

In addition, knowledge of the chemical composition and physical attributes (texture, surface roughness, soil water content, and size distribution) of the soil particles is required to estimate dust production (Zhang et al. 1997).

Dust storms occur only when the surface wind exceeds a certain threshold, which is related to surface roughness, particle size, and soil humidity. Several dust production models have been developed based on field observations, wind tunnel measurement and assumptions. For example, Wang et al. (2000) developed a dust deflation model for northern China, based on the long-term observations of hundreds of weather stations over northern China, which allowed the simulation of online dust emissions. Shao (2002) developed a model to predict dust emission based directly on the physical characteristics of the particles. The uncertainties of estimation of dust emission are quite large and the key mechanisms of dust production need to be determined. Human activity can also increase or decrease the motion of dust particles by disturbing the land surface. Tegen and Fung (1995) estimated that up to 50% of dust aerosols could be classified as anthropogenic as they came from disturbed soil surfaces. This should be considered in future deflation modules when a way is found to separate dust production as a result of human activities from that produced by natural processes.

The duration of dust in the atmosphere depends on the size of the suspended particles. Large particles are removed from the atmosphere quickly by gravity; whereas sub-micron size particles may be transported for several weeks. Many transport models have been applied to investigate the transport of Asian dusts on regional and global scales. Figure 9.2 shows some key features of the long-range transport of Asian dust. The dust is transported horizontally over long distances with a multi-layered profile, one layer near the ground, and the other in the middle troposphere. The high concentrations of dust transported over the coast are usually between 500 and 1500 m high (Lin 2001).

Physical and chemical characteristics of dust particle may change during transport by absorbing pollutants or sea salt in their passage over polluted industrial areas or over the sea (Gong et al. 2003). Evidence of sulfate and nitrate coatings on dust has been observed in the Beijing and Qingdao regions and these have been found to change the hygroscopic nature of the dust (Orsini et al. 2003). When dust encounters fresh pollutants, these heterogeneous reactions can lead to a series of complex responses of the photochemical system, which also affect the radiative influence of dust. In addition, these reactions can alter the chemical-size

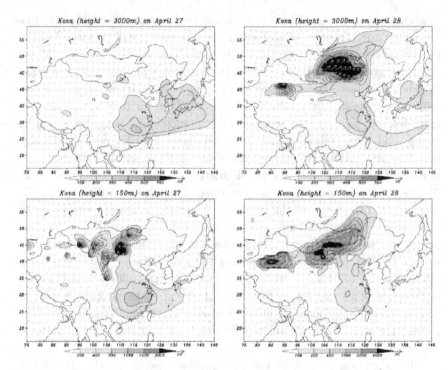

Fig. 9.2. Distribution of simulated Asian dust (Kosa) concentration (μg m^{-3}) at heights of 150 and 3000 m in East Asia on 27 April (left) and 28 April (right), 2000. (Revised from Lin 2001.)

distribution of the aerosol. Under heavy dust loadings, these reactions can lead to >20% of the sulfate and >70% of the nitrate being associated with the coarse fraction (Tang et al. 2004).

3. Transport of Dust Particles to the Pacific Ocean

In the 1960s and 1970s, deep-sea cores revealed that aeolian dust was a major source of marine deposition. Observations taken during the Sea-Air Exchange over the Pacific Ocean program (Duce et al. 1980) and the Atmosphere Ocean Chemistry Experiment over the Atlantic ocean (Riley et al. 1989) indicated that for several elements, such as Pb, Al, V, Mn, Zn, certain hydrocarbons, and various synthesized organic compounds, long-range transport and fall out from the atmosphere was as important for deposition as discharge from rivers. Merrill et al. (1985), by analyzing mineral aerosols, showed that the Asian dust carried by the Westerly could reach islands at lower latitudes.

The Asian plateau is the only area where frequent dust storms occur in the middle latitude (35°–45°N) of the northern hemisphere. As a result of these dust storms, mineral particles and particles mixed with pollutants are transported eastward over the North Pacific, especially in spring (Jaffe et al. 1999). Their interaction is expected to lead to increased solubility of iron and other nutrients, which will impact on oceanic productivity when deposited to the ocean surface. Bishop et al. (2002) reported a near doubling of biomass in the ocean mixed layer over a two-week period after the passage of a dust storm, and attributed this to a biotic response to iron in the dust.

More than two years of observations over the northern Yellow Sea showed that the deposition of dust particles in the China Seas was greater than that to the central North Pacific Ocean (Zhang et al. 1993). Most of the dust aerosol derived from continental Northeastern Asia was deposited in the Yellow and East China Seas. Observations at dust sampling sites on Pacific islands have shown that 480×10^{12} g of dust is transported from Asia to the North Pacific Ocean each year. This is greater than the transport of dust from the Sahara desert to the North Atlantic Ocean (220×10^{12} g per year; Duce et al. 1991).

Uematsu et al. (2003) analyzed the spatial and temporal variation of the mineral aerosol concentration and its total deposition flux over the western North Pacific region with a regional dust transport model. They found that dry deposition accounted for more than 60% of the total deposition of Asian mineral dust from March 1994 to February 1995. The annual deposition fluxes decreased rapidly from $21 \mathrm{\,g\,m^{-2}\,yr^{-1}}$ in the coastal area to $0.8 \mathrm{\,g\,m^{-2}\,yr^{-1}}$ over the open ocean. Recent studies show that 2.4 Tg of dust was deposited on the ocean surface during 5–15 April and 8.7 Tg was deposited during all of March and April, 2001. With a measured dust composition of 4% Fe at Zhenbeitai, 0.1 Tg Fe was deposited in the dust during the dust storm and 0.35 Tg Fe was deposited during the spring of 2001 (Seinfeld et al. 2004).

Dust particles transported over long distances contain not only crustal elements such as iron, aluminum, silicon, and calcium, but also various anthropogenic pollutants and organic compounds. Analysis of dust for a single particle base determined the molecular form of nitrate and sulfate present (Masahiko and Tanaka 1994). Nitrate was heterogeneously formed on the mineral dust particles and was present in the air parcels as they were transported over Japan. Both crustal and anthropogenic elements displayed high concentrations in the peak dust season (spring) and low concentrations in the summer (Yasushi et al. 1999). Size-segregated aerosols and composition measured at Mauna Loa Observatory from 1993 through 1996 indicated that long-range transport episodes were characterized by anthropogenic aerosols mixed with Asian dust and Asian pollution with relatively small amounts of soil (Kevin et al. 1999).

Iron in surface water is the limiting factor for primary productivity in certain ocean regions. Recent studies show that nitrogen fixation in oceans is closely related to the soluble iron (Chus et al. 2003), and iron and aluminum concentrations were found to increase synchronously along the southern portion of transects between California and Hawaii in spring 2001 (Johnson et al. 2003). Experiments involving the artificial addition of iron in many areas of the ocean have demonstrated that iron in the atmosphere has a limiting and crucial effect on the primary productivity in certain ocean regions (Hall and Safi 2001). *In situ* optical measurements provide evidence that the presence of the submicron Sahara dust result in abnormal color of waters of the Mediterranean Sea (Claustre et al. 2002).

About half the dust derived from Asian deserts is transported and deposited to Chinese seas and the remote North Pacific Ocean. During the long-range transport of the dust, the percentage of Fe(II) in it increases gradually, and then Fe(II) may lead to a significant increase of the primary productivity in certain ocean regions, activate marine organisms, and affect the euphotic layer and photosynthesis (Zhuang 2003). On the other hand, pollution products and heavy metals adsorbed by dust aerosols during long-range transport may damage marine organisms and ecosystems. The impacts of dust transport and deposition are far from straightforward.

4. Impacts of Dust Particles on Climate and Environment

About 1.0–5.0 Gt dust particles are introduced into the atmosphere each year (Duce 1995), and the dust content is highly variable in both time and space; however, the dust definitely affects climate. Dust particles change optical depth, alter direct climatic forcing by scattering and absorbing solar short-wave radiation and earth's long-wave radiation (Takemura et al. 2000). Uncertainties associated with these changes are considerable and may vary between $-0.6\,\mathrm{W}$ and $0.4\,\mathrm{W\,m^{-2}}$ (Penner et al. 2001).

Because dust reacts with sulfur dioxide and nitric acid and becomes more hygroscopic, it may play a role in regional cloud formation. Cloud nucleation by dust particles produces indirect climatic effects through the alteration of cloud reflection and radiation by altering cloud microphysics

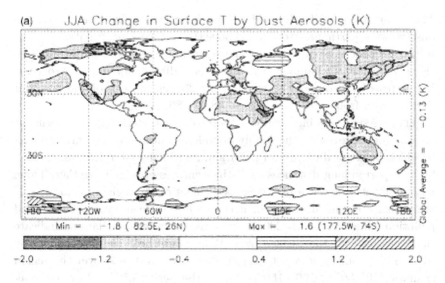

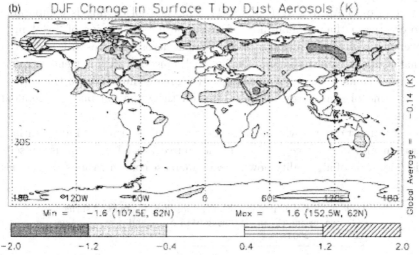

Fig. 9.3. Change in surface temperature (K) by soil dust aerosols for (a) summer and (b) winter (Miller and Tengen 1998).

and the production or suppression of rainfall. The effect of radiative forcing by soil aerosols upon climate is calculated with atmospheric general circulation models (Miller and Tegen 1998) indicating that surface temperature is reduced on the order of 1 K, typically in regions where deep convection is absent (Fig. 9.3). It is also suggested that dust can make considerable perturbation to Indian and African monsoon rainfall.

The ability to quantify these effects has been limited by a lack of critical observations, particularly of layers above the surface. Uncertainties relating to the frequencies of dust storm occurrence in China over the last 50 years (Fang et al. 2001) hamper studying this radiation forcing due to dust. Estimates of the number of strong dust storms in each of the last four decades vary from 5, 8, 13, 14, 23 to 48, 68, 89, 47, 36.

The Asian Pacific regional aerosol characterization experiment measurements revealed the highly complex structure of the atmosphere, in which layers of dust, urban pollution, and biomass-burning smoke may be transported long distances as distinct entities or mixed together. These measurements allowed a first-time assessment of the regional climatic and atmospheric chemical effects of a continental-scale mixture of dust and pollution (Huebert et al. 2003). Radiative transfer calculations indicate that aerosols cool the surface by $14\,\mathrm{W\,m^{-2}}$ and heat the atmosphere by $11\,\mathrm{W\,m^{-2}}$, resulting in a net climate forcing of $-3\,\mathrm{W\,m^{-2}}$ over the region spanning 20°–50°N, 100°–150°E during the period of the dust outbreak from 5 to 15 April 2001 (Table 9.1). This aerosol cooling is comparable to or greater than human-induced greenhouse gas warming of 2–$3\,\mathrm{W\,m^{-2}}$ over vast regions downwind of anthropogenic aerosol sources. It is suggested that radiative flux reductions during such episodes are sufficient to cause regional climate change by exerting a far greater influence on the surface and atmospheric energy budgets than on the radiation budget (Seinfeld et al. 2004).

The most direct environmental effect of a dust storm is to induce surface erosion and destroy vegetation in the source area. Dust storms cause a deterioration of air quality downstream, increase concentrations of particles

Table 9.1. Calculated top of the atmosphere (TOA) and surface radiative forcing ($\mathrm{W\,m^{-2}}$) during 5–15 April 2001 over east Asia (20°–50°N, 2004).

	Surface	Atmosphere	TOA
Dust	−9.3	3.8	−5.5
Sulfate	−3.6	0.3	−3.3
Organic carbon	−3.9	1.7	−2.2
Black carbon	−4.1	4.5	0.4
Sea salt	−0.4	0.0	−0.4
Internal mixture	−2.2	3.5	1.3
Thermal IR	3.0	−2.3	0.7
Total forcing (clear sky)	−20.5	11.5	−9.0
Total forcing (with clouds)	−14.0	11.0	−3.0

that may induce diseases of the respiratory organs and cardiovascular system, endangering human health. In addition, deposited dust particles could change soil acidity and nutrient supply, and thus could influence crops and other plants. On the other hand, neutralization by alkaline dust particles may increase the pH value of rainwater by 0.8–2.5 in northern China, by 0.5–0.8 in South Korea, and by 0.2–0.5 in Japan. This may explain why acid rain seldom occurs in northern China even though it has very high sulfur emissions.

5. Conclusions

No other region on the Earth is as large and diverse source of aerosols (and trace gases) as the Asian continent (Seinfeld et al. 2004). In spring, when frontal activity in Asia is the most prevalent, industrial pollution, biomass burning, and mineral dust outflows produce an extraordinarily complex regional aerosol mix, composed of inorganic compounds (such as sulfates, nitrates, and sea-salts), organic carbon, black carbon, and mineral dust.

Although many studies on Asian dust have been investigated through surface network, satellite measurements, Lidar observations and transport models, and large international field experiments such as the Asian Pacific Regional Aerosol Characterization Experiment and the Aeolian Dust and its Impacts on Climate Experiment, the collected data on the chemical, physical, and optical properties of Asian dust and pollution aerosols is far from enough to classify their impacts on climate change.

Several important questions relevant to the chemical and climatic effects of Asian aerosol outflow remain to be answered. For example:

- The height of the dust cloud, the impacts on particulate transport and deposition to the ocean?
- To what extent are dust and pollution particles internally mixed during transport, and what impact does this have on their optical and hygroscopic properties?
- To what extent does dust serve as a reaction surface for pollutant gases, and what are their heterogeneous reaction rates?
- How good are the chemical transport models and will they be good enough when coupled with climate models to assess climate forcing?

Clearly, further work is needed to quantify the influence of dust on the photochemical and biogeochemical cycles in east Asia. Continuous, comprehensive, remote and *in situ* measurements of trace gases and aerosols, combined with state of the art models are needed to quantify and accurately predict the anthropogenic alteration of atmospheric composition and its impact on marine environment and climate.

Literature Cited

Bishop, J. K. B., R. E. Davis and J. T. Sherman. 2002. Robotic observations of dust storm enhancement of carbon biomass in the North Pacific. *Science* 298:817–820.

Chus S., S. Elliott and M. E. Maltrud. 2003. Global eddy permitting simulations of surface ocean nitrogen, iron, sulfur cycling. *Chemosphere* 50:223–235.

Claustre, H., A. Morel, S. B. Hooker, M. Babin, D. Antoine, K. Oubelkheir, A. Bricaud, K. Leblanc, B. Quéguiner and S. Maritorena. 2002. Is desert dust making oligotrophic waters greener? *Geophysical Research Letters* 29:107.

Duce, R. 1995. Distribution and fluxes of mineral aerosol. In *Aerosol Forcing of Climate*, edited by R. J. Charlson and J. Heintzenberg, Chichester: John Wiley Press.

Duce, R. A., C. K. Unni, B. J. Ray, J. M. Prospero and J. T. Merrill. 1980. Long-range atmospheric transport of soil dust from Asia to the tropical North Pacific: Temporal variability. *Science* 209:1522–1524.

Duce, R. A., P. S. Liss, J. T. Merrill, E. L. Atlas, P. Buat-Menard, B. B. Hicks, J. M. Miller, J. M. Prospero, R. Arimoto, T. M. Church, W. Ellis, J. N. Galloway, L. Hansen, T. D. Jickells, A. H. Knap, K. H. Reinhardt, B. Schneider, A. Soudine, J. J. Tokos, S. Tsunogai, R. Wollast and M. Zhou. 1991. The atmospheric input of trace species to the world ocean. *Global Biogeochemical Cycles* 5:193–259.

Fang, Z. Y., Y. G. Zhang, X. J. Zheng and Y. C. Cao. 2001. Methods and primary results of remote monitoring dust storm using meteorological satellites. *Quaternary Sciences* 21:48–55.

Gong S. L., X. Y. Zhang, T. L. Zhao, I. G. Mc Kendry, D. A. Jaffe and N. M. Lu. 2003. Characterization of soil dust aerosol in China and its transport and distribution during 2001 ACE-Asia: 2. Model simulation and validation. *Journal of Geophysical Research* 108:4262, doi:10.1029/2002JD002633.

Hall, J. A. and K. Safi. 2001. The impact of *in situ* Fe fertilization on the microbial food web in the Southern Ocean. *Oceanography* 48:2591–2613.

Huebert, B. J., T. Bates, P. B. Russell, G. Y. Shi, Y.-J. Kim, K. Kawamura, G. Carmichael and T. Nakajima. 2003. An overview of ACE-Asia: Strategies for quantifying the relationships between Asian aerosols and their climatic impacts. *Journal of Geophysical Research* 108(D23):8633, doi:10.1029/2003JD003550.

Jaffe, D., T. Anderson, D. Covert, R. Kotchenruther, B. Trost, J. Danielson, W. Simpson, T. Berntsen, S. Karlsdottir, D. Blake, J. Harris, G. Carmichael and I. Uno. 1999. Transport of Asian Air Pollution to North America. *Geophysical Research Letters* 26:711–714.

Johnson, K. S., V. A. Elrod, S. E. Fitzwater, J. N. Plant, F. P. Chavez, S. J. Tanner, R. M. Gordon, D. L. Westphal, K. D. Perry, J. F. Wu and D. M. Karl. 2003. Surface ocean-lower atmosphere interactions in the Northeast Pacific Ocean Gyre: Aerosols, iron and the ecosystem response. *Global Biogeochemical Cycles* 17:1063, 10.1029/2002GB002004.
Kevin, D. P. T. A. Cahill, R. C. Schnell and J. M. Harris. 1999. Long-range transport of anthropogenic aerosols to the National Oceanic and Atmospheric Administration baseline station at Mauna Loa Observatory, Hawaii. *Journal of Geophysical Research* 14(D15):18521–18533.
Lin, T. H. 2001. Long-range transport of yellow sand to Taiwan in Spring 2000: Observed evidence and simulation. *Atmospheric Environment* 35:5873–5882.
Masahiko, Y. and H. Tanaka. 1994. Aircraft observation of aerosols in the free marine troposphere over the North Pacific Ocean: Particle chemistry in relation to air mass origin. *Journal of Geophysical Research* 99(D3):5353–5377.
Merrill, J. M., R. Bleck and L. Avila. 1985. Modeling atmospheric transport to the Marshall islands. *Journal of Geophysical Research* 90:12927–12936.
Mikami, M., O. Abe and M. Du. 2002. The impact of Aeolian dust on climate: Sino-Japanese cooperative project ADEC. *Journal of Arid Land Studies* 11:211–222.
Miller, R. L. and I. Tegen. 1998. Climatic response to soil dust aerosols. *Journal of Climate* 11:3247–3267.
Orsini, D. A., Y. Ma, A. Sullivan, B. Sierau, K. Baumann and R. J. Weber. 2003. Refinements to the Particle-Into-Liquid Sampler (PILS) for ground and airborne measurements of water soluble aerosol composition. *Atmospheric Environment* 37:1243–1259.
Penner, J. E., M. Andreae, H. Annegarn, L. Barrie, J. Feichter, D. Hegg, A. Jayaraman, R. Leaitch, D. Murphy, J. Nganga and G. Pitari. 2001. Aerosols, their direct and indirect effects. Pp. 313 in *Climate Change 2001: The Scientific Basis: Contribution of Working Group I to the Third Assessment Report of the Intergovernmental Panel on Climate Change*, edited by J. T. Houghton, Y. Ding, D. J. Griggs, M. Noguer, P. J. van der Linden, X. Dai, K. Maskell and C. A. Johnson. Cambridge: Cambridge University Press.
Riley, J. P., R. Chester and R. A. Duce. 1989. SEAREX: The Sea/Air Exchange Program. Pp. 404 in *Chemical Oceanography*. New York: Academic Press.
Schneider, B., N. W. Tindale and R. A. Duce. 1990. Dry deposition of Asian mineral dust over the central North Pacific. *Journal of Geophysical Research* 95(D7):9873–9878.
Seinfeld, J. H., G. R. Carmichael, R. Arimoto, W. C. Conant, F. J. Brecthtel, T. S. Bates, T. A. Cahill, A. D. Clarke, S. J. Doherty, P. J. Flatau, B. J. Huebert, J. Kim, K. M. Markowicz, P. K. Quinn, L. M. Russell, P. B. Russell, A. Shimizu, Y. Shinozuka, C. H. Song, Y. Tang, I. Uno, A. M. Vogelmann, R. J. Weber, J.-H. Woo and X. Y. Zhang. 2004. ACE-Asia: Regional climatic and atmospheric chemical effects of Asian dust and pollution. *Bulletin of American Meteorological Society* 85:367–378, doi: 10.1175/BAMS-85-3-367.

Shao, Y. P. 2002. Northeast Asian dust storms: Real-time numerical prediction and validation. *Journal of Geophysical Research* 108(D22):4691, doi:10.1029/2003JD003667.

Shi, G. Y., X. Y. Zhang and L. M. Wang. 2002. *Scientific Outline of International Research Project on Dust Storm*. China: Chinese Academy of Sciences.

Sugimoto, N., I. Uno, M. Nishikawa, A. Shimizu, I. Matsui, X. Dong, Y. Chen and H. Quan. 2003. Record heavy Asian dust in Beijing in 2002: Observations and model analysis of recent events. *Geophysical Research Letters* 30:1640, doi:10.1029/2002GL016349.

Takemura, T., H. Okamoto, Y. Maruyama, A. Numaguti, A. Higurashi and T. Nakajima. 2000. Global three-dimensional simulation of aerosol optical thickness distribution of various origins. *Journal of Geophysical Research* 105:17853–17873.

Tegen, I. and I. Fung. 1995. Contribution to the atmospheric mineral aerosol load from land surface modification. *Journal of Geophysical Research* 100:18707–18726.

Tang, Y., G. R. Carmichael, G. Kurata, I. Uno, R. J. Weber, C.-H. Song, S. K. Guttikunda, J.-H. Woo, D. G. Streets, C. Wei, A. D. Clarke, B. Huebert and T. L. Anderson. 2004. Impacts of dust on regional tropospheric chemistry during the ACE-Asia experiment: A model study with observations. *Journal of Geophysical Research* 109(D):19S21, doi:10.1029/2003JD003806.

Uematsu, M., Z. F. Wang and I. Uno. 2003. Atmospheric input of mineral dust to the western North Pacific regional based on direct measurements and a regional chemical transport model. *Geophysical Research Letters* 30:1342, 10.1029/2002GL016645.

Uno, I., G. R. Carmichael, D. G. Streets, Y. Tang, J. J. Yienger, S. Satake, Z. Wang, J. H. Woo, S. Guttikunda, M. Uematsu, K. Matsumoto, H. Tanimoto, K. Yoshioka and T. Iida. 2003. Regional chemical weather forecasting system CFORS: Model descriptions and analysis of surface observations at Japanese island stations during the ACE-Asia experiment. *Journal of Geophysical Research* 108(D23):8668, doi:10.1029/2002JD002845.

Wang Z. F., H. Ueda and M. Y. Huang. 2000. A deflation module for use in modeling long-range transport of yellow sand over East Asia. *Journal of Geophysical Research* 105(D22):26947–26960.

Wang, Z. F., H. Akimoto and I. Uno. 2002. Neutralization of soil aerosol and its impact on the distribution of acid rain over East Asia: Observations and model results. *Journal of Geophysical Research* 107, 10.1029/2001JD001040.

Yasushi, N., S. Tanaka and S. J. Santosa. 1999. A study on the concentration, distribution and behavior of metals in atmospheric particulate matter over the North Pacific Ocean by using inductively coupled plasma mass spectrometry equipped with laser ablation. *Journal of Geophysical Research* 104(D21):26859–26866.

Zhang, D. 1982. History of 'muddy rain' phenomenon in China. *Science in China* 27:294–297.

Zhang, J., S. M. Liu and X. Lu. 1993. Characterizing Asian wind-dust transport to the Northwest Pacific Ocean: Direct measurements of dust flux for two years. *Tellus* 45B:335–345.

Zhang, X. Y., R. Arimoto and Z. S. An. 1997. Dust emission from Chinese desert sources linked to variations in atmospheric circulation. *Journal of Geophysical Research* 102(D23):28041–28047.

Zhang, X. Y., S. L. Gong, Z. X. Shen, F. M. Mei, X. X. Xi, L. C. Liu, Z. J. Zhou, D. Wang, Y. Q. Wang and Y. Cheng. 2003. Characterization of soil dust aerosol in China and its transport/distribution during 2001 ACE-Asia. 1. Network Observations. *Journal of Geophysical Research* 108(D9), doi:10.1029/2002JD002632.

Zou, X. K. and P. M. Zhai. 2004. Relationship between vegetation coverage and spring dust storms over northern China. *Journal of Geophysical Research* 109, doi:10.1029/2003JD003913.

Zhuang, G. S. 2003. Time brings a great change to the world — dust storm, aerosol and global biogeochemical cycles in China. *Science in China* 6:38–42.

Part III

LAND-USE CHANGE

Chapter 10

LAND USE AND LAND COVER CHANGE IN EAST ASIA AND ITS POTENTIAL IMPACTS ON MONSOON CLIMATE

CONGBIN FU and SHUYU WANG

1. Introduction

The type of land cover of an area can significantly affect the impact of solar variation on climate by changing surface characteristics such as albedo, stomatal resistance, and surface roughness. Changes in management practices within land use and land cover lead to changes in terrestrial carbon stocks and fluxes, and directly contribute to the carbon dioxide concentration of the atmosphere. By causing change and responding to impacts, land use/land cover change is linked in complex and interactive ways to other global environmental changes, as well as human actions at multiple spatial and temporal scales. Land use/land cover change occurs at a local scale, and can have significant impacts on the regional environment, for example, on local air quality, the lowering of groundwater tables, soil quality and fertility that control the productivity of agriculture, the character of the urban landscape, local weather, the occurrence and spread of infectious disease (IPCC 2001), and other aspects of human health and welfare. Land use/land cover change is the key to addressing regional economic development and societal vulnerability, and it is integral to local and regional policy, resource management and development issues. It is therefore widely recognized that better scientific understanding of land cover/land use change is required for carbon trading, food supply, water and air quality and for assessing the potential impacts on climate.

Numerous studies point out that due to population growth and social and economic developments worldwide, land use and land cover, including vegetation type, vegetation fraction and soil properties, have been greatly modified (Ramankutty and Foley 1998). Major land cover types like grassland, woodland, and forest which have been converted to agricultural land to raise crops and cattle, now account for one-third of the world's surface. Modeling studies indicate that large-scale deforestation in humid tropical areas of South America, Africa, and Southeast Asia has a clear impact on local climate (Zhang et al. 1996; Hahmann and Dickinson 1997). In addition, some significant extra-tropical impacts have been identified (Sud et al. 1996). Observations show that replacing tropical forest with degraded pasture can reduce evaporation and increase surface temperature, and the effect can be reproduced by most climate models.

As one of the regions that is undergoing rapid economic and social developments, Asia has been greatly transformed by land use/land cover change, and the regional environment has been subjected to stresses such as salinization, desertification, deforestation, soil erosion, and water and air pollution. Increasing industrialization in the region, which depends on the availability of land and water resources, will further increase the risk of environmental degradation and damage to local ecological systems. In addition, the need to produce food and supply clean water for the growing number of people in the region will further focus attention on changes in land use and land cover (Ojima 2000).

2. Land Use and Land Cover

In view of the impacts of land use and land cover change on regional climate and environment, and the important role it may play in future climate change and sustainable development, it is critical to have an accurate understanding of past land use practices, current land use and cover patterns, and projections of future land use and cover. The combination of climate and land use change may have more profound effects on the habitability of the planet than either acting alone. While land use change is often a driver of environmental and climate change, changing climate can in turn affect land use and land cover. Climate variability alters land use practices differently in different parts of the world, highlighting differences in societal vulnerability and resilience. Providing a scientific understanding of the process of land use change, and the impacts of different land use decisions, and how they will be affected by a changing climate and increasing climate variability is a priority area for research.

2.1. Land cover in the Holocene period

Land use/land cover changes in the Asian region have been a major factor in ecosystem changes for a long time. Agricultural and livestock developments have been recorded for millennia in this region. During the early Holocene, which is around 9,000 years ago, the central Asian steppe and forest-steppe zone seemed to extend further north than at present due to a warmer climate. In northern and central China, based on various pollen cores, Winkler and Wang (1993) suggests that a forest cover returned and that the vegetation zones were shifted slightly northwards relative to the present, due to a warmer climate. Vegetation also extended westwards into the present arid belt due to greater rainfall, although at that stage the peak Holocene warmth and moisture had not been reached.

During the Holocene warm period, which is about 6,000 years ago, the air temperature in East China was about 2.5°C higher than the current climate, and early agriculture began to appear in the region. At that time the yearly total precipitation was about 300 mm in north and northwest China, 100–200 mm more than the current value. With higher temperatures and greater precipitation, the cropped land and proportion of cultivable land increased, and the northern boundary of paddy fields was about 2–3° further north at 35°N than at present. As a result, the dryland boundary extended northward by the same amount into Inter Mongolia and Gansu and Qinghai Province in West China.

2.2. Land use/land cover change in the past 300 years

Over the past 300 years, in an ever accelerating process, much of the arable land in areas with long-standing agricultural civilizations, such as China, India, the Middle East, and Europe, has been modified by humans. Even in areas without large agricultural populations, humans used fire to modify vegetation. Over the last three centuries, nearly 20% of the world's forests and woodlands have disappeared, and the total global area of forests and woodlands has diminished by 12 million km^2. Grasslands and pastures declined by 5.6 million km^2 (8%), although many grasslands have been converted to pastures), and croplands have increased by 12 million km^2 (466%). Such large changes in land cover can have important consequences such as changes in climate, modification of the global cycles of carbon, nitrogen, and water, and increased rates of extinction and biological invasion. In the monsoon area of south and southeast Asia, the most important change in land use over the last 100-year period was the conversion of 107×10^6 ha of forest/woodland to categories with lower biomass. Land thus transformed accounted for 13.5% of the total area of the south and southeast Asia (Richards 1990; Flint and Richards 1994).

2.3. Land cover in north China for the next 50 years

Through the statistical analysis of social, economic, ecological, and environmental information and satellite data, and the use of different climate scenarios, a high resolution dataset of land use change for next 10–50 years has been established. The dataset will be helpful in simulating environmental change in north China, evaluating the water resource and food productivity, and organizing reforestation and reduction of deserts.

3. Potential Impact of Land Use Changes

Land surface conditions affect the dynamics and thermodynamics of the atmosphere through the control of energy and water exchanges between the land surface and the atmosphere, and therefore affect the regional and global circulation patterns, as well as weather and climate. Using climate models to help understand the impacts of land use and land cover change on climate has been one of the major challenges for climatologists (Xue et al. 1996; Bonan 1997). The role of land surface ranges from purely physical influences, for example, the aerodynamic drag on the atmosphere or the role of soil characteristics in controlling soil moisture and runoff, to some major biological influences, such as the response of leaf stomata to environmental changes and biogeochemical cycling. However, quantitative evaluation of climate changes induced by land use and land cover changes is difficult because of the lack of observed land cover datasets and the highly nonlinear land atmosphere interactions.

Both observational as well as theoretical studies have shown that large scale land cover changes, such as deforestation, reforestation, over cultivation, and over grazing of grassland, have resulted in either deterioration or amelioration of the regional climate and environment (Wei and Fu 1998; Pielke 2001). Changes to the land surface resulting from climate change or increased CO_2 concentration are likely to become more important over the mid to long term. For example, the extension of the growing season in high latitudes will probably result in increases in biomass density, biogeochemical cycling rates, photosynthesis, respiration and fire frequency in the northern forests, leading to significant changes in albedo, evapotranspiration, hydrology, and the carbon balance of the zone.

In the mid and high latitudes the essential role of vegetation in the surface water balance through transpiration has been demonstrated by the analysis over various regions using contour plots from the European Center for Medium range Weather Forecasting model. This model runs to 7 days and provides data at 24 h intervals for sea level pressure, 850 mb winds and temperatures and 500 mb heights. Reanalysis of the data revealed the time space characteristics of the relationship between precipitation, water vapor transportation in and out of the region which is composed of advection and evapotranspiration. In the humid tropics, both advection and evapotranspiration contribute nearly equally to precipitation, or advection is the main source for precipitation. In the mid to high latitudes of the continent, in contrast, evapotranspiration is a major source for precipitation, although the amounts of precipitation and evapotranspiration are far smaller than those in the tropics. The large contribution of evapotranspiration to precipitation in the mid and high latitudes strongly suggests the essential role of vegetation in the surface water balance through transpiration.

Simulation work has indicated that the progressive cultivation of large areas in the East and Midwest USA over the last century has induced a regional cooling of the order of 1–2°C due to enhanced evapotranspiration rates and increased winter albedo (Bonan 1997). Analysis of the interaction between Sahelian vegetation and rainfall suggests that the persistent rainfall anomaly observed there in the 1970s and 1980s could be related to land surface changes (Claussen 1997; Xue 1997). All these studies indicate that large scale land use changes can lead to significant regional climatic impacts.

The Asian Monsoon is known as a land–atmosphere–ocean coupled system. The land–ocean heating contrast between the Tibetan Plateau and the Indian Ocean produce strong southwesterly monsoon flow across the equator, which transports huge amounts of water vapor over south, southeast and East Asia (Li and Huang, Chapter 7 this volume). Land use and land cover changes can also affect regional hydrological cycles. Zhang et al. (2003) studied the feedback between vegetation cover and summer precipitation, which is complex and uncertain. Analysis of the relationship between the satellite derived normalized difference vegetation index and precipitation indicated that summer precipitation has a positive response to increased vegetation cover changes in most regions of China. Spatial analyses of correlations between the vegetation index over each sensitive

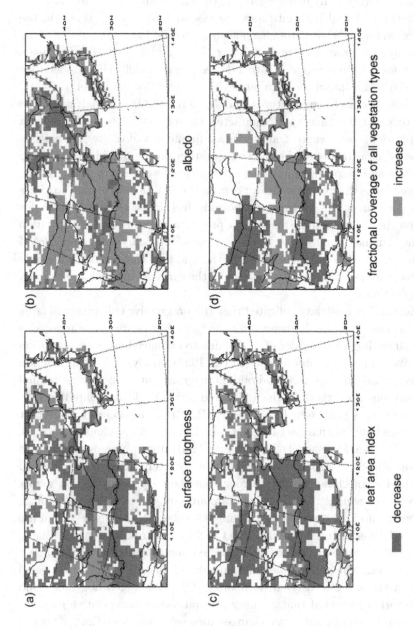

Fig. 10.1. Changes of four physical parameters of land-surface from potential to current vegetation cover.

region and precipitation at 160 stations suggest that vegetation cover strongly affects summer precipitation not only over the sensitive region, but also over other regions, especially the downstream region along the monsoon flow.

In the past 3000 years, more than 60% of the region has been affected by conversion of various categories of natural vegetation into farmland, conversion of grassland into semidesert, and widespread land degradation. Such human induced land cover changes result in significant changes of surface dynamic parameters, such as albedo, surface roughness, leaf area index, and fractional vegetation coverage. Using a regional climate model, Fu (2003) showed that by altering the complex exchanges of water and energy from surface to atmosphere, the changes in land cover result in significant changes to the East Asian monsoon. These include the changes of main surface characters. Surface roughness and leaf area index decrease (black areas of Figs. 10.1(a) and 10.1(b)) over central China where the forests have been turned into farmland, northwest China and some parts of Mongolia where the grassland has been turned into semidesert or desert, and the region over southwest China where the evergreen broad leaf forest has been turned into mixed forest or shrub.

The increase of surface albedo occurs over the regions where the natural vegetation covers were significantly destroyed as shown in Fig. 10.1(c). The total fractional vegetation coverage is higher in farmland area than in natural forests, but it is lower in the area of semidesert and desert in comparison with grassland and the area of mixed forests in comparison with evergreen broadleaf forests (Fig. 10.1(d)). Changes of these surface parameters modify the energy and water budgets between the land surface and the atmosphere and result in the changes of atmospheric circulation. The weakening of the summer monsoon is shown by the northerly anomalous flow in Fig. 10.2, which would be against the development of summer monsoon circulation and the moisture transport northward and the development of convective activities, resulting in a drier atmosphere over most of the domain. The changes of summer total precipitation (Fig. 10.3) show the main dry areas are located in northern and southeast China.

These also result in the reduction of all the other components of surface water balance such as evaporation, runoff, and soil water content. The consequent diminution of northward and inland moisture transfer may be a significant factor in explaining the decreasing atmospheric humidity and soil water content, and thus the trend in aridification observed in many parts of the region, particularly over northern China during the last 3000 years.

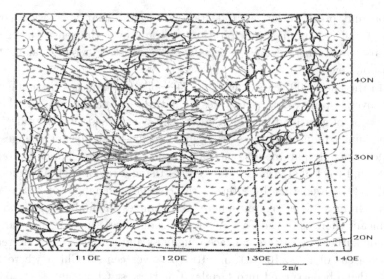

Fig. 10.2. Changes of summer monsoon circulation over East Asia under two vegetation covers (current minus potential) during summer (June, July, August).

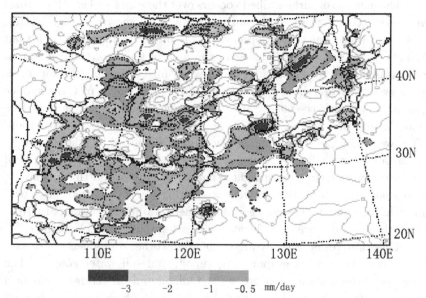

Fig. 10.3. Summer total precipitation changes in East Asia under two vegetation covers (current minus potential) during summer (June, July, August).

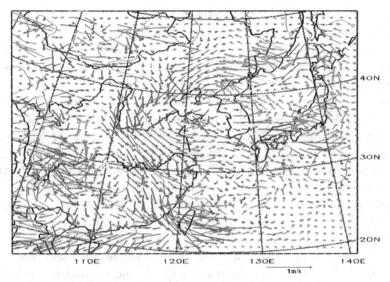

Fig. 10.4. Changes of winter monsoon circulation over East Asia under two vegetation covers (current minus potential) during winter (December, January, February).

Figure 10.4 shows that the winter monsoon over East Asia becomes stronger under the deterioration of natural vegetation cover by having the stronger anomalous northerly flow, which means the dry and cold air mass from the inland area would be brought down to all regions of East Asia. This would result in changes of surface climate, such as reduction of atmospheric humidity and precipitation, mostly in the southern part of the region, and colder temperatures over almost the whole region.

Simulation of the climate impacts of desertification on climate in Inter Mongolia (Xue 1996; Fu et al. 1996), and direct measurements, show that desertification leads to higher surface air temperature and less precipitation and evaporation. Zheng et al. (2002) pointed out that, due to the combined effects of desertification in North China and deforestation in South China, flooding in the Yangtze River Basin and drought in North China might be getting worse. In West China degradation of forest, pasture and crop land, due to irrational land use, has resulted in severe storms and expanding of the desert. Wang et al. (2001) investigated the impacts of land use and land cover changes on the structure of the summer monsoon system in East Asia. The simulation suggests that the East Asian summer monsoon would be enhanced because of the large scale land use and land cover changes in western China, which is favorable to water vapor transfer from the adjacent oceanic areas to the continent of China. The humidity increase in China can be attributed partly to evapotranspiration, but mostly to water vapor transfer by the enhanced monsoon circulation. The integrated impact of land use and land cover variations on temperature, wind speed, and humidity is not only limited to the west part of China, but to adjacent areas where land use and land cover does not vary. In a similar study, Wang and Hao (2003) identified the benefits of constructing the 3N (northeast, north and northwest) Shelterbelt in China, which is the largest reforestation project in the world, and investigated its climate impacts by using a mesoscale meteorological model. The simulation showed that construction of the shelterbelt and the corresponding land use and land cover variations would change the roughness length significantly. The simulation suggests that the regional climate in North China would be improved because of decreased wind speed, increased air humidity and precipitation. Two multi-year simulations, one with current land use and the other with potential vegetation cover (Gao et al. 2003) suggest that land use change over China would decrease annual precipitation over northwest China, a region with a prevalence of arid and semiarid areas, increase mean annual surface air temperature over some areas, and decrease the temperature along coastal areas. Summer mean daily maximum temperature would increase in many locations, while winter mean daily minimum temperature would decrease in East China and increase in northwest China. Soil moisture would decrease significantly across China. The results indicate that the same land use change may cause different climate effects in different regions depending on the surrounding environment and climate characteristics.

Simulation studies of the impacts of land cover changes on regional summer climate over east Asia, using two types of land cover for the cool/wet and warm/dry climate (Suh and Lee 2004) suggest that land–atmosphere–land cover change interaction is nearly independent of climate regime (in particular, surface wind, latent and sensible heat fluxes, and thus surface temperature). However, the simulations suggested that the impact of land cover change on precipitation is dependent not only on climate regime, but also on time of the year. The resulting impacts were significant in June but slowly decreased as summer advanced. The study also suggested that impacts would be more pronounced in warm and dry years than in cool and wet years. Relatively weak impacts of land cover change in cool and wet climates can be related to a reduction in net solar radiation at the surface and evaporation because of an increase in cloud cover.

4. Remaining Questions

Although much progress has been made concerning land use and land cover change on local, regional, and global climate, uncertainty still exists due to the following factors:

1. At present, only limited global datasets for land surface are available and these need to be further improved. A comprehensive land use/land cover dataset, providing a global time series of vegetation and soil parameters over the last two centuries at General Circulation Model resolution, would be a very useful tool to separate land use change impacts on regional climate from global scale warming effects.
2. In order to understand the impacts of land use and land cover change, better efforts and co-operation are required to improve the understanding of the interrelationships and dynamic feedbacks between land use/land cover change and carbon budget, atmospheric chemistry, water resources, and climate variability. The challenge will be to use contemporary impacts of land use and land cover change to calibrate impacts on ecosystem goods and services; biogeochemical, water, and energy cycles; and climate processes. These investigations must be undertaken on multiple scales so that the full dimensions of the perturbations of environmental processes can be determined.
3. For both historical analyses and future projections, there is a need for interactive vegetation models that can simulate changes in vegetation parameters and carbon cycle variables in response to climate change. Furthermore, development of coupled climate land use/cover models, taking socioeconomic factors into consideration, should be accelerated. Simulation of climate land use/cover feedbacks will require advancement of current understanding of multiple stress processes at local to global scales. Validation of the interacting climate land use effects for specific regions of the globe will be particularly challenging. International cooperation will be needed to optimize the currently existing and emerging observational networks.

Literature Cited

Bonan, G. B. 1997. Effects of land use on the climate of the United States. *Climate Change* 37:449–486.

Charney, J. G. 1975. Dynamics of deserts and drought in the Sahel. *Quarterly Journal of the Royal Meteorological Society* 101:193–202.

Claussen, M. 1997. Modeling bio-geophysical feedback in the African and Indian monsoon region. *Climate Dynamics* 13:247–257.

Flint, E. P. and J. F. Richards. 1994. Historic land use and carbon estimates for south and southeast Asia 1880–1980, Dataset NDP-046 CDIAC/ORNL. Oak Ridge Laboratory, Oak Ridge.

Fu, C. B., H. L. Wei, W. Z. Zheng and B. K. Su. 1996. Sensitive experiment on the vegetation cover in mainland China by a meso-scale model. In *Global Change and Future Environment of China*. Beijing: China Meteorological Press.

Fu, C. B. 2003. Potential impacts of human-induced land cover change on East Asia monsoon. *Global and Planetary Change* 37:219–229.

Gao, X. J., Y. Luo, W. T. Lin, Z. C. Zhao and F. Giogri. 2003. Simulation of effects of land use change on climate by a regional climate model. *Advances in Atmospheric Sciences* 20:583–592.

Hahmann, A. N. and R. E. Dickinson. 1997. RCCM2-BATS model over tropical South America. Applications to tropical deforestation. *Journal of Climate* 10:1944–1964.

IPCC (Intergovernmental Panel on Climate Change) 2001. *Impacts, Adaptation and Vulnerability — Contribution of Working Group II to the IPCC Third Assessment Report*, edited by J. J. McCarthy, O. F. Canziani, N. A. Leary, D. J. Dokken and K. S. White. Cambridge: Cambridge University Press.

Ojima, D. 2000. *Land Use/Land Cover Change in Temperate East Asia: Current Status and Future Trend*. Washington D.C.: International START Secretariat.

Pan, Z., E. Takle, M. Segal and R. Arritt. 1999. Simulation of potential impacts of man-made land cover change on U.S. summer climate under various synoptic regimes. *Journal of Geophysical Research* 104:6515–6528.

Pielke, R. A. 2001. Influence of the spatial distribution of vegetation and soils on the prediction of cumulus convective rainfall. *Reviews of Geography* 39:151–177.

Ramankutty, N. and J. A. Foley. 1998. Characterizing patterns of global land use: An analysis of global croplands data. *Global Biogeochemical Cycles* 12:667–685.

Suh, M. S. and D. K. Lee. 2004. Impacts of land use cover changes on surface climate over East Asia for extreme climate cases using RegCM2. *Journal of Geophysical Research* 109:D02108.

Sud, Y. C., K. Yang and G. K. Walker. 1996. Impact of *in situ* deforestation in Amazonia on the regional climate: General circulation model simulation study. *Journal of Geophysical Research* 101(D3):7095–7109.

Wang, H. J., H. F. Zhang and A. Pitman. 2001. Numerical simulations on the effect of present day (1700–1990) land use change to climate in China. Pp. 143–164 in *West development and biosystem construction*. Beijing: China Forestry Press (In Chinese).

Wang, H. L. and Z. Hao. 2003. A simulation study on the eco-environmental effects of 3N shelterbelt in North China. *Global and Planetary Change*. 37:231–246.

Wei, H. L. and C. B. Fu. 1998. Study of the sensitivity of a regional model in response to land covers change over northern China. *Hydrological Processes* 12:2249–2265.

Winkler, M. G. and P. K. Wang. 1993. The late-quaternary vegetation and climate of China. Pp. 221–261 in *Global climates since the last glacial maximum*, edited by H. E. Wright Jr., J. E. Kutzbach, J. Webb III, W. F. Ruddiman, F. A. Street-Perrott and P. J. Bartlein. Minneapolis: University of Minnesota Press.

Xue, Y. K. 1996. The impact of desertification in the Mongolian and Inner-Mongolian grassland on the regional climate. *Journal of Climate* 6:2173–2189.

Xue, Y. K. 1997. Biosphere feedback on regional climate in tropical north Africa. *Quarterly Journal of the Royal Meteorological Society* 123:1483–1515.

Xue, Y., M. J. Fennessy and P. J. Sellers. 1996. Impact of vegetation properties on U.S. summer weather prediction. *Journal of Geophysical Research* 101:7419–7430.

Zhang, J. Y., W. J. Dong, C. B. Fu and L. Y. Wu. 2003. The influence of vegetation cover on summer precipitation in China, a statistical analysis of NDVI and climate change. *Advances in Atmospheric Sciences* 20:1002–1006.

Zhang, H., A. Henderson-Sellers and K. Mc Guffie. 1996. Impacts of tropical deforestation. Part I: Process analysis of local climatic change. *Journal of Climate* 9:1497–1517.

Zheng, Y. Q., Y. F. Qian, M. Q. Miao and G. Yu. 2002. The effects of vegetation change on regional climate I: Simulation results. *Acta Meteorologica Sinica* 60:1–16 (In Chinese).

Chapter 11

THE TERRESTRIAL CARBON BUDGET IN EAST ASIA: HUMAN AND NATURAL IMPACTS

HANQIN TIAN, JIYUAN LIU, JERRY MELILLO, MINGLIANG LIU, DAVID KICKLIGHTER, XIAODONG YAN and SHUFEN PAN

1. Introduction

Carbon sequestration by terrestrial ecosystems is considered to play an important role in today's global carbon budget (Schimel et al. 2001). There is now a long list of studies that point to the existence of an extra-tropical Northern Hemisphere sink for anthropogenic CO_2. These studies are of various kinds including inventories of land-cover change (Brown et al. 1996), atmospheric CO_2 and O_2 monitoring (Keeling et al. 1996), isotopic analyses (Rayner et al. 1999), and ecosystem process modeling (Tian et al. 2003). However, there is a great uncertainty about the magnitude and geographical location of this sink (Prentice et al. 2001). A recent review of the evidence has confirmed that both North America and Eurasia served as carbon sinks during the 1990s (Prentice et al. 2001), suggesting a net sink of 1–2.5 $Pg C yr^{-1}$ that is distributed relatively evenly between North America and Eurasia. The magnitude and mechanisms underlying the carbon sink in North America have received considerable attention (Pacala et al. 2001).

The Eurasian sink has received less attention. We do not yet understand how the terrestrial carbon budget in East Asia is balanced (Houghton and Hackler 2003). Preliminary analyses indicate that the temperate and boreal forests of East Asia are making important contributions to the carbon sink in Eurasia (Myneni et al. 2001). Since the early 1980s, China has implemented many large-scale afforestation projects that have led to an increase in forest cover of about 30 million hectares (Streets et al. 2001). One recent study indicated that reductions in deforestation combined with afforestation activities have led to a net carbon storage of about

0.10 Pg C yr^{-1} during the period 1900–2000 (Streets et al. 2001). A similar study by Fang et al. (2001) reported that the net carbon storage rate for all forests in China over a similar period (1989–1998) averaged about 0.03 Pg C yr^{-1}. These forest inventory-based estimates of net carbon flux do not identify the mechanisms responsible for the carbon sink. From both scientific and policy perspectives, however, it is of critical importance to understand the mechanisms responsible for the terrestrial carbon sink.

The land ecosystems in East Asia have been intensively disturbed or managed by human activities for thousands of years, and are now involved in rapid economic, social, and environmental changes. Monsoon climate and land use have been suggested as two major factors that control net primary production and carbon storage in the ecosystems of East Asia (Tian et al. 2003). The purpose of this chapter is to synthesize the state-of-the-art understanding of (1) recent pattern of land-use change in East Asia, (2) the terrestrial carbon budget in East Asia, with a specific reference to China, the world's third largest country, (3) the role of human activity and natural processes on terrestrial carbon budget, and (4) the uncertainty and needs for future research.

2. Mechanisms Responsible for the Terrestrial Carbon Sink

Several mechanisms were proposed to explain the terrestrial carbon uptake that is required to balance land use emissions (Schimel et al. 2001). These mechanisms include (1) physiological or metabolic factors, including elevated concentrations of atmospheric carbon dioxide (carbon dioxide fertilization), increased availability of nutrients (e.g., nitrogen fertilization), and changes in temperature and rainfall, any of which could increase growth rates in forests, and (2) disturbance and recovery mechanisms, including both natural disturbances and the direct effects of changes in land use and management (Houghton 2003). These mechanisms can control the net carbon exchange between the atmosphere and terrestrial ecosystems through influencing rates of photosynthesis, respiration, growth, and decay as well as carbon storage in terrestrial ecosystems.

The annual net carbon exchange of terrestrial ecosystems with the atmosphere can be described by the equation (Tian et al. 2003):

$$\text{NCE} = \text{NPP} - R_\text{H} - E_\text{NAD} - E_\text{AD} - E_\text{P},$$

where NCE is the annual net carbon exchange, NPP is the net primary production, R_H is the heterotrophic respiration (i.e., decomposition), E_NAD represents the emissions associated with non-anthropogenic disturbance (e.g., lightning fires, insect infestations), E_{AD} represents the emissions from anthropogenic disturbance (e.g., deforestation), and E_P represents the decomposition of products harvested from ecosystems for use by humans and associated livestock. A negative net carbon exchange indicates that terrestrial ecosystems are a source of atmospheric carbon dioxide (CO_2), whereas a positive net carbon exchange indicates a terrestrial sink.

The fluxes NPP and R_H are influenced by spatial and temporal variations in natural environmental conditions such as atmospheric CO_2 concentration, climate, and nitrogen availability. Human activities modify these fluxes further by (1) influencing the availability of water and nitrogen to the plants, (2) breaking up soil aggregates during cultivation or construction to expose more soil organic matter to oxidation, (3) managing the community composition and amount of biomass present on a site, and (4) introducing air pollutants that can either enhance or reduce plant growth. In addition, NPP and R_H of a site will change with time as vegetation regrows after a human or natural disturbance.

The magnitude of the carbon loss from a disturbance depends on the intensity and frequency of the disturbance. Disturbances are episodic events that cause immediate carbon losses from an ecosystem, but may also change environmental conditions at a site so that additional carbon is lost from the ecosystem over a longer time period.

Human activities also directly influence terrestrial carbon storage by the production and redistribution of agricultural products, paper, and other wood products. The carbon stored in these products is returned to the atmosphere at a variety of rates that are related to the consumption of food, the burning and decomposition of trash, and the deterioration of wooden structures in buildings, bridges, etc. As a result of trade, many of these products may decompose or deteriorate in places far removed from the site where the biomass was created. Thus, regional terrestrial carbon sinks may be overestimated if this product flux of carbon back to the atmosphere is not considered properly.

3. Recent Pattern of Land-Use Change

Since the 1980s, China has the fastest growing economy in the world, but faces pressure from a growing human population. There is an increasing demand for land to provide food and services so that the large-scale land transformation across China has inevitably occurred in recent decades (Brown 1995). Data quality and reliability, however, is one of the greatest problems in generating a clear picture of China's carbon cycle. Essential information to assess cross-cutting issues, such as the size of the cropland, woodland, and urban areas, was not known or not reported correctly for decades (Tian et al. 2003). There are large discrepancies among estimates on the state and change of land use in China (Keto et al. 2000). To reduce uncertainty in the estimates of cropland, woodland, and urban areas, Liu and his colleagues have developed land-cover datasets at a resolution of 30 m for the entire nation from Landsat TM imagery for three years (1990, 1995, and 2000) using approximately 500 Landsat TM scenes per year. They have also developed land-cover datasets at a resolution of 1:1 M from a combination of landsat MSS imagery and aerial photographs for 1980.

For the entire nation, Liu et al. (2003) estimated that, in the year 2000, the area of cropland was about 141.14 M ha, with the paddy field of 35.7 M ha and the dry cropland of 105.5 M ha (Table 11.1). For the period from late 1980s to late 1990s, China's cropland increased about 2.99 M ha or 2.17%. Of this increase, dry cropland increased about 2.85 M ha or 2.78% and paddy field increased about 0.14 M ha or 0.4%. This net change in cropland area represents an imbalance between loss and gain in cropland areas.

About 3.2 M ha of croplands were converted to other land uses, including 1.5 M ha to built-up land. The conversion of cropland to woodland is about 0.52 M ha or 16% and the conversion to grassland is about 0.64 M ha or 20.2%. However, there are 6.2 M ha of cropland converted from other land uses, which is much larger than cropland area lost. As a result, the cropland area appeared to increase during this period.

Woodland area was about 226.7 M ha in 1999/2000, decreasing 1.1 M ha or 0.48% in the past decade (Table 11.1). This has resulted from the imbalance between a loss of 2.7 M ha and a gain of 1.6 M ha in woodland

Table 11.1. Changes in land use in China during 1990–2000 derived from Landsat TM data (Unit: million ha; based on Liu et al. 2003).

Land cover types	1990	2000	Change	% Change
Cropland	138.1	141.1	3.0	2.17
Paddy	35.5	35.7	0.2	0.56
Dryland	102.6	105.5	2.9	2.83
Woodland	227.8	226.7	−1.1	−0.48
Forests	138.3	137.6	−0.7	−0.51
Grassland	306.3	302.9	−3.4	−1.11
Water body	32.7	32.9	0.2	0.61
Built-up	44.7	46.4	1.7	3.80
Urban	3.3	4.1	0.8	24.24
Unused land	200.5	200.1	−0.4	−0.20

areas. It was estimated that 64% of the loss was converted to cropland and 29.8% of the loss converted to grassland. Liu et al. (2003) estimated that forest areas were about 137.6 M ha in 1999/2000. The classification of forests in Liu et al. (2003), which is natural and planted forests with canopy coverage greater than 30%, is comparable to the definition of forest in the National Forest Resource Inventory database, but the estimation of forest areas by Liu et al. (2003) is higher than that reported in that database (Table 11.1). The definition of forest has been changed in the fourth forest inventory, which could lead to a bias while comparing it to previous inventory data.

Landsat TM data shows a net loss of 3.4 M ha of grassland area in the past decade, or 1.12% during this period (Table 11.1). Landsat-based estimate indicates that the total area of built-up land was about 46.43 M ha, with urban area of 4.15 M ha. The built-up land increased 1.8 M ha or 3.94%, but the urban area showed the largest percent increase, 24.78%. The increase of built-up land was converted primarily from cropland (85.38%).

Data on city growth and area in China are unreliable. According to National Bureau of Statistics of China, at the end of 2000, urban area was 1,622.1 thousand ha, which is considerably lower than our Landsat TM-based estimate of 4,145.4 thousand ha. This large difference is because we accounted for all developed urban areas.

4. Estimates of Terrestrial Carbon Storage

Three methods have been used to estimate carbon pools and fluxes in terrestrial ecosystems of East Asia: (1) the forest and soil inventory approach (e.g. Wang et al. 2003), (2) book-keeping models (e.g., Houghton and Hackler 2003), and (3) ecosystem models (e.g. Tian et al. 2003).

4.1. Terrestrial carbon storage

To assess the effects of human activities on carbon storage in terrestrial ecosystems of East Asia, we need information on the amount of carbon stored prior to human disturbance. Tian et al. (2003) estimated that the potential carbon storage in natural ecosystems in East Asia is about 195 Pg C, with 73 Pg C stored in vegetation and 122 Pg C stored in soils. Terrestrial ecosystems in East Asia stored about 155 Pg C in the 1990s. Thus, potential loss of carbon from terrestrial ecosystems in East Asia due to land-use change has been about 40 Pg C, including 29 Pg C from vegetation and 11 Pg C from soils. Simulations with the Terrestrial Ecosystem Model indicate that, from 1860 to 1990, land-use change caused a loss of 12.9 Pg C. Thus, there had been a larger carbon loss before 1860 than after 1860. From 1860 to 1980, simulation indicates that total carbon storage decreased continually (Fig. 11.1). However, after 1980, terrestrial ecosystems in East Asia began to sequester atmospheric CO_2, which might be related to large-scale afforestation activity in this region. Changes in vegetation carbon reflected temporal pattern of total carbon storage. Soil organic carbon remained relatively stable from 1860 to 1950, but decreased from 1950 to 1980, and showed a slight increase since 1980.

Estimates of potential total carbon storage in natural ecosystems in China were similar, varying from 155 to 160 Pg C (Peng and Apps 1997; Ni 2001; Houghton and Hackler 2003). However, there was greater variation in estimates of the distribution between vegetation and soils, 35–58 Pg C for vegetation and 100–120 Pg C for soils. For contemporary carbon storage, modeled results suggest a total of 103–117 Pg C in 2000, with 12–25 Pg C in vegetation and 91–102 Pg C in soils (Houghton and Hackler 2003; Tian et al. 2003; Cao et al. 2003).

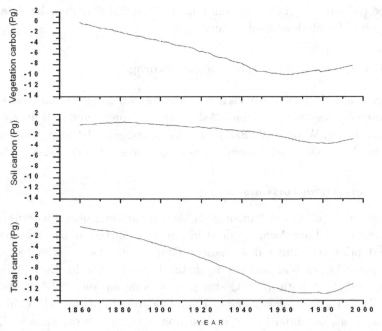

Fig. 11.1. Changes in terrestrial carbon storage since 1860 in East Asia as estimated by the Terrestrial Ecosystem Model.

Several studies have used the inventory-based method to estimate contemporary carbon storage in forests, grasslands, and soils in China, but none of these studies has provided an estimate of contemporary carbon storage for all ecosystems in China. There are large differences among various estimates of carbon storage, varying from 3.26 to 7.0 Pg C in forests (Fang et al. 2001; Wang et al. 2001), from 1.23 to 3.06 Pg C in grasslands (Ni 2001), and from 77.4 to 92.4 Pg C in soils (Wang et al. 2003; Wu et al. 2003). Wang et al. (2001) attribute their lower value to the use of different conversion factors (stem volume to total biomass) for different age classes within each species. Fang et al. (2001) used a single conversion factor for all age classes within a species. All estimates of soil carbon storage were based on the second national soil survey of China.

4.2. Terrestrial carbon budget

The forest inventory data estimated that forests in China represent a net sink of atmospheric CO_2 of 0.021–0.112 Pg C yr^{-1} from the 1980s to present. Estimates vary, in part, because they pertain to different years and often include only partial accounting of carbon. Fang et al. (2001) reported an average release of carbon (0.022 Pg C yr^{-1}) for the years 1949–1980 and an average uptake (0.021 Pg C yr^{-1}) for the years 1980–1998. Their estimates are based on changes in forest biomass only; they did not include soils or changes in wood products. The same is true for the estimate of Chinese forests from Goodale et al. (2002), including soils as well as living vegetation. Zhou (2000) calculated a net forest uptake of 0.048 Pg C yr^{-1} for the period 1989–1993. Streets et al. (2001) reported an annual sink of 0.098 Pg C in 1990 and of 0.112 Pg C in 2000.

4.3. Variation in net carbon exchange

The annual net carbon exchange between the atmosphere and terrestrial ecosystems in East Asia showed substantial interannual and decadal variations for the time period 1860–2000 according to the Terrestrial Ecosystem Model simulations (Tian et al. 2003). Over the period 1980–2000, in China, the nationwide net carbon exchange ranged from a source of 0.75 Pg C yr^{-1} in 1999 to a sink of 0.58 Pg C yr^{-1} in 1990. Variations in the net carbon exchange across years are coupled tightly to variations in climate, particularly precipitation. Annual net carbon exchange for the period 1980–2000 was significantly correlated with annual precipitation. In a specific year, the effect of drought or wet climate on carbon storage could be much larger than that of either increasing CO_2 or land-use change. Precipitation control on the net carbon exchange has also been found by empirical analysis by Yang and Wang (2000), which is especially true in monsoon regions such as East Asia. In any year over the period 1980–2000, net carbon exchange can be very large in one location but very small or negative in another location because of the spatial heterogeneity of vegetation, soils, and climate. Figure 11.2 shows quite different spatial

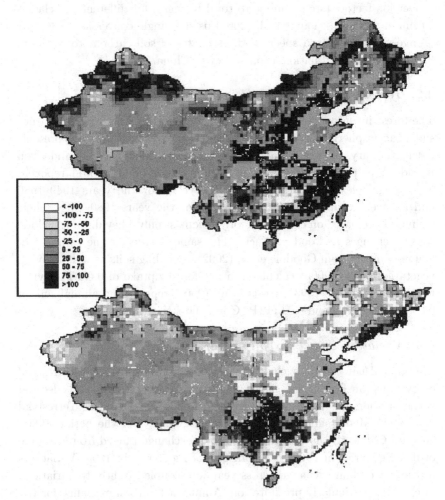

Fig. 11.2. Net carbon exchange between the atmosphere and terrestrial ecosystems for 1998 (top) and 1999 ($g\,C\,m^{-2}\,yr^{-1}$) as estimated by the Terrestrial Ecosystem Model (black means CO_2 uptake, bright means CO_2 emission).

patterns of carbon source and sink in 1998 and 1999. Locations with large positive annual net carbon exchange often receive a large amount of precipitation. In contrast, locations with negative net carbon exchange often receive little precipitation. Year-to-year changes in the spatial pattern of net carbon exchange are mostly caused by changes in the spatial pattern of precipitation, which can change dramatically with monsoon events. In strong monsoon years, which bring rainfall to much of the monsoon region, most ecosystems act as a sink of carbon from the atmosphere. Conversely, in weak monsoon years, which bring hot, dry weather to much of monsoon Asia, most ecosystems act as a source of carbon to the atmosphere. However, increasing atmospheric CO_2 generally leads to increased carbon storage over the entire region.

5. Effects of Land Use, Climate, and Carbon Dioxide on Budgets

Quantifying the relative contribution of human activity and natural processes on the terrestrial carbon budget is of critical importance to carbon cycle science as well as policy making, but scientifically and technically difficult.

5.1. Land-use change

The contribution of land-use change to the terrestrial carbon budget has traditionally been estimated with book-keeping models (e.g., Houghton and Hackler 2003) that balance deforestation and forest regrowth over time, assuming generic time-dependent functions for carbon gains and losses in different ecosystem types. Recently, Tian et al. (2003) also accounted for the effect of spatial and temporal variability in climate and CO_2 on forest regrowth in the simulation of carbon dynamics of monsoon Asia. For China, during the 1990s, the effect of land-use change on net carbon exchange varied from a small source of 0.008 Pg C per year (Houghton and Hackler 2003) to a small sink of 0.015 Pg C yr^{-1} (Tian et al. 2003). For East Asia, during the 1990s, land-use change resulted in a carbon uptake of 0.086 Pg C yr^{-1} by terrestrial ecosystems in East Asia (Table 11.2).

5.2. Carbon dioxide and climate

Besides influencing the recovery of disturbed ecosystems, carbon dioxide and climate also influence terrestrial carbon uptake in undisturbed ecosystems. Physiological or metabolic factors responsible for terrestrial

Table 11.2. Relative contribution of land use, climate, and CO_2 to carbon storage in China and East Asia as estimated by the Terrestrial Ecosystem Model (Pg C yr^{-1}).

	China		East Asia	
	1980s	1990s	1980s	1990s
Land use	−0.103	0.015	0.030	0.086
Climate	0.060	−0.035	0.031	0.058
CO_2	0.091	0.096	0.088	0.096
Total	0.084	0.058	0.149	0.240

carbon uptake were not included in the analysis of Houghton and Hackler (2003). The net carbon storage associated with physiological mechanisms has been independently estimated in a number of different process-based model analyses (e.g., Cao et al. 2003). Tian et al. (2003) used the Terrestrial Ecosystem Model to simultaneously consider the effects of land-use change and physiological mechanisms. Their analysis indicated that the combination of climate, CO_2, and land use resulted in a carbon uptake of 0.149 Pg C yr^{-1} in East Asia, with a net carbon sink of 0.138 Pg C yr^{-1} in China for the 1980s (Table 11.2). Recent estimates with the same model in conjunction with high resolution land use data indicate that the combination of climate, CO_2, and land use led to a carbon sink of 0.084 Pg C yr^{-1} for the 1980s and 0.058 Pg C yr^{-1} for the 1990s. Changes in land use resulted in a carbon source of 0.103 Pg C yr^{-1} for the 1980s, but led to a carbon sink of 0.015 Pg C yr^{-1} for the 1990s. Climate variability led to a carbon uptake by ecosystem in the 1980s, but resulted in a carbon loss from ecosystems in the 1990s. The effect of CO_2 fertilization contributes about 0.09 Pg C yr^{-1} to the terrestrial carbon uptake for the period 1980s and 1990s. Using another ecosystem model, Cao et al. (2003) also suggested that, for the period 1981–2000, climate variability and CO_2 led to a carbon uptake of 0.07 Pg C yr^{-1}. These recent sinks are similar to the estimates by forest inventory analysis (Fang et al. 2001; Streets et al. 2001).

6. Needs for Future Research

We have shown the importance of considering spatial and temporal variability in the effects of land-use change, CO_2 fertilization, and climate variability on carbon dynamics in East Asia. However, our analyses have been limited by uncertainties related to the representation of environmental factors by spatially explicit datasets and by our current inability to simulate the effects of some potentially important factors on the regional carbon budget of East Asia. For example, there are large discrepancies among estimates of cropland area in China (Tian et al. 2003). This uncertainty in cropland area influences our estimate of regional carbon fluxes.

The land-use analysis in the current process-based ecosystem models is incomplete because it only considered the effects on the regional carbon budget associated with row crop agriculture (Tian et al. 2003). It did not consider (1) the conversion of forests to pastures, (2) possible changes in harvest and regrowth cycles within managed forests, and (3) urbanization and desertification. To address these issues, spatially explicit datasets of the historical distribution of pastures, forest harvest, urban areas and the extent of desertification need to be developed to improve estimates of regional carbon budgets. In addition, the locations, severity, and timing of forest fires need to be considered. During the second half of the 20th century, there were several large fires in the forest area of North China including the famous "Black Dragon" fire in 1987 that burnt an estimate of 10 million ha (Wang et al. 1996). More information from field experiments is required to develop algorithms that better simulate the effects of these other land-use changes and fire disturbance on the regional carbon budget.

The fate of cleared lands is an important factor that affects carbon fluxes and storage (Tian et al. 2003). While reforestation and plantation establishment can lead to carbon sequestration (Brown et al. 1996), urbanization, desertification, and other land degradation could cause long-term carbon loss to the atmosphere. To reduce uncertainty in regional carbon budgets, more work needs to be done to better account for these factors in future studies.

Besides CO_2 fertilization, recent reviews of the global carbon budget also indicate that terrestrial carbon storage could also be affected by nitrogen deposition and other changes in atmospheric chemistry (Prentice et al. 2001) which are not considered in this study. In recent decades, the rapid industrialization in monsoon Asia has resulted in increased anthropogenic nitrogen deposition and tropospheric ozone levels. Nitrogen deposition should enhance terrestrial carbon storage (Holland et al. 1997), but it is also possible that chronic high inputs of nitrogen may cause terrestrial ecosystems to lose carbon (Aber et al. 1993). This nitrogen deposition may have substantial interactions with increasing atmospheric CO_2 on the regrowth of forests after cropland abandonment to influence carbon sequestration in terrestrial ecosystems. On the other hand, both field and modeling studies have shown that ozone decreases crop yield

(Chameides et al. 1999) and forest productivity (McLaughlin and Percy 2000). In addition, atmospheric aerosols and regional haze may also reduce crop productivity (Cohan et al. 2002). To better understand the carbon dynamics in East Asia, future studies should take these atmospheric chemistry factors into account.

In summary, there is an important need to understand three linked questions, (1) how have carbon fluxes and storage changed in East Asia in the past 300 years?, (2) how will carbon fluxes and storage change in East Asia in the next 50–100 years?, and (3) what mechanisms have had major effects on changes in these fluxes and storage. We need to consider the relative roles of: (a) climate variability, (b) changes in land cover and use, (c) changes in fire disturbance, (d) changes in the chemistry of precipitation (particularly nitrogen), and (e) changes in the composition of the atmosphere (carbon dioxide, ozone). Such research will need to be done in a fully collaborative fashion with a team of scientists from both within and outside East Asia. We see the initiation of this partnership in science as an effort to develop a common understanding of perhaps the most important issue facing human kind in the 21st century, our disruption of the global carbon cycle.

Literature Cited

Aber, J. D., A. Magill, R. Boone, J. Melillo, P. Steudler and R. Bowden. 1993. Plant and soil responses to chronic nitrogen additions at the Harvard Forest, Massachusetts. *Ecological Applications* 3:156–166.

Brown, L. 1995. *Who will feed China?: Wake-up call for a small planet.* New York: W. W. Norton and Company.

Brown, S. A., J. Sathaye, M. Cannell and P. Kauppi. 1996. Management of forests for mitigation of greenhouse gas emissions. Pp. 773–797 in *Climate change 1995: Impacts, adaptions and mitigation of climate change,* edited by R. T. Watson, M. C. Zinyowera and R. H. Moss. Cambridge: Cambridge University Press.

Chameides, W. L., X. Li, X. Tang, X. Zhou, C. Luo, C. S. Kiang, J. St. John, R. D. Saylor, S. C. Liu, K. S. Lam, T. Wang and F. Giorgi. 1999. Is ozone pollution affecting crop yields in China? *Geophysical Research Letters* 26:867–870.

Cao, M. K., S. D. Prince, K. Li, B. Tao, J. Small and X. Shao. 2003. Response of terrestrial carbon uptake to climate interannual variability in China. *Global Change Biology* 9:536–546.

Cohan, D. S., J. Xu, R. Greenwald, M. H. Bergin and W. L. Chameides. 2002. Impact of atmospheric aerosol light scattering and absorption on terrestrial net primary productivity. *Global Biogeochemical Cycles* 16:1090, doi:10.1029/2001GB001441.

Fang, J., A. Chen, C. Peng, S. Zhao and L. Ci. 2001. Changes in forest biomass carbon storage in China between 1949 and 1998. *Science* 292:2320–2322.

Goodale, C. L., M. J. Apps, R. A. Birdsey, C. B. Field, L. S. Heath, R. A. Houghton, J. C. Jenkins, G. H. Kohlmaier, W. A. Kurz, S. Liu, G. J. Nabuurs, S. Nilsson and A. Z. Shvidenko. 2002. Forest carbon sinks in the Northern Hemisphere. *Ecological Applications* 12:891–899.

Holland, E., B. Braswell, J.-F. Lamarque, A. Townsend, J. Sulzman, J.-F. Muller, F. Dentener, G. Brasseur, H. Levy II, J. E. Penner and G.-J. Roelofs. 1997. Variations in the predicted spatial distribution of atmospheric nitrogen deposition and their impact on carbon uptake by terrestrial ecosystems. *Journal of Geophysical Research* 102(D13): 15849–15866.

Houghton, R. A. 2003. Why are estimates of the terrestrial carbon balance so different? *Global Change Biology* 9:500–509.

Houghton R. A. and J. L. Hackler. 2003. Sources and sinks of carbon from land-use change in China, *Global Biogeochemical Cycles* 17:1034, doi:10.10 29/2002GB00 1970.

Keto, K. C., R. K. Kaufmann and C. E. Woodcock. 2000. Landsat reveals China's farmland reserves, but they're vanishing fast, *Nature* 406:121.

Keeling, R. F., S. Piper and M. Heimann. 1996. Global and hemispheric CO_2 sinks deduced from changes in atmospheric O_2 concentration. *Nature* 381:218–221.

Li, K., S. Q. Wang and M. K. Cao. 2003. Carbon storage in vegetation and soils of China. *Science in China (D)* 33:72–80.

Liu, J., Z. X. Zhang, D. F. Zhuang, Y. Wang, W. Zhou, S. Zhang, R. Li, N. Jiang and S. Wu. 2003. A study on the spatial-temporal dynamics changes of land-use and driving forces analyses of China in the 1990s. *Geographic Research* 22:1–12.

McLaughlin, S. and K. Percy. 2000. Forest health in North America: Some perspectives on actual and potential roles of climate and air pollution. *Water, Air and Soil Pollution* 116:151–197.

Myneni, R. B., J. Dong, C. J. Tucker, R. K. Kaufmann, P. E. Kauppi, J. Liski, L. Zhou, V. Alexeyev and M. K. Hughes. 2001. A large carbon sink in the woody biomass of northern forests. *Proceedings of the National Academy of Sciences USA* 98:14784–14789.

Ni, J. 2001. Carbon storage in terrestrial ecosystems of China: Estimates at different spatial resolutions and their responses to climate change. *Climatic Change* 49:339–358.

Pacala, S. W., G. C. Hurtt, D. Baker, P. Peylin, R. A. Houghton, R. A. Birdsey, L. Heath, E. T. Sundquist, R. F. Stallard, P. Ciais, P. Moorcroft, J. P. Caspersen, E. Shevliakova, B. Moore, G. Kohlmaier, E. Holland, M. Gloor, M. E. Harmon, S.-M. Fan, J. L. Sarmiento, C. L. Goodale, D. Schimel and C. B. Field. 2001. Consistent land- and atmosphere-based U.S. carbon sink estimates. *Science* 292:2316–2320.

Peng, C. and M. J. Apps. 1997. Contribution of China to the global carbon cycle since the last glacial maximum. *Tellus* 49B:393–408.

Prentice, I. C., G. D. Farquhar, M. J. R. Fasham, M. L. Goulden, M. Heimann, V. J. Jaramillo, H. S. Kheshgi, C. Le Quéré, R. J. Scholes and D. W. R. Wallace. 2001. The carbon cycle and atmospheric carbon dioxide. Pp. 183–237, in *Climate change 2001: The scientific basis*, third assessment report of the Intergovernmental Panel on Climate Change, edited by J. T. Houghton, Y. Ding, D. J. Griggs, M. Noguer, P. J. van der Linden, X. Dai, K. Maskell and C. A. Johnson. Cambridge: Cambridge University Press.

Rayner, P. J., I. G. Enting, R. J. Francey and R. Langenfelds. 1999. Reconstructing the recent carbon cycle from atmospheric CO_2, $d^{13}C$ and O_2/N_2 observations. *Tellus* 51B:213–232.

Schimel, D. S., J. I. House, K. A. Hibbard, P. Bousquet, P. Ciais, P. Peylin, B. H. Braswell, M. J. Apps, D. Baker, A. Bondeau, J. Canadell, G. Churkina, W. Cramer, A. S. Denning, C. B. Field, P. Friedlingstein, C. Goodale, M. Heimann, R. A. Houghton, J. M. Melillo, B. Moore III, D. Murdiyarso, I. Noble, S. W. Pacala, I. C. Prentice, M. R. Raupach, P. J. Rayner, R. J. Scholes, W. L. Steffen and C. Wirth. 2001. Recent patterns and mechanisms of carbon exchange by terrestrial ecosystems. *Nature* 414:169–172.

Streets, D. G., K. Jiang, X. Hu, J. E. Sinton, X. Q. Zhang, D. Xu, M. Z. Jacobson and J. E. Hansen. 2001. Recent reductions in China's greenhouse gas emissions. *Science* 294:1835–1836.

Tian, H., J. M. Melillo, D. W. Kicklighter, S. Pan, J. Liu, A. D. McGuire and B. Moore III. 2003. Regional carbon dynamics in monsoon Asia and its implications to the global carbon cycle. *Global and Planetary Change* 37:201–217.

Wang, S. Q., H. Q. Tian, J. Y. Liu and S. Pan. 2003. Pattern and change of soil organic carbon storage in China: 1960s–1980s. *Tellus* (B) 55:416–427.

Wang, X. K., Z. W. Feng and Z. Y. Ouyang. 2001. The impact of human disturbance on vegetative carbon storage in forest ecosystems in China. *Forest Ecology and Management* 148:117–123.

Wang X. K., Z. W. Feng and Y. H. Zhuang. 1996. Forest fires in China: Carbon dioxide emission to the atmosphere. Pp. 764–770 in *Biomass burning and global change*, edited by J. Levine. Boston: MIT Press.

Wu, H., Z. Guo and C. Peng. 2003. Distribution and storage of soil organic carbon in China. *Global Biogeochemical Cycles* 17:1048, doi:10.1029/2001GB001844.

Yang, X. and M. Wang. 2000. Monsoon ecosystem control on atmospheric CO_2 interannual variability: Inferred from a significant positive correlation between year-to-year changes in land precipitation and atmospheric CO_2 growth rate. *Geophysical Research Letters* 27:1671–1674.

Zhou, Y. 2000. Carbon storage and budget of major Chinese forest types. *Acta Phytoecologica Sinica* 24:518–522.

Chapter 12

IMPACTS OF AGRICULTURE ON THE NITROGEN CYCLE

ZHENGQIN XIONG, JOHN FRENEY, ARVIN MOSIER, ZHAOLIANG ZHU, YOUN LEE and KAZUYUKI YAGI

1. Introduction

Agriculture in East Asian countries, especially China, is highly intensive and with the aid of new technology and high input of fertilizer and pesticide is both very productive and meets the demands for food by the rapidly increasing population. Unfortunately, it is becoming apparent that the changing agriculture is having adverse effects, and the sustainability of land resources and the environment have become matters for concern (Zhu et al. 2005). This chapter documents the use of increased nitrogen to improve food supply, considers the effects of changing agricultural systems on the nitrogen cycle, and discusses approaches to reduce the impact of nitrogen on the environment.

2. Agricultural Patterns

The type and intensity of agricultural practice varies tremendously between and within countries of the region and as a result the effect on the nitrogen cycle can be small or large and regional or global.

2.1. China

Even though the population of China almost doubled between 1961 and 2004 (0.67 billion to 1.32 billion) sufficient food has been produced in the region to sustain the increased population. During the period 1961–2005 total grain production increased from 110 Tg yr^{-1} to 429 Tg yr^{-1} (Fig. 12.1) and total meat production from 2.5 Tg yr^{-1} to 78 Tg yr^{-1} (FAO 2007).

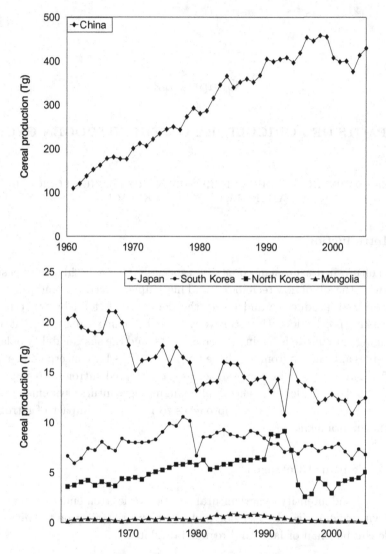

Fig. 12.1. Cereal production in East Asia between 1961 and 2005.

Paddy rice (*Oryza sativa* L.) is the major grain crop in China, grown mainly in the Yangtze River valley and southern China, and on the Yunnan-Guizhou Plateau. Maize (*Zea mays* L.) is grown in the provinces of northeastern, northern, and southwestern China, while wheat is planted throughout China, but mainly on the North China Plain. The large increase in food production was made possible by the introduction of high yielding varieties of rice, maize, and wheat (*Triticum aestivum L.*), intensification of cropping (double and sometimes triple cropping), irrigation, and the increased input of nitrogen (Zhu and Chen 2002).

The input of nitrogen may come from animal wastes, increased biological fixation, mineralization of crop residues, increased mineralization through cultivation of organic soils and fertilizer nitrogen (Smil 1999). Since 1961 the methods used to provide plant available nitrogen have changed dramatically. In 1961, of the nitrogen applied to agricultural fields in China, 13.5% came from fertilizer, 43.9% from animal wastes, 10.8% from biological nitrogen fixation, and 31.8% from reutilization of crop residue. By 1998, 70.9% of the nitrogen came from fertilizer nitrogen, 13.8% came from animal wastes, and 9.9% was fixed biologically (Mosier and Zhu 2000; Zhu and Chen 2002). In 1961, 0.54 Tg fertilizer N yr^{-1} was used to grow food and fiber for humans, and feed for animals. This has grown steadily over the intervening years to 28 Tg in 2005 (IFA 2007; Fig. 12.2). Zhu and Chen (2002) showed that there was a very strong positive relationship between annual food production and annual fertilizer nitrogen consumption.

Because of the increasing population and changing dietary preferences for meat, milk, and eggs, the animal population in China has also increased dramatically. The total number of animals in China has increased from 272 million in 1961 to 1,013 million in 2005. Many of these animals graze on the 300 million ha of grasslands, distributed mainly throughout Inner Mongolia, the basin between the Tianshan and Altai mountains in Xinjiang and on the Qinghai-Tibet plateau, and others are raised on feed lots. In 2005, livestock included large numbers of poultry (5.4 billion), pigs (489 million), goats (196 million), sheep (171 million), cattle (115 million), buffalo (23 million), and horses (7.6 million). The animals modify the physiology and nitrogen uptake characteristics of the pasture by grazing, change the physical characteristics of the soil by treading, and deposit excreta directly onto the pasture (Jarvis et al. 1995). Animals do not utilize the nitrogen they ingest efficiently; on an average only 10.5% of the nitrogen in grass, silage, or other feedstuff is converted into milk, meat, eggs or wool and the remainder is excreted in dung and urine (Van der Hoek 1998). In this chapter we used the default values given in Mosier et al. (1998) to determine the excretion for each animal type and summed the values to get total excretion. The results show that nitrogen excreted by livestock in China has increased from 6.1 Tg N yr^{-1} in 1961 to 21.9 Tg N yr^{-1} in 2005.

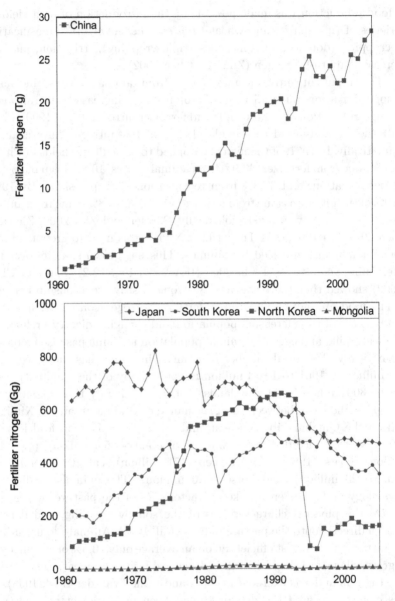

Fig. 12.2. Fertilizer nitrogen consumption in East Asia between 1961 and 2005.

2.2. Japan

The most striking feature of Japanese agriculture is the decrease in arable land. This has decreased from 5.7 million hectares in 1961 to 4.4 million hectares in 2005 (FAO 2007). This is due mainly to the decrease in cereal production area, including rice, wheat, and barley (*Hordeum vulgare L.*). However, the land is intensively cultivated, and rice paddies occupy most of the lowland area of the countryside. The terraces and lower slopes are planted with wheat and barley in the autumn and with sweet potatoes (*Ipomoea batatus (L.) Lam.*), vegetables, and upland rice in the summer. While the population of Japan has increased by a third between 1961 and 2004 (95–128 million), cereal grain production has almost halved in that time (20–12 Tg yr^{-1}; Fig. 12.1; FAO 2007). The reduction in cereal grain production appears to be the result of the decrease in available farm land, because the efficiency of fertilizer nitrogen use has been maintained at about 27 kg grain produced per kg nitrogen added over the whole period (Fig. 12.3). Production per hectare has been maintained at high levels through the use of technically advanced fertilizers and farm machinery (Mishima 2002; Kumazawa 2002). Fertilizer nitrogen use in Japan in 1961 was 633 Gg N yr^{-1}; it increased and fluctuated about 700 Gg N yr^{-1}

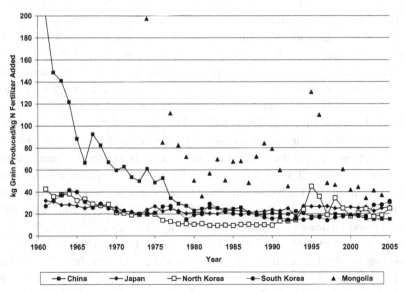

Fig. 12.3. The ratio of cereal production to total fertilizer nitrogen consumption for East Asian countries.

until 1983, after which it decreased steadily to the present day usage of ~460 Gg N yr^{-1} (Fig. 12.2) due to better management practices and concern for the environment (Yagi and Minami 2005).

While cereal grain production decreased over the last 40 years, meat production has increased dramatically, from 0.69 Tg yr^{-1} in 1961 to 3.1 Tg yr^{-1} in 2005. Consequently the number of animals in Japan increased from 7.9 million in 1961 to 14.1 million in 2005 (FAO 2007). Milk cows are numerous in Hokkaido, but they are also raised in Tohoku, and Kanto areas. Beef cattle are mostly concentrated in south western Japan. In addition to the cattle, 265 million chickens and 10 million pigs were raised in Japan in 2005. As a result of the extra animals the amount of nitrogen excreted in faeces and urine increased from 300 Gg N yr^{-1} in 1961 to 560 Gg N yr^{-1} in 2005. Much of this nitrogen is derived from imported grain (FAO 2007).

2.3. South Korea

South Korea's terrain is made up of partially forested mountain ranges separated by a deep, narrow valleys, and most agricultural activity takes place in the cultivated plains along the coasts in the west and south where the climate is temperate. South Korea's arable land comprises 16.7% of its land area, and permanent crops are grown on only 12% of the arable land (FAO 2007). Crop production is focused on rice (6.4 Tg), barley (0.26 Tg), maize, wheat, and millet. The population increased from 25.7 million in 1961 to 48 million in 2004, but cereal grain production (5.9 Tg yr^{-1} to 6.8 Tg yr^{-1}; Fig. 12.1) did not keep pace with population increase. This was primarily caused by the enormous growth of urban areas which led to a decrease of arable land (down from 2.03 million ha in 1961 to 1.64 million hectares in 2005; FAO 2007). The result of these developments is that by 2003 South Korea produced only 31% of the cereal grain it needed. However, vegetable production (especially in plastic greenhouses) has increased from 3.8 Tg in 1970 to 12 Tg in 2005. Fertilizer nitrogen use increased from 217 Gg N yr^{-1} in 1961 to 507 Gg N yr^{-1} in 1997, but then decreased rapidly to 354 Gg N yr^{-1} in 2005 (Fig. 12.2; FAO 2007). Because of increased incomes and changing preferences for food, meat consumption increased from 0.11 to 1.7 Tg yr^{-1} in the period 1961–2005 (FAO 2007). This has resulted in an increase in the animal population from 2.6 million to 11.9 million, and the need to import 8.3 Tg of cereal (mainly corn) in 2003 to feed the animals (Lee, personal communication). The livestock in 2005 included 9 million pigs and 2.3 million cattle, along with 110 million chickens. Nitrogen excretion from the animals increased from 74 Gg N yr^{-1} in 1961 to 270 Gg N yr^{-1} in 2005.

2.4. North Korea

North Korea's sparse agricultural resources limit agricultural growth. Climate, terrain, and soil conditions are unfavorable for farming. The major portion of the country is mountainous, and only about 22% of the land is arable (FAO 2007). Farming is concentrated in the flatlands of the four west coast provinces, where a longer growing season, level land, good soil, and adequate rainfall permit intensive cultivation of crops. The population increased from 11.7 million in 1961 to 23 million in 2004. The area of arable land increased from 2.15 million hectares in 1961 to 2.8 million hectares in 2005, as a result of reclaiming tidal land and terracing hill slopes (FAO 2007). At present about 590,000 ha are used for rice, 497,000 ha for maize, and 135,000 ha for wheat and barley (FAO 2007). In 1961, cereal grain production was 3.6 Tg yr^{-1} and this had increased to 5 Tg yr^{-1} by 2005 because of increased land area and fertilizer nitrogen. Fertilizer nitrogen use increased from 84 Gg N yr^{-1} in 1961 to 659 Gg N yr^{-1} in 1991. However, in the early 1990s, there was a shortage of fertilizer because of a lack of foreign exchange, and consumption of fertilizer nitrogen fell rapidly to 72 Gg N yr^{-1} in 1996 (Fig. 12.2). Current use is of the order of 160 Gg N (IFA personal communication).

Meat consumption has also increased in North Korea, from 86 Gg yr^{-1} in 1961 to 249 Gg yr^{-1} in 2005 and the number of livestock has increased from 2 to 6.7 million. This has increased excretion of nitrogen in faeces and urine from 55 to 120 Gg yr^{-1}.

2.5. Mongolia

In Mongolia, towering mountain ranges, plateaus, and the Gobi desert cover much of the country. Almost 90% of the land area is pasture or desert wasteland, of varying usefulness. Less than 1% is arable and 9% is forested (FAO 2007), temperatures are usually very cold or very hot (mean annual temperature is 4°C), and the little rainfall (average 250 mm) occurs in a few summer storms (Chuluun and Ojima 2002). Economic activity, traditionally, has been based on agriculture and breeding of livestock. In 1961 the human population of Mongolia was 984,000, and in 44 years this has increased to 2.6 million. During this period grain production decreased from 124 to 75 Gg yr^{-1}. No fertilizer nitrogen was recorded as having been used until 1970, when 0.1 Gg N was used. Fertilizer nitrogen use increased to 14 Gg in 1987 and then decreased to 4 Gg in 2003 (FAO 2007). The decrease in agricultural production in the late 1990s and early 2000s was probably due to a combination of widespread opposition to privatization, and adverse effects of weather.

Rangeland is the main source of forage for livestock, which increased from 23 million in 1961 to 28 million in 2005. During this period meat consumption increased from 166 to 194 Gg yr^{-1}, and excretion of nitrogen from animals increased from 430 Gg yr^{-1} in 1961 to 500 Gg yr^{-1} in 2005.

3. Fate of Added Nitrogen

As outlined above, in 2005 as a direct result of population growth, 27.1 Tg of nitrogen was applied to soils in East Asia to stimulate plant growth, and 23.4 Tg N was excreted from animals in the form of urine and faeces. It is well known that fertilizer nitrogen is not used efficiently and that much of the nitrogen applied can be lost from the plant–soil system by denitrification, ammonia volatilization, leaching, run-off, and by wind and water erosion (Cai 1997; Yan et al. 2003; Zhu 1997; Zhu and Chen 2002). The nitrogen excreted by animals can also be lost in the same way, but the mode and rate of loss varies depending on whether the nitrogen is excreted in urine or faeces (Misselbrook et al. 1998).

Using the methodology developed for calculating nitrous oxide emissions (IPCC 1997), we estimated gaseous emissions of nitrous oxide and ammonia, and leaching and run-off losses of nitrogen for the regions of East Asia. Whenever possible, we used East Asia specific parameter values, but when these were unavailable, the IPCC (1997) defaults were used.

As expected, because of the large amount of fertilizer nitrogen used and the large excretion of animal nitrogen, most of the nitrogen lost from the agroecosystems in East Asia came from China (Table 12.1). Emissions of nitrous oxide from the region totaled 1,111 Gg N yr^{-1}, with 94.4% emanating from China, 2.6% from Japan, and 1.4% from South Korea. On average, 52% was emitted directly, 28% came from animal systems, and 20% resulted indirectly as a result of ammonia and nitric oxide emissions and deposition. However, in Mongolia the bulk of the emissions (80%) came from animal production. Bujidmaa and Enkhman (2005) report that an additional 0.1 Gg N yr^{-1} is emitted to the atmosphere in Mongolia

Table 12.1. Loss of nitrogen (Gg N yr^{-1}) from East Asian agroecosystems to the environment in 2004.

	Nitrous oxide emission	Ammonia volatilization*	Leaching and runoff*
China	1,049	9,419	2,601
Japan	29	233	52
South Korea	16	127	39
North Korea	8	56	15
Mongolia	9	82	5
Total	1,111	9,917	2,712

*The values shown for ammonia volatilization and leaching and runoff are total estimated losses and do not account for local NH$_x$ and NO$_y$ deposition.

from biomass burning in cooking stoves. According to Zhu et al. (2005), East Asia contributes 47.6% of the total Asian emission of nitrous oxide.

Our calculations suggest that 9.9 Tg N as ammonia was emitted into the atmosphere in 2004 as a result of agricultural activities, and that China was the source of ~95% of it. In China 4.5 Tg of the ammonia nitrogen came from the application of fertilizer and 4.9 Tg from human and livestock excreta. East Asia is responsible for about 32% of the global ammonia emissions from animals and fertilizer, calculated to be 30.7 Tg (Bouwman et al. 1997).

Another 2.7 Tg N yr^{-1} is lost from East Asia by leaching and run-off to aquatic systems, and Zhu et al. (2005) estimate that 72% of this nitrogen is derived from application of fertilizer. Most of the leaching and runoff loss occurred in China and little nitrogen was lost by this pathway in Mongolia (Table 12.1).

The emission of nitrous oxide from East Asia is about 18% of the total emitted from global agriculture (6.3 Tg yr^{-1}; Mosier et al. 1998), which means that the East Asian emission has a significant effect on global climate. On the other hand, in addition to long-range transport of NH$_x$ aerosols, it is anticipated that much of the ammonia emitted into the atmosphere will be deposited nearby, and along with the nitrogen lost by leaching and run-off, affect, from a national perspective, the local environment (Howarth et al. 1996; Mosier, 2001).

4. Environmental Problems

Agriculture in East Asia has impacted the nitrogen cycle in a variety of ways ranging from too little nitrogen to produce food or to maintain soil organic matter levels, loss of surface soil, with its indigenous nitrogen by wind erosion, as a result of over grazing and cultivation, to an excess of nitrogen and the resulting contamination of water and air.

4.1. China

The transport of dissolved inorganic nitrogen in the three major rivers of China (Yangtze, Yellow, and Zhu) to estuaries increased by 0.775, 0.055, and 0.145 Tg per year, respectively, between 1980 and 1989 (Duan et al. 2000), and Shen et al. (2003) observed that 2,849 x 10^3 Gg total nitrogen, and $1,746 \times 10^3$ Gg dissolved inorganic nitrogen were discharged at the mouth of the Yangtze river in 1998. The annual transport of dissolved inorganic nitrogen in the Yangtze is strongly correlated with the consumption of fertilizer nitrogen, population and economic level in the river's catchment (Chen et al. 1999; Duan et al. 2000).

In recent years, algal blooms have been found to occur frequently in some lakes in China, and eutrophication has become a serious environmental problem. For example, Sun and Zhang (2000) reported that 61% of the 28 lakes investigated were significantly eutrophied. Research indicated that 7–35% of the total N load in the water body came from agricultural land (Jin 1995). Most of the eutrophic lakes are distributed in the middle and lower reaches of the Yangtze River and Yungui Plateau (Jin et al. 1995). Xing et al. (2001) concluded, from the high ammonium content, the low nitrate content, and the high $\delta^{15}NH_4^+$ value in the surface water of the Taihu Basin of the Yangtze Valley, that the source of the nitrogen in rivers and lakes was sewage from the cities, excrement from the rural population, and fish feeding, rather than fertilizer nitrogen.

Xing et al. (2001) also investigated the source of nitrogen in well water from the same region and found that 28% of the wells investigated had a nitrate concentration greater than the World Health Organization limit of 10 mg N L^{-1}. They also found that in wells, protected from domestic sewage and animal excreta, the nitrate levels were high during the wheat growing season and concluded that the nitrogen in well water came from fertilized farmland.

Brown (2001) reported that the combination of overploughing and overgrazing are creating a dust bowl in northwest China. Huge flocks of sheep and goats strip the land of its protective vegetation, then the strong winds of late winter and early spring can remove millions of tons of topsoil in a single day (Zhu and Liu 1989). The frequency and the intensity of dust storms in Northern China have increased from eight events in the 1960s to 14 events in the 1980s, and to more than 55 events during 2000–2002 (Liang 2002). According to the National Soil Erosion Survey in 2002, the area eroded by wind erosion in China is about $191 \times 10^4 \, \text{km}^2$ (Xu et al. 2002). The worst affected areas are Xinjiang (920,726 km^2), Inner Mongolia (594,607 km^2), Gansu (141,969 km^2), and Qinghai (128,972 km^2). The dust storms result in top soil loss, removal of nutrients, air pollution, massive deterioration of the country's agricultural resources and environment, reduction in food production, and economic loss (Wang Xiaobin and Cai Dianxiong, personal communication; Zhang et al. 2002; Lian 2002). Soils which have been strongly degraded by wind erosion have lost 66% of their organic matter and 73% of their total nitrogen (Zhao et al. 2002). It has been estimated that 5,000 Tg surface soil, with a mean total nitrogen concentration of 0.76 g kg^{-1}, is removed each year, resulting in the loss of 3.8 Tg nitrogen (Xi et al. 1990).

4.2. Japan

In Japan, nitrogen pollution is caused by excessive use of nitrogen fertilizer in tea plantations (661 kg N ha^{-1}) and vegetable fields (e.g. tomato, 243 kg N ha^{-1}). High rates of nitrogen are also applied to rusk (576 kg N ha^{-1}), tobacco (*Nicotiana tabacum L.*) (280 kg N ha^{-1}), maize (200 kg N ha^{-1}), and sugarcane (*Saccharum officinarum L.*) (226 kg N ha^{-1}), but the rate of application for rice is reasonable at 73 kg ha^{-1} (IFA 2002). Much of the nitrogen applied is not absorbed by the plants, and in many cases, except for rice and leguminous crops, excessive fertilizer applications depressed plant growth and yields (Nishio 2001). Mishima (2001) indicated that national average recovery of nitrogen applied as fertilizer or animal manure ranged from 0.28 to 0.35 kg kg^{-1} during 1980 to 1997. A survey of ground water conducted by the Ministry of the Environment in 1999, showed that 5.8% of tested wells and 4.7% of drinking water wells exceeded the environmental standard of 10 mg N L^{-1} (Kumazawa 2002).

More important, however, is the massive influx of nitrogen into Japan in food and feedstuffs. Nitrogen imported into Japan in 1997 in the form of food and feedstuff amounted to 1,212 Gg N, which is considerably greater than the 495 Gg of fertilizer nitrogen used (Yagi and Minami 2005). Much of the nitrogen from food and feedstuffs is finally excreted by people and livestock and results in contamination of the environment.

4.3. South Korea

While fertilizer nitrogen use has been reduced significantly since the mid 1990s, the use of animal manure as fertilizer has increased. Most of the animal manure is processed as compost before use in the field with the result that ammonia is volatilized during composting. Ammonia loss has not been measured directly, but changes in the N/P ratio of the material before and after composting suggests that about 40% of the manure nitrogen was lost during handling and processing (Lee personal communication). The results suggest that in 2002, 95 Gg N was lost as ammonia during composting of animal manure, compared with the 127 Gg lost as a result of fertilizer application (Table 12.1).

Although farmers now use less fertilizer nitrogen for rice production; down from 154 kg ha^{-1} in 1999 to 108 kg ha^{-1} in 2004, the rate of application of nitrogen for vegetables (202 kg N ha^{-1}) has increased well above that used for cereal production (134 kg N ha^{-1}). With the rapid growth in vegetable production and the high rate of application of fertilizer nitrogen environmental problems have increased. Vegetable production in the east highland slope area has resulted in erosion of soil and loss of nitrogen to the Han River, the water supply source for Seoul (Lee personal communication).

4.4. North Korea

In recent years, agricultural production in North Korea has decreased due to shortage of inputs and drought, and the food supply situation has caused international concern (FAO 2003). The average rate of nitrogen application for the two main crops, rice (81 kg N ha^{-1}) and maize (58 kg N ha^{-1}), is well below the recommended rates required for high yields (IFA 2002). In order to overcome the shortage of fertilizers, the government is seeking alternative sources of nutrients, and recognizes the need to improve the efficiency of nitrogen (FAO 2003).

4.5. Mongolia

Deforestation and efforts to increase grain and hay production in Mongolia by plowing up more virgin land have increased soil erosion from wind and rain. Most recently, with the rapid growth of newly privatized herds, overgrazing in selected areas is a concern. This has been documented in the National Plan of Action to Combat Desertification in Mongolia (1998).

According to this report, about 30% of total land is currently considered degraded.

Simulated grazing studies (Ojima et al. 1999; Chuluun and Ojima 1999) indicated that heavy summer grazing would result in a 15% soil carbon loss, and continuous year long grazing at a stocking rate of 4 sheep-month ha^{-1} would decrease soil organic carbon about 12% compared with grazing of 1 sheep per month per ha. As the nitrogen content of these soils is closely related to the carbon content ($r = 0.988$; C/N=9.4; Ojima et al. 1999; Chuluun and Ojima 1999) grazing will reduce the total nitrogen content by the same amount. Intensification of grazing can also reduce nitrogen inputs from soil algae populations. The field surveys and modeling analysis indicate that these ecosystems are best suited to nomadic grazing management and not for intensive sedentary grazing systems, or to intensive cropping management systems.

5. Approaches to Reduce Environmental Impacts of Lost Nitrogen

Nitrogen is indispensable for maintaining the recently achieved food self-efficiency in East Asia. Many areas still do not use enough fertilizer to maximize crop yields and the regional variability is large (compare North Korea and China). The average rate of nitrogen application for rice production in China was 180 kg ha^{-1}, which is markedly higher than the world average and among the highest average national nitrogen rates for rice in the world (Buresh et al. 2004). In the Taihu Basin 300 kg N ha^{-1} is applied to the rice crop and 250 kg N ha^{-1} is applied to wheat. The results for efficiency of use of fertilizer nitrogen in China (Fig. 12.3) suggest that the rate of application is too high. The ratio of cereal production to nitrogen fertilizer consumption can be considered a crude indicator of nitrogen use efficiency at the national scale (Dobermann and Cassman 2004). In general, large values for this ratio occur in low input systems that use little fertilizer nitrogen, as was typical of China in 1961 and Mongolia, and small values occur in high input systems.

In order to keep pace with a still-rapidly increasing human population, fertilizer use will continue to grow as food production does (Zhu et al. 2005). Sustainable development in food production is especially critical for the region because the needs for food are urgent. As arable land is restricted, production can only be increased by increasing the efficiency of the farming system. Fertilizer nitrogen use efficiency must be improved, and more use needs to be made of the large amount of manure nitrogen excreted in the region. Fertilizer nitrogen loss and its impact on the environment must be reduced. Nitrogen use efficiency depends on the conditions which influence crop growth (e.g., precipitation, temperature, nutrient uptake), ammonia volatilization, denitrification and leaching rates. In areas of high input agriculture, it has been assumed that fertilizer use efficiency could be improved by 20% (Mosier et al. 2002). If we can improve fertilizer use efficiency in East Asia to the same extent then fertilizer nitrogen use in the region can be reduced.

Improved efficiency of fertilizer use can be attained through improved management, using current technology. Some management options have proven useful including optimizing the rate and timing of nitrogen application and deep placement of fertilizer nitrogen into soil. It has also been demonstrated that balanced fertilizer use not only improves agronomic efficiency, but also reduces the accumulation of mineral nitrogen in soil and nitrogen loss (Mosier et al. 2004).

Policies and regulations should be developed and implemented, so as to encourage and facilitate the extension of current technology. Strong extension services with a high regard for protection of the environment are required for there is a need to balance the benefits derived from applications of nitrogen with the associated environmental costs (Zhu et al. 2005). To help meet future food needs, governments should consider promoting non-cereal food crops such as potato and vegetables which utilize land, water, and sunlight more efficiently and yield 5–10 times more per hectare than cereals (Ahmed 1996).

Literature Cited

Ahmed, S. 1996. Agriculture–fertilizer interface in the Asian and Pacific Region: Issues of growth, sustainability, and vulnerability. Food and Fertilizer Technology Center, Extension Bulletin. Taipei: Food and Fertilizer Technology Center.

Bouwman, A. F., D. S. Lee, W. A. H. Asman, F. J. Dentener, K. W. Van der Hoek and J. G. J. Olivier. 1997. A global high-resolution emission inventory for ammonia. *Global Biogeochemical Cycles* 11:561–587.

Brown, L. R. 2001. Dust bowl threatening China's future, in *Worldwatch Institute Report*, edited by L. R. Brown, C. Flavin, H. French et al., 23 May, 2001. New York: W. W. Norton & Company.

Bujidmaa, B. and S. Enkhman. 2005. The nitrogen monitoring and emission in Mongolia. *Science in China* 721–724, in Proceedings of the 3rd International Nitrogen Conference. Monmouth Junction, New Jersey, USA: Science Press.

Buresh R., S. B. Peng, J. L. Huang, J. C. Yang, G. H. Wang, X. H. Zhong and Y. B. Zou. 2004. Rice systems in China with high nitrogen inputs. Pp 143–153 in *Agriculture and the nitrogen cycle*, edited by A. R. Mosier, J. K. Syers and J. R. Freney. Washington, DC: Island Press.

Cai, G. X. 1997. Ammonia volatilization. Pp. 193–213 in *Nitrogen in soils of China*, edited by Z. Zhu, Q. Wen and J. R. Freney. Dordrecht, The Netherlands: Kluwer Academic Publishers.

Chen, J. S., X. M. Gao, X. H. Xia and D. W. He. 1999. Nitrogen pollution of Yangtze River. *Environmental Science* 18:289–293 (in Chinese).

Chuluun, T. and D. Ojima. 1999. Climate and grazing sensitivity of the Mongolian rangeland ecosystem. http://www.nrel.colostate.edu/projects/lutea/chuluun_poster.htm

Chuluun, T. and D. Ojima. 2002. Land use change and carbon cycle in arid and semi-arid lands of East and Central Asia. *Science in China* (Series C) 45:48–54.

Dobermann, A. and K. G. Cassman. 2004. Environmental dimensions of fertilizer nitrogen: What can be done to increase nitrogen use efficiency and ensure global food security? Pp. 261–278 in *Agriculture and the nitrogen cycle*, edited by A. R. Mosier, J. K. Syers and J. R. Freney. Washington, DC: Island Press.

Duan, S., S. Zhang and H. Huang. 2000. Transport of dissolved inorganic nitrogen from the major rivers to estuaries in China. *Nutrient Cycling in Agroecosystems* 57:13–22.

FAO (Food and Agriculture Organization). 2003. *Fertilizer use by crop in the Democratic People's Republic of Korea*. First version. Rome, Italy: FAO.

FAO (Food and Agriculture Organization). 2007. *FAOSTAT Database Collections*. Rome, Italy: FAO. http://www.apps.fao.org

Howarth, R. W., G. Billen, D. Swaney, A. Townsend, N. Jaworski, K. Lajtha, J. A. Downing, R. Elmgren, N. Caraco, T. Jordan, F. Berendse, J. Freney, V. Kudeyarov, P. Murdoch and Z. Zhu. 1996. Regional nitrogen budgets and riverine N & P fluxes for the drainages to the North Atlantic Ocean: Natural and human influences. *Biogeochemistry* 35:75–139.

IFA (International Fertilizer Industry Association). 2002. *Fertilizer use by crops*. 5th ed., Rome, Italy: IFA, IFDC, IPI, PPI, FAO. http://www.fertilizer.org/ifa/statistics/crops/fubc5ed.pdf

IFA (International Fertilizer Industry Association). 2007. http://www.fertilizer.org/ifa/

IPCC (Intergovernmental Panel on Climate Change). 1997. *Revised 1996 IPCC guidelines for national greenhouse gas inventories*. Paris, France: The Organization for Economic Cooperative Development.

Jarvis, S. C., D. Scholefield and B. Pain. 1995. Nitrogen cycling in grazing systems. Pp. 381–419 in *Nitrogen fertilization in the environment*, edited by P. E. Bacon. New York: Marcel Dekker Inc.

Jin, X. C. 1995. Eutrophication of the lakes in China. Pp. 267–322 in *The environment of lakes in China*, edited by X. C. Jin et al. Oceanology Publishers (in Chinese).

Jin, X., S. Liu, Z. Zhang et al. 1995. *Lakes in China — Research of their environment.* China Ocean Press.

Kumazawa, K. 2002. Nitrogen fertilization and nitrate pollution in groundwater in Japan: Present status and measures for sustainable agriculture. *Nutrient Cycling in Agroecosystems* 63:129–137.

Lian, L. 2002. Characteristics of dust storm distribution and their causes. *Ziran Zazhi* 24:335–338 (in Chinese).

Liang, Q. 2002. Dust storm coming again. *Meteorology* 3:2–6.

Mishima, S. 2001. Recent trend of nitrogen flow associated with agricultural production in Japan. *Soil Science and Plant Nutrition* 47:157–166.

Mishima, S. 2002. The recent trend of agricultural nitrogen flow in Japan and improvement plans. *Nutrient Cycling in Agroecosystems* 63:151–163.

Misselbrook, T. H., D. R. Chadwick, B. F. Pain and D. M. Headon. 1998. Dietary manipulation as a means of decreasing N losses and methane emissions and improving herbage N uptake following application of pig slurry to grassland. *Journal of Agricultural Science* 130:183–191.

Mosier A. R. 2001. Exchange of gaseous nitrogen compounds between terrestrial systems and the atmosphere. Pp. 291–309 in *Nitrogen in the environment: Sources, Problems, and management*, edited by R. F. Follett and J. L. Hatfield. Amsterdam: Elsevier Science B.V.

Mosier A. R. and Z. L. Zhu. 2000. Changes in patterns of fertilizer nitrogen use in Asia and its consequences for N_2O emissions from agricultural systems. *Nutrient Cycling in Agroecosystems* 57:107–117.

Mosier, A. R., J. K. Syers and J. R. Freney. 2004. *Agriculture and the nitrogen cycle: Assessing the impacts of fertilizer use on food production and the environment.* Washington, DC: Island Press.

Mosier, A., C. Kroeze, C. Nevison, O. Oenema, S. Seitzinger and O. van Cleemput. 1998. Closing the global N_2O budget: Nitrous oxide emissions through the agricultural nitrogen cycle. *Nutrient Cycling in Agroecosystems* 52:225–248.

Mosier, A. R., M. A. Bleken, P. Chaiwanakupt, E. C. Ellis, J. R. Freney, R. B. Howarth, P. A. Matson, K. Minami, R. Naylor, K. N. Weeks and Z. Zhu. 2002. Policy implications of human-accelerated nitrogen cycling. *Biogeochemistry* 57/58:477–516.

Nishio, M. 2001. Analysis of the actual state of nitrogen application in arable farming in Japan. *Japanese Journal of Soil Science and Plant Nutrition* 72: 513–521 (in Japanese).

Ojima, D. S., L. Tieszen, T. Chuluun, J. Belnap, J. Dodd and Z. Z. Chen. 1999. Factors influencing production systems and soil carbon of the Mongolian Steppe. http://www.nrel.colostate.edu/projects/lutea/dennis_poster.htm

Shen, Z. L., Q. Liu, S. M. Zhang, H. Miao and P. Zhang. 2003. A nitrogen budget of the Changjiang River catchment. *Ambio* 32:65–69.

Smil, V. 1999. Nitrogen in crop production: An account of global flows. *Global Biogeochemical Cycles* 13:647–662.

Sun, S. and C. Zhang. (2000). Nitrogen distribution in the lakes and lacustrine of China. *Nutrient Cycling in Agroecosystems* 57:23–31.
Van der Hoek, K. W. 1998. Nitrogen efficiency in global animal production. *Environmental Pollution* 102:127–132.
Xi, C. F. et al. 1990. *Soils of China*. Beijing: China Agriculture Publishing House.
Xing, G., Y. Cao, S. Shi, G. Sun, L. Du and J. Zhu. 2001. N pollution sources and denitrification in waterbodies in Taihu Lake region. *Science in China* (Series B) 44:304–314.
Xu, F., S. Guo and Z. Zhang. 2002. Soil erosion in China based on the 2000 national remote sensing survey. *Journal of Geographical Sciences* 12:435–443.
Yagi, K. and K. Minami. 2005. Challenges of reducing excess nitrogen in Japanese agroecosystems. *Science in China* Ser. C Life Sciences 48:928–936.
Yan, X., H. Akimoto and T. Ohara. 2003. Estimation of nitrous oxide, nitric oxide and ammonia emissions from croplands in East, Southeast and South Asia, *Global Change Biology* 9:1080–1096.
Zhang, Q., X. Zhao, Y. Zhang and L. Li. 2002. Preliminary study of sand-dust storm disaster and countermeasures in China. *Chinese Geographical Science* 12:9–12.
Zhao, H., X. Zhao, T. Zhang and R. Zhou. 2002. Changes of soil environment and its effects on crop productivity in desertification process of sandy farmlands. *Journal of Soil and Water Conservation* 16:1–4.
Zhu, Z. 1997. Fate and management of fertilizer nitrogen in agro-ecosystems. Pp. 239–279 in *Nitrogen in soils of China*, edited by Z. Zhu, Q. Wen and J. R. Freney. Dordrecht, The Netherlands: Kluwer Academic Publishers.
Zhu, Z. L. and D. L. Chen. 2002. Nitrogen fertilizer use in China-contributions to food production, impacts on the environment and best management strategies. *Nutrient Cycling in Agroecosystems* 63:117–127.
Zhu, Z. and S. Liu. 1989. *Desertification and its Control in China*. Beijing, China: Science Press.
Zhu, Z. L., Z. Q. Xiong and G. X. Xing. 2005. Impacts of population growth and economic development on the nitrogen cycle in Asia. Science in China Ser. C Life Sciences 48:729–737.

Chapter 13

IMPACTS OF LAND USE ON STRUCTURE AND FUNCTION OF TERRESTRIAL ECOSYSTEMS AND BIODIVERSITY

KANEHIRO KITAYAMA, GUANGSHENG ZHOU and KEPING MA

1. Introduction

A web of organisms is maintained through energy (carbon) and mineral flows in a given ecosystem. Energy and mineral flows include a grazing chain that starts from live plant parts (biomass) and a detritus chain that starts from dead plant parts (necromass). In either case, plants provide dependent organisms with food resources. The magnitude of primary productivity (i.e., the major ecosystem function), theoretically, defines the number of trophic levels and the complexity of food web (i.e., ecosystem structure). Dependent organisms, on the other hand, maintain plant populations through pollination and mineral recycling. Ecosystem structure and function are thus intimately related to each other through the web of organisms. These ecosystem processes are the basis for the services that an ecosystem provides including the dynamics of carbon.

Our attempt is to broadly review the observational and model research on how climate change, land use, and land-use change impact the terrestrial ecosystems and biodiversity in the East Asian monsoon region. As vascular plants are the major primary producers, which influence the biodiversity of dependent organisms, we confine our discussion to the richness and composition of vascular plant species.

2. Overview of Terrestrial Ecosystems in East Asia

The region through Japan, Korea, and eastern parts of China is characterized by a moist climate during summer, which is the growing season for plants. Therefore, at a biome scale, moisture is not severely limiting terrestrial vegetation. The western part of East Asia is, however, arid. This is because the moisture-laden air, which flows from the Indian subcontinent, is interrupted by the Himalayas. The orographic uplift of the moisture-laden

Fig. 13.1. Zonobiomes of East and Southeast Asia (after Walter 1979). (I) Equatorial with diurnal climate; (II) tropical with summer rains; (III) subtropical-arid; (IV) winter rain and summer drought; (V) warm-temperate (maritime); (VI) typical temperate with a short period of frost; (VII) arid-temperate with a cold winter; (VIII) cold-temperate (boreal); (IX) Arctic.

air dumps much of the rain in India, and the lee side (Tibet) becomes arid. The Tibetan Plateau becomes the source of high heat and reinforces the monsoon circulation.

Soils are Entisols and Inceptisols mostly of young volcanic origin throughout Japan and Korea. Southeastern China has a large extent of more weathered Udults, which are characterized by subsurface horizons of clay accumulation and low base supply. The Tibetan Plateau includes permafrost and rugged mountains devoid of soils.

Figure 13.1 illustrates the climatically defined biomes of East Asia (after Walter 1979). A zonation of thermally defined biomes can be seen from south to north (equatorial to warm-temperate to typical temperate with a short period of frost). Interior of China is characterized by the biome of arid temperate with a cold winter.

Altitudinal vegetation belts can be defined on the horizontal plane of East Asia. Ohsawa (1995) demonstrated the vegetation belts of maritime East Asia in an altitudinal–latitudinal plane (Fig. 13.2). Evergreen broad-leaved forests are replaced with deciduous broad-leaved forests and evergreen coniferous forests at middle latitudes. He explains the distribution of these biomes based on two thermal variables: warmth index (WI, sum of

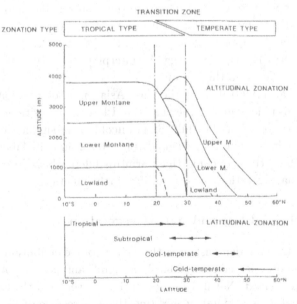

Fig. 13.2. Latitudinal–altitudinal template of the vegetation of maritime East Asia (after Ohsawa 1995).

monthly mean temperatures which exceed 5°C month^{-1}), and the coldest month temperature. The distribution of forest biomes (either altitudinal or latitudinal) is controlled by WI 15. The limit of evergreen broad-leaved (but not deciduous) forests is, however, controlled by the coldest month mean temperature of −1°C. Therefore, the forests in maritime East Asia are controlled by temperature.

Western East Asia demonstrates arid biomes in the lee of the Tibetan Plateau, namely in upslope sequence the montane shrub–grass tussock or shrub–steppe, montane fruticose desert, and montane drawf suffruticose desert (Zhang et al. 1996). The contrast of biome distributions in eastern (maritime) versus western East Asia suggests that controlling factors for biomes are different: temperature sum and/or coldest temperature control biomes in the east, while temperature as well as moisture control biomes in the west (Zhang et al. 1996). Zhang et al. (1996) obtained the following climatic equations for the Tibetan Plateau region:

$$BT = 46.18 - 0.44965L - 0.13627G - 0.0037006H,$$
$$PET = 2727.5 - 26.486L - 8.086G - 0.218432H,$$
$$PER = 10^{**}(4.7241 + 0.041754L - 0.05083G - 0.00040665H),$$

where L, G, and H denote latitude, longitude, and altitude of each site. BT is biotemperature (i.e., temperature that controls biome distributions, comparable to WI by definition), PET is potential evapotranspiration, and PER is potential evapotranspiration rate. PET is more strongly responding to latitude than BT is, suggesting the interplays of orographic effects and monsoon circulation in this region.

Fang et al. (1996) divided the East Asian monsoon region into three macro-topographic regions: the Tibetan Plateau, transitional mountain area, and eastern plains and low hills. Siberian cold air mass influences the eastern plains and low hills, while it is prevented from entering into the Tibetan Plateau. Fang et al. (1996), thus, concludes that warmth is different between east and west, and altitudinal vegetation zones are uplifted in the Tibetan side.

3. Current Land Use and Actual Vegetation

Although Fig. 13.1 demonstrates the continuous distributions of biomes (which are potential climatic zones), a significant portion of East Asia had been converted. Natural forests in Japan were mostly converted to plantations of high economic values (mostly evergreen conifers) in the past four decades. In China the major land-use changes occurred in the past several centuries (Fu and Wang, Chapter 12, this volume), unlike currently rapidly changing regions such as the Amazon. Agricultural practices in East Asia are the intensive subsistence tillage of rice crop from Japan, though Korea, to the middle and south of China. The intensive subsistence tillage of non-rice crops is practiced in north East Asia. Rudimental sedentary cultivation is practiced in some mountainous regions of China, and nomadic herding in interior China.

4. Impacts of Land Use and Climate Change on Ecosystems and Biodiversity

"Cultural scenery" which consists of rice fields admixed with substituted vegetation managed by humans actually characterizes much of East Asia. A relatively long history of land-use change, and the long existence of substituted vegetation make the ecosystem assessment of land-use change difficult. Moreover, the synergy with climate change makes the assessment even harder. Table 13.1 lists ecological stress factors arising from land use or land-use change, which are considered important in East Asia. All these factors may override climate-change effects and their consequences may have to be assessed as the synergy with climate change.

Another complexity is the interaction with biodiversity itself. Ecological factors affect the physiology of all organisms. Consequences of the factors are transmitted through the chain of biotic organization from tissues, organs, individuals, the web of species interactions, and lastly to ecosystem. Different species differ in the response to a stress factor. Differential responses may change the ranking of species in competitive ability. Scientists need to take all of these aspects into consideration when assessing the consequence of ecosystem change and biodiversity alteration. Furthermore, it is difficult to extrapolate the expected ecosystem consequence to a landscape process. Numerous observational and experimental studies deal with ecological processes at specific sites but few studies have successfully extrapolated how the local processes are transmitted to a landscape or a biome level.

With these constraints, ecologists in the site-based projects in East Asia have taken "the-space-for-the-time approach" by making use of natural environmental gradients. Several research sites are established along a spatial environmental gradient with standardized methods of monitoring applied, and temporal changes are inferred from spatial patterns obtained

Table 13.1. Factors arising from land use and land-use change, which can affect terrestrial ecosystems in East Asia.

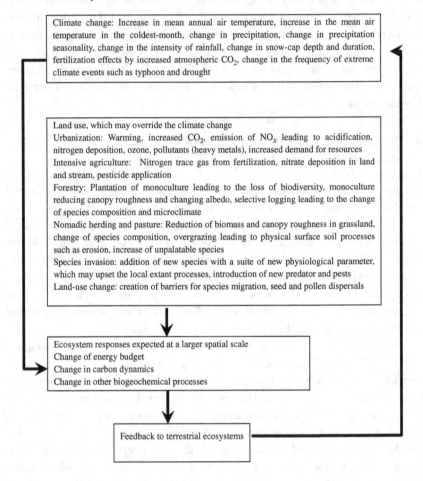

along the gradient. A major emphasis was placed on such projects in the 1990s and later and the results of these studies are presented below.

4.1. *Northeast and East China: Transect analyses*

Terrestrial ecosystems play a critical role in modulating the global carbon cycle. Human activities are disrupting terrestrial ecosystems that directly affect ecosystem functions. To understand how terrestrial ecosystems will respond to global change and to reduce the uncertainty of the carbon

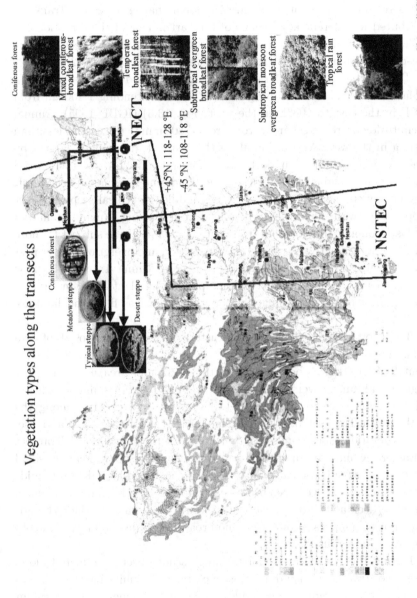

Fig. 13.3. Vegetation types along the Northeast China Transect (NECT) and the North–South Transect of Eastern China (NSTEC).

budget in monsoon Asia, the global change terrestrial transects suggested by the International Geosphere — Biosphere Program — were established in this region. At present, there are 15 global change terrestrial transects distributed in different key regions of the earth. Two of these are located in China. They are the water-driven Northeast China Transect and the heat-driven North–South Transect of Eastern China (Zhou 2002).

The Northeast China Transect, established in 1994, was mainly a rainfall-driven transect centering at 43.5°N, and extending 1,400 km from 132°E in the east to 108°E in the west (Fig. 13.3) (IGBP 1995). Annual precipitation decreased sharply from around 800 mm in the east to less than 100 mm in the west. Vegetation along the transect, determined largely by rainfall gradient, shifts from dark conifer forest, conifer broad-leaved mixed forest, deciduous broad-leaved forest, woodlands and shrublands in the east, via meadow steppe and cropland in the middle, to typical steppe, desert steppe in the west. Due to high pressures from growing human population and fast economic development, the transect area has been subject to more and more human activities, including cultivation and overgrazing over past years. A general question guiding the Northeast China Transect is "how does water availability influence the composition of plant functional types, soil organic matter, net primary production, trace gas flux, and land-use distribution?"

The North–South Transect of Eastern China is composed of two parts: 110°–120°E, 15°–40°N, and 118°–128°E, 40°–57°N. It covers about 2,889,100 km^2, about 30.1% of the total land area in China. The geographical range covers 1,323 counties from 25 provinces, municipalities, and autonomous regions (Fig. 13.3). The main driving forces along the North–South Transect of Eastern China are temperature and land use. Eight vegetation types from south to north along the North–South Transect of Eastern China are mountain tropical rain forest, tropical seasonal rain forest, low subtropical monsoon evergreen broad-leaved forest, mid-latitude subtropical evergreen broad-leaved forest, high-latitude subtropical evergreen/deciduous broad-leaved forest, warm temperate deciduous broad-leaved forest, temperate coniferous and broad-leaved mixed forest, and cold temperate coniferous forest.

During the last decade, global-change studies were conducted along these two transects including the experiments simulating the effects of precipitation, temperature and doubled CO_2, long-term ecosystem observation and field survey with sampling at 25 km intervals as well as modeling (Zhou 2002).

The spatial characteristics of plant species, such as its geographical distribution, pattern of frequency, and pattern of dominance are indicative of responses to an environmental gradient and climatic changes. Plant species may move at different rates and in different directions as the result of climate change, and may have arisen from species-specific tolerances to the climate variables. Thus, plant migrations in coming decades may result in changing community structure as spreading populations outrun their slower-migration competitors, pollinators, seed dispersers or natural enemies and they encounter new ones as the population frontier extends to new areas (Wiegand et al. 1998). Therefore, detecting the change of spatial characteristics of plant species on environmental gradients would be helpful to study the composition of terrestrial ecosystems at a large scale under global climate change.

In the analysis of the spatial characteristics of 16 tree species along the Northeast China Transect from 287 permanent plots in 1986 and 1994 (Chen et al. 2002), it was found that species changed individualistically but not in association. For instance, the spatial correlation between species changed during eight years, such as those between *Larix olgensis* and *Betula platyphylla*, and *Acer mono* and *Ulmus* spp. Distribution areas, the size of patch, frequency, and dominance of species all changed individualistically. *Betula costata* and *Phellodendron amurense* spread most quickly toward west and east, respectively. All tolerant tree species extended their frontiers and all intolerant tree species retracted their frontiers except *Betula platyphylla*. Although *Quercus mongolica*, *Acer mono*, and *Larix olgensis* still dominated the vegetation along the Northeast China Transect, it seemed that environmental change and habitat loss would favor the population of *Betula costata*, *Juglans mandshurica*, *Ulmus* spp., and *Fraxinus rhynchophylla* but disfavor the population of *Populus davidiana*, *Betula platyphylla*, *Betula dahurica*, *Tilia* spp., and *Larix olgensis*.

The changes of five main plant communities in the Songnen Plain along a grazing-disturbance gradient showed that grazing disturbed the grassland environment through gnawing and treading of livestock, and resulted in the changes in species composition and population's dominance. The species richness in the communities gradually decreased with aggravating grazing intensity from saline-moist meadow steppes to typical steppes (Yang et al. 1999).

The plant diversity of nine main grassland communities along the grazing-disturbance gradient on the Northeast China Transect showed that Shannon index (an index of species diversity) was highest in moderate grazing or heavy grazing stages, and decreased in the following sequence: moderate grazing (heavy grazing) > heavy grazing (moderate grazing) > light grazing > over-grazing. Species evenness changed disproportionately more than species richness did, indicating that the competitive relationships among co-existing species were changed by grazing. The life-form diversity also changed along the grazing-disturbance gradient. Plant diversity of the steppe region on the Northeast China Transect decreased in the following order: meadow steppe > typical steppe > desert steppe > alkaline meadow, and species richness had higher contribution rate to the change of species diversity than did species evenness (Yang et al. 2001).

Bai et al. (2000) pointed out that the moisture and temperature conditions in the Xilin River basin, Inner Mongolia, have significant influence on plant species diversity and productivity of grassland communities. The species richness, diversity, and productivity of plant communities have positive relationships with precipitation, elevation, soil organic carbon and total nitrogen content, and negative relationships with mean annual temperature, $\geq 10°C$ cumulative temperature, and aridity. Soil organic carbon and aridity are the main factors controlling plant species richness and productivity of grassland communities. Moreover, rainfall and its seasonal distribution have different effects on community productivity. Annual changes and seasonal distribution of precipitation led to the fluctuation of the primary productivity of *Stipa krylovii* community. Positive effects of ten-day precipitation from early January to early April and from the end of June to the end of August were found on the primary productivity. These seasons were critical to plant growth in terms of water requirement, and the effect was greater in the first period than in the second period. The reversed pattern was found in the effect of ten-day precipitation from the middle of April to the middle of June. Some degrees of water stress may be necessary to promote the growth of roots and carbohydrate reserves (Bai 1999).

Zhou et al. (2002) discussed the relations of plant species number, soil carbon and nitrogen, and above-ground biomass with a precipitation gradient and interactions with land-use practices (grassland fencing, mowing, and grazing) on the basis of data from the west part of the Northeast China Transect. The results indicated that the above-ground biomass of grassland communities has a linear relationship with precipitation under three land-use practices. Plant species number, soil organic carbon, and soil total nitrogen have linear relationships with precipitation under fencing and mowing, while the relationships are nonlinear under grazing. Plant species number, soil carbon and total soil nitrogen have strong linear relationships with the above-ground biomass under both fencing and mowing, while they seem to have nonlinear relationships under grazing. Land-use practices along the precipitation gradient result not only in change of grassland communities but also in qualitative changes of their structure and function. Thus, grasslands are more vulnerable to the changes in climate under mowing than under grazing. At a given precipitation level, number of plant species, above-ground biomass, and soil organic carbon are higher under low to medium intensity of human activities (mowing and grazing).

The responses of vegetation on the North–South Transect of Eastern China to global change, under different climatic scenarios, were obtained by subjecting modified models to different climatic scenarios derived from seven general circulation models (Yu et al. 2002). The simulations indicated that there were significant increases in the area of deciduous broad-leaved forests but evergreen broad-leaved forests being unchanged under all seven scenarios. The distributions of all the other natural vegetation decreased. Differential responses between deciduous versus evergreen broad-leaved forests are due to the climatic conditions of the extant vegetation. Under contemporary climate and atmospheric CO_2 concentration, deciduous broad-leaved forests in the northern part of the North–South Transect are more likely droughty than evergreen broad-leaved forests in southern China. With similar increases in temperature, a higher precipitation increase would lead to a greater increase in area for the deciduous broad-leaved forests because more water is available for assimilation at higher precipitation. The increases in area of deciduous broad-leaved forests were low for two scenarios because the greater increases in temperature for these two scenarios increased evapotranspiration and hence reduced the water use efficiency of deciduous trees. This effect of increasing precipitation and temperature on evergreen broad-leaved forests is less evident because this vegetation type is mainly located in the southern area where the baseline precipitation is much higher than in the north.

Tsunekawa et al. (1996b) predicted vegetation changes in entire China with two scenarios of climate change using a multinominal logit model: scenario A, a 2°C rise in mean annual air temperature and a 20% rise in annual precipitation; scenario B, a 4°C rise in mean annual air temperature and a 20% rise in annual precipitation. They assumed an even climate change in entire China. They concluded that (1) the conifer forest distributed in northeast China will be lost from China, (2) the broad-leaved forest distributed in east China will shift northward by around 3 degrees in latitude for scenario A, and 5 degrees for scenario B, and (3) the desert region in west China will expand and steppe and savanna will decrease especially in scenario B. The latter prediction is in line with Zhang et al. (1996).

4.1.1. Tibetan Plateau

Zhang et al. (1996) have modeled the influences of climate change on the vegetation of the Tibetan Plateau by using an improved Holdridge life zone classification scheme, a permafrost model and a net primary productivity model. They assumed a 4°C rise in mean annual air temperature and a 10% increase in annual precipitation over the entire Tibetan Plateau. They predict a shift of altitudinal vegetation zones and a shrinkage of the permafrost region; these effects add to the desertification which is spreading in the region due to the massive land-use change.

4.1.2. Japan and maritime East Asia

The project "Terrestrial Ecosystem of Monsoon Asia" was conducted as one of the core projects of Global Change and Terrestrial Ecosystems to predict the effects of elevated CO_2 and climate change on the distribution and structure of forests in monsoon Asia, and to determine the associated feedback effects to the global carbon cycle. The research strategy of the project was based on the environmental gradient concept along a transect in monsoon Asia (Fig. 13.2) from boreal forests in Siberia and Hokkaido, through cool and warm temperate forests in mainland Japan and eastern China, to tropical rain forests in Southeast Asia. This transect included two high priority areas of Global Change and Terrestrial Ecosystems: boreal forests, which were expected to change significantly because of increased temperature due to increasing concentrations of greenhouse gases in the atmosphere, and tropical rain forests, which were endangered by the rapid change in land use due to deforestation and high population pressure. A particular emphasis was placed on the linkage between physiological processes of foliage canopy and landscape-scale processes of plant demography and plant community dynamics, where plant individual processes were integrated from physiology, which in turn were projected to geographic patterns.

The above-ground biomass of the forests in the region significantly correlated with cumulative air temperature during the growing season of trees. Kitayama and Aiba (2002) studied the interplays of soil nutrients and temperature in rain-forest ecosystems (i.e., the ecosystems that are not limited by moisture). The patterns of net primary productivity and decomposition in relation to air temperature were similar in soil-phosphorus rich versus poor forests. Underlying mechanisms were that nutrient-use efficient species (particularly phosphorus) in photosynthesis replaced inefficient species in cooler and phosphorus-poor forests. This clearly indicated the interaction of species diversity, climate, and nutrients.

In another project, Tsunekawa et al. (1996a) predicted the changes in Japanese vegetation using a multinominal logit model. Japanese vegetation was divided into 1 km × 1 km grids. They concluded that vegetation types classified on a scale equivalent to biomes or climatic belts will differ by 23% in grid cell for a 1°C rise in mean annual air temperature, by 44% for 2°C, and by 62% for 3°C.

"Conventional" studies use a statistical examination of a suite of climatic parameters with the distribution of extant vegetation types. Prediction will be conducted by changing the values of explanatory parameters. By contrast, Ishigami et al. (2001) used the process-based model, BIOME3 (Haxeltine and Prentice 1996) to predict the change of potential natural vegetation using 100 km^2 mesh climate data developed by a Japanese institute for the year 2050. In this model, the combination of climate and edaphic data (parameters) in a given grid cell is used to estimate the maximum net primary productivity, and the functional plant type (life form) which can achieve that maximum net primary productivity is assigned to the grid cell. According to their conclusions, cool-temperate deciduous broad-leaved forests shift northward and occupy the majority of the northernmost Hokkaido Island, which now more-or-less lacks such vegetation. This model does not take into consideration the migration speed of plant species, nor current land use types. It predicts merely "potential" vegetation, which assumes that any human land-use impacts cease. Harasawa and Nishioka (2003) compiled studies on global warming impacts on Japanese ecosystems.

4.1.3. Species migration and land use

Effects of land use on terrestrial ecosystems are manifested not only through biogeochemical processes but also through creating a barrier for migrating species. Kohyama and Shigesada (1995) took a unique approach to ecosystem-change evaluation by incorporating the bottom-up processes of competition for light among tree populations using a size-structure model. They predicted the effects of global warming on the migration of three vegetation zones along a latitudinal gradient, which was not limited by moisture, such as the one in maritime East Asia. When they shifted climate parameters with the speed of 700 km per 100 years, surprisingly there was a time lag of more than 2,000 years for trees to reach a new climatically equilibrated condition. This time lag was derived from extremely slow colonization of migrating tree seedlings at each ecotone due to the competition with "current" vegetation. This study points out the importance of bottom-up ecological processes of constituting trees. Paleoecological evidence actually suggests that migration speeds were often slow and time lags existed during the post-glacial era (Davis 1989). Another piece of empirical evidence for the problem of migration speed was obtained by Matsui et al. (2004), who studied the climatic controls of the distribution of Japanese beech, Fagus crenata, the dominant species of cool-temperate deciduous forests.

5. Conclusions

Excellent infrastructure and networks of field-observation sites exist in East Asia. Capacity to continue quality monitoring is still high in this region. One of the strengths is the ability to measure photosynthesis or carbon exchange at a leaf to a stand level, and to model forest structure as an integration of functional plant types. However, the measurement and modeling of soil processes have been less emphasized. Considering biota, climate, and soils together form biogeochemical processes of a terrestrial ecosystem, and considering East Asia has a unique combination of soil types, soil and related biogeochemical processes need to be emphasized in future studies. The participation of soil scientists and biogeochemists is strongly encouraged.

Furthermore, plant-species responses to the same climate change are individualistic due to physiological differences, and thus realistic prediction of the effects of climate, and land-use changes depends on the knowledge of species physiology. Those sites, which were established in previous projects, should continue to be used to detect ecosystem and biodiversity changes, to provide physiological parameters to evaluate the feedback to earth and atmospheric processes, and to suggest adaptation policies for human societies.

Literature Cited

Bai, Y. 1999. Influence of seasonal distribution of precipitation on primary productivity of Stipa krylovii community. *Acta Phytoecologica Sinica* 23:155–160.

Bai, Y., L. Li, Q. Wang, L. Zhang, Y. Zhang and Z. Chen. 2000. Changes in plant species diversity and productivity along gradients of precipitation and elevation in the Xilin River Basin, Inner Mongolia. *Acta Phytoecologica Sinica* 24:667–673.

Chen, X., G. Zhou and X. Zhang. 2002. Spatial characteristics and change for tree species along the North East China Transect (NECT). *Plant Ecology* 64:65–74.

Davis, M. B. 1989. Lags in vegetation response to greenhouse warming. *Climate Change* 15:75–82.

Fang, J.-Y., M. Ohsawa and T. Kira. 1996. Vertical vegetation zones along 30°N latitude in humid East Asia. *Vegetatio* 126:135–149.

Harasawa, H. and S. Nishioka. 2003. *Global warming and Japan* Tokyo: Kokon-Shoin Publisher (in Japanese).

Haxeltine, A. and I. C. Prentice. 1996. BIOME3: An equilibrium terrestrial biosphere model based on ecophysiological constraints, resource availability, and competition among plant functional types. *Global Biogeochemical Cycles* 10:693–709.

IGBP. 1995. Spatial extrapolation and modeling on IGBP transects. *Global Change Report* 36:15–20.

Ishigami, Y., Y. Shimizu and K. Omasa. 2001. Prediction of changes in vegetation distribution in Japan using process based model and GCM data. Proceedings of LUCC Symposium 2001.

Kitayama, K. and S. Aiba. 2002. Ecosystem structure and productivity of tropical rain forests along altitudinal gradients with contrasting soil P pools on Mount Kinabalu, Borneo. *Journal of Ecology* 90:37–51.

Kohyama, T. and N. Shigesada. 1995. A size-distribution-based model of forest dynamics along a latitudinal environmental gradient. *Vegetatio* 121:117–126.

Matsui T., Yagihashi T., Nakaya T., Tanaka N. and Taoda H. 2004. Climatic controls on distribution of Fagus crenata forests in Japan. *Journal of Vegetation Science* 15:57–66.

Ohsawa, M. 1995. Latitudinal comparison of altitudinal changes in forest structure, leaf type, and species richness in humid monsoon Asia. *Vegetatio* 121:3–10.

Tsunekawa, A., H. Ikeguchi and K. Omasa. 1996a. Prediction of Japanese potential vegetation distribution in response to climate change. Pp. 57–65 in *Climate change and plants in East Asia,* edited by K. Omasa, K. Kai, H. Taoda, Z. Uchijima and M. Yoshino. Tokyo: Springer.

Tsunekawa, A., X. Zhang, G. Zhou and Omasa K. 1996b. Climatic change and its impacts on the vegetation distribution in China. Pp. 67–84 in *Climate change and plants in East Asia,* edited by K. Omasa, K. Kai, H. Taoda, Z. Uchijima and M. Yoshino. Tokyo: Springer.

Walter, H. 1979. *Vegetation of the earth.* New York: Springer.
Wiegand, T., K. A. Moloney and S. J. Milton. 1998. Population dynamics, disturbance, and pattern evolution: Identifying the fundamental scales of organization in an model ecosystem. *The American Naturalist* 152:321-337.
Yang, L., M. Han and J. Li. 2001. The plant diversity change of grassland communities along grazing disturbance gradient in Northeast China Transect. *Acta Phytoecologica Sinica* 25:110-114.
Yang, L., J. Li and Y. Yang. 1999. β-diversity of grassland communities along gradient of grazing disturbance. *Chinese Journal of Applied Ecology* 10:442-446.
Yu, M., Q. Gao and X. Zhang. 2002. Responses of vegetation structure and primary production of a forest transect in eastern China to global change. *Global Ecology and Biogeography* 11:223-236.
Zhang, X., D. Yng, G. Zhou, C. Liu and J. Zhang. 1996. Model expectation of impacts of global climate change on biomes of the Tibetan Plateau. Pp. 25-38 in *Climate change and plants in East Asia,* edited by K. Omasa, K. Kai, H. Taoda, Z. Uchijima and M. Yoshino. Tokyo: Springer.
Zhou, G., Y. Wang and S. Wang. 2002. Responses of grassland ecosystems to precipitation and land use along the Northeast China Transect. *Journal of Vegetation Science* 13:361-368.
Zhou, G. 2002. *Northeast China transect and global change — aridification, human activities and ecosystems.* Beijing: Meteorological Press.

Chapter 14

IMPACT OF LAND USE IN SEMIARID ASIA ON ARIDIFICATION AND DESERTIFICATION

DENNIS OJIMA, WEN JIE DONG, ZHUGUO MA,
JEFF HICKE and TOM RILEY

1. Introduction

The temperate East Asian semiarid lands have undergone dramatic land use and climatic changes during its past 100 years. The region is noted for its rapid increase in surface temperature, approximately a 1°C increase during the past 50 years (Jones and Moberg 2003). The recent changes in climate and socioeconomic factors affecting land use in the region has led to cropland abandonment, destocking of certain rangelands and increased stocking of others, degradation of soils due to salinization and desertification, and damage to wetlands due to modifications of water regime. Carbon stores and fluxes throughout the region have been modified by changes in land use over the decades.

The modifications of the social–economic situation in the region have been a primary agent of land-use change during the past century, and especially during the past 50 years. Interactions between and among policies, human responses, and the Earth system function cannot be decoupled. The semiarid ecosystems in the East Asian region present unique research challenges to assess the effect of land cover and land-use changes on carbon biogeochemistry and climate feedbacks. The main focus of this chapter is to describe the general nature of changes in climate and land use intensity on the land resource and how these may affect dust aerosol fluxes, water vapor exchange, and carbon dioxide (CO_2) exchange of these ecosystems. The region of study extends from the Mongolian plateau in Mongolia to the arid regions of China. This region experiences moderate influences from the Asian monsoon (Fu and Wang, Chapter 10, this volume; Li and Huang, Chapter 5, this volume).

2. Regional Characteristics

The steppe of East Asia is large in areal extent, has long land use history, and is situated in a geographical region where large climate warming trends are observed. The semiarid region of Mongolia and northern and western China which experiences annual precipitation between 180 and 400 mm per year with 70–80% of the rainfall occurring between May and August. The region has a continental climate with the mean annual temperature ranging from $-10°C$ in the north and high elevation regions to $9°C$ in the southern portions of the desert steppes. Half of this land has precipitation of less than $300\,\text{mm}\,\text{yr}^{-1}$ (Fig. 14.1(a)).

The inter-annual variability in rainfall in this region is extremely high (Yasunari et al., Chapter 6, this volume), especially in the western region where the coefficient of variation commonly observed is around 30% (Fig. 14.1(b)). The temperature in this region is typified by cold winters (mean monthly January temperatures ranging from $-6°C$ to $-30°C$) and warm summers (mean monthly July temperatures ranging from $+2°C$ to $24°C$). The strong aridity of the region results in high rates of water loss as estimated by the potential evapotranspiration rates which can exceed annual precipitation by more than 500%. Steppe soils are classified as mollisols with soil organic matter ranging from 1.5% to 5.5%. Although these soils are generally fertile, water is often limiting to plant productivity. The soils are vulnerable to wind erosion and to land degradation resulting from overgrazing and cropland conversion. Rangeland is the main source of forage for livestock, which are kept on natural pasture year-round. Mongolian rangelands occupy approximately 125.8 million hectares or about 80% of the total area of the country.

The potential of this region to alter the global carbon cycle is not well understood, but it seems to be largely dependent on land use dynamics affecting the source–sink relationship in these rich steppe soils. Export of carbon, other nutrients, and soils from the region due to dust export may also alter regional and global climate. The land use dynamics themselves are different from those presently observed in tropical countries undergoing development, or in temperate countries whose land use systems are in relative equilibrium. For example, the simultaneous occurrence of cropland abandonment in Mongolia with increase in livestock numbers presents a

Fig. 14.1. Long-term mean (1950-2000) climate data (New et al. 2002) for the semiarid and arid region. (a) Mean annual precipitation (°C). (b) Coefficient of variation for mean annual precipitation.

complex matrix of land cover dynamics radically different from the mainly unidirectional process of deforestation observed in many tropical forests. Large scale intensification efforts took place in the 1950s through 1990 with large tracts of land converted to cropland, livestock production was made more sedentary, and collectives for agricultural production were organized throughout the region. During the past decade, most of the land use management in these countries has adjusted to changes in market availability, levels of government support, and land tenure policies. These countries are among the most vulnerable in terms of their economic, political, and environmental systems.

The climate warming experienced by Eurasia during the past 50 years (Jones and Moberg 2003) has accelerated the speed at which glaciers are receding and contributing to the change in hydrological cycle in the region. The temperature increase and overall enhanced aridity of the region have contributed to the massive dust inputs from the semiarid and arid regions (Wang and Ostubo 2002).

Given these changes in land use and grazing systems, we anticipate that degraded rangelands and abandoned croplands will alter carbon cycling in the region. Recovery of degraded rangelands represents a potentially important sink for the global carbon cycle (Ojima et al. 1993; Conant et al. 2001; Follett et al. 2001). Carbon inversion studies have indicated that the Eurasian region can modify the global terrestrial biospheric sink–source relationship (Bousquet et al. 2000). Their analyses indicate that in certain years the region can be a sink as large as 1 Pg per year or a source of equal magnitude. Based on research concerning organic matter reserves and net primary productivity, Asian rangelands have the potential to sequester considerable amounts of carbon, especially if conditions and productivity of overgrazed and desert areas could be improved through better management practices and reclamation efforts (Ojima et al. 1993; Gilmanov 1997). Overall, rangelands may play an important role in sequestering atmospheric CO_2 (Allen-Diaz et al. 1996; Hungate et al. 1997).

The recent changes in the physical climate and factors affecting land use decisions in the region (i.e., transition or the political-economic state) has led to changes in cropland abandonment, destocking of certain rangelands and increased stocking of others, degradation of soils due to salinization and desertification, and damage to wetlands due to modifications of water regime and industrial development. The region is a key source of large dust plumes observed from March through May and has regional C cycle and climate implications (Bousquet et al. 2000; Uno et al. 2003). The interaction between climate and land-use changes will determine how land cover and carbon biogeochemistry will respond to these physical and social–economic caused modifications, and may affect both long-term and short-term carbon storage/fluxes and sustainability of these Asian ecosystems.

3. Land Use

The steppe and the North China Plain ecosystems constitute one of the most productive agricultural regions of the world. The ecosystems found in these regions are sensitive to climate changes and capable of inducing changes in regional climate systems (Fu and Wei 1993; Ojima et al. 1998). These lands have supported human cultures for several millennia and have undergone several climatic and socioeconomic changes during that time. The ecological and cultural history of Western China and the Mongolian steppe are long and rich. Current changes provide an opportunity to understand the interaction between the human and ecological systems.

Multi-year field observations in the Xilin river basin have shown large interannual variation of climate and aboveground biomass (Xiao et al. 1996). Century model results indicated the grassland soil carbon is sensitive to climate change (Xiao et al. 1995). Soil carbon storage in the ecosystems are much higher than expected given the annual precipitation (Xiao et al. 1995, Chuluun and Ojima 2002), however grazing or cropping these soils may result in 30–50% losses of soil carbon (Ojima et al. 1998).

In these semiarid regions of semiarid ecosystems, nomadic pastoralism has been the dominant agronomic activity for many centuries. Recent changes in cultural, political, and economic factors have caused changes in how the pastoral systems operate within the region. These systems encompass a range of grazing patterns (i.e., frequency, intensity of grazing, and the types of animals), and have incorporated new breeding stocks that are potentially not suitable for certain climate regimes (e.g., drought conditions of the Gobi desert, cold hardiness against severe winter storms in the Mongolian steppe region). These changes in pastoral management have altered the nomadic patterns of the region.

Land degradation by overgrazing has already taken place. It was reported in the National Plan of Action to Combat Desertification in Mongolia (1998) that about 30% of the total land is under degradation. Consequently, rangeland degradation in both size and intensity will have a direct impact on ecosystem carbon cycling and regional climate. Pasture productivity is dependent on the physical conditions (e.g., climate, availability of water resources, snow cover distribution) as well as pasture management.

The land use systems have undergone significant changes during the 1990s and into the 21st Century. Socioeconomic liberalization has altered market accessibility, agricultural infrastructure, and land tenure policies affecting land use in the region. Large-scale wheat farming was not economically or environmentally sustainable, and resulted in significant depletion of soil carbon. Since 1990, many of these croplands in Mongolia have been abandoned. The gradual recovery of soil carbon on the once-cultivated rangelands represents a sink for atmospheric carbon as productive rangelands are re-established.

A comparative study of culture and environment in Inner Asia conducted by Humphrey and Sneath (1999) found that pasture degradation was associated with the loss of mobility in pastoral systems. Pasture degradation was more severe at the research sites, on the steppes of Eastern Russia, Buyatia, and Chita Oblast, where pastoralists were limited in their livestock movements than at research sites, in Mongolia, Tuva (Russia), Inner Mongolia, and Xinjiang (China), where movement was not as restricted. Thus, they concluded that mobile pastoralism still remains a viable and useful technique in the modern age.

4. Climate Variability and Land Use Patterns

The inherent climate variability influences land use patterns. In regions with more than 250 mm annual precipitation herders tend to be more sedentary and localized in their movements and land use patterns. Government policies to increase the level of productivity from these grasslands have altered both the land use and the demographic patterns. Grassland conversion to dryland and irrigated cropland has taken place in various parts of the region. For instance, in Inner Mongolia agricultural policies in these areas have converted the most productive grassland sites to farm cropland and improved pasture (usually a monoculture) (Zhang 1992). These policies not only reduced the amount of rangeland available for range livestock production, but they have also increased grazing intensity, and often on less fertile grazing lands (Sheehy 1992). Desert regions, degraded lands, and croplands contribute to carbon and soil losses due to over use and mis-management.

The satellite derived vegetation index, normalized difference vegetation index, data provides a long-term (1982-2002) regional observation of greenness of the land surface (Tucker et al. 2001). Four regions within Mongolia were analyzed to investigate the temporal behavior of net primary productivity of the vegetation, including an area in the far west; a region around the capital, Ulaanbaatar; a region just north of Ulaanbaatar; and an area in the east. The satellite points estimating plant productivity within each of these regions were aggregated based on the global land cover classification scheme (Loveland et al. 2000). The regions were dominated by one or two biomes, typically grasslands, but additional biomes are reported as well.

Mean plant productivity was highest around Ulaanbaatar, where grassland values exceeded $200\,\text{g}\,\text{C}\,\text{m}^{-2}\,\text{yr}^{-1}$. Plant productivity in the eastern and northern regions was about $150\,\text{g}\,\text{C}\,\text{m}^{-2}\,\text{yr}^{-1}$, while in the western region, the plant productivity was less than $50\,\text{g}\,\text{C}\,\text{m}^{-2}\,\text{yr}^{-1}$. We computed trends in annual production from 1982 to 2002. The plant productivity decreased in the major biomes of the eastern region and Ulaanbaatar, was near zero in the northern region, and was slightly positive in the western region. Interannual plant productivity variability is high, ranging from 25% to 40% of the mean, with the exception of the western region. In this region, plant productivity is nearly constant through the first 10 years (i.e., from 1982 through 1992), but increases substantially from 1992 to 2002. This increase is likely associated with greater precipitation.

Analysis of rangeland recovery in this region of Asia indicates that as growing season increases, there is more carbon uptake into soils, suggesting that there is carbon sequestration potential. Improved grazing management practices may prove useful for the Mongolian steppe situation, which is undergoing large increases in livestock numbers. Implementation of improved grazing practices can lead to sustainable carbon storage in rangeland ecosystems, and offset some of the negative impacts of climate warming by conserving soil moisture.

During spring the drier areas of the region are key sources of large dust plumes which has led to health concerns in the Asian population, and deposited dust into the Pacific Ocean and onto the North American continent. These changes in land use and climate interactions have altered carbon stores and fluxes from extensive areas in the region, and have potentially affected the feedback to the climate system (Mahowald and Luo 2003).

The Gobi and Taklimakan deserts have been identified as key sources of dust emissions. During the past three decades, the frequency of dust events in some areas of Asia has increased in association with the expansion of land degradation (Wang and Ostubo 2002). Earth observations associated with various earth orbiting satellites (e.g., TOMS, NOAA-7, and now MODIS) have indicated areas of intense activity east of the Caspian Sea in the temperate region of Eurasia. The spring-time dust fluxes are carried aloft into the eastern areas of China, Korea, and Japan, and one event was even documented to deposit dust in the Alps (Grousset et al. 2003). These dust events have major climatological effects on the radiative forcing of the atmosphere and nutrient export (Uno et al. 2003; Zhang et al. 2003). In addition, investigations have begun on the adsorption of soot and elemental carbon onto the dust particles and their effects on radiative properties of the dust (Chuang et al. 2003; Qian et al., Chapter 8, this volume).

5. Conclusions

Dramatic changes in land use have occurred in this region of East Asia during the last several decades. The extent of grassland conversion into croplands and grassland degradation is large due to increased human population and political reforms of pastoral systems. Rangeland ecosystems of this region are vulnerable to environmental and political shocks, and response of pastoral systems to these shocks have varied across the region. Indeed, this part of the world has experienced many perturbations in the recent decades, such as, the collapse of the livestock sector in some states of central Asia, expansion of livestock in China and Mongolia, increased cropland conversion from pasture, multiple winter storms of 1999–2001 in Mongolia, and recent intensive dust storm events.

However, this region has great potential for rangeland improvements and carbon sequestration, if appropriate land management practices are adopted. Traditional pasture management with greater mobility should be incorporated in rural development policies. For instance, traditional pastoral networks should be encouraged with a new land reform policy in Mongolia. The traditional pastoral networks emerged as dissipative structures during the past in areas of limited natural resources (water and soil organic matter) and highly variable environment conditions. These strategies improved the resilience and sustainability of grazing lands in Mongolia and surrounding pastoral regions.

The sensitivity of these ecosystems to climate variability and change are tightly coupled to the periodicity and intensity of monsoonal rainfall. The manner in which these precipitation patterns are altered by land use and by other global changes are difficult to evaluate. However, it is evident that any increase in the variability or decline in rainfall will greatly reduce the ability of people in the region to maintain their livestock trade.

Literature Cited

Allen-Diaz, B., T. Benning, F. Bryant, B. Campbell, J. duToit, K. Galvin, E. Holland, L. Joyce, A. K. Knapp, P. Matson, R. Miller, D. Ojima, W. Polley, T. Seastedt, A. Suarez, T. Svejcar and C. Wessman. 1996. Rangelands in a changing climate: Impacts, adaptations and mitigation. Pp. 131–158 in *Climate change 1995. Impacts, adaptations and mitigation of climate change. Scientific-Technical Analyses. Contribution of Working Group II to the Second Assessment Report of the Intergovernmental Panel on Climate Change*, edited by R. T. Watson, M. C. Zinyowera and R. H. Moss. New York: Cambridge University Press.

Bousquet, P., P. Peylin, P. Ciais, C. Le Quéré, P. Friedlingstein and P. P. Tans. 2000. Regional changes in carbon dioxide fluxes of land and oceans since 1980. *Science* 290:1342–1346.

Chuang, P. Y., R. Duvall, M. Bae, A. Jefferson, J. Schauer, H. Yang, J. Yu and J. Kim. 2003. Observations of elemental carbon and absorption during ACE-Asia and implications for aerosol radiative properties and climate forcing. *Journal of Geophysical Research-Atmospheres* 108(D23):Art. No. 8634.

Chuluun, T. and D. S. Ojima. 2002. Land use change and carbon cycle in arid and semi-arid lands of East and Central Asia. *Science in China. Series C* 45:48–54.

Conant R. T., K. Paustian and E. T. Elliott. 2001. Grassland management and conversion into grassland: Effects on soil carbon. *Ecological Applications* 11:343–355.

Follett, R. F., J. M. Kimble and R. Lal. 2001. The potential of US grazing lands to sequester soil carbon. Pp. 401–130 in *The potential of US grazing lands to sequester soil carbon and mitigate the greenhouse effects*, edited by R. F. Follett, J. M. Kimble and R. Lal. Boca Raton, FL: Lewis Publishers of CRC Press.

Fu, C. and H. Wei. 1993. Study on sensitivity of meso-scale model in response to land cover classification over China. *Eos Supplement* (October 26, 1993). Page 172.

Gilmanov, T. G. 1997. Ecology of rangelands of Central Asia and modeling of their primary productivity. Pp. 147–178 in *Central Asia Livestock Regional Assessment Workshop*. Tashkent, Uzbekistan, February 27–March 1, 1996. Davis, CA: Small Ruminant Collaborative Research Support Program-US-Agency for International Development.

Grousset, F. E., P. Ginoux, A. Bory and P. E. Biscaye. 2003. Case study of a Chinese dust plume reaching the French Alps. *Geophysical Research Letters* 30:Art. No. 1277.

Humphrey, C. and D. Sneath. 1999. *The end of nomadism?: Society, state and the environment in inner Asia*. Durham, North Carolina: Duke University Press.

Hungate, B. A., E. A. Holland, R. B. Jackson, F. S. Chapin III, H. A. Mooney and C. B. Field. 1997. The fate of carbon in grasslands under carbon dioxide enrichment. *Nature* 388:576–579.
Jones, P. D. and A. Moberg. 2003. Hemispheric and large-scale surface air temperature variations: An extensive revision and an update to 2001. *Journal of Climate* 16:206–223.
Loveland, T. R., B. C. Reed, J. F. Brown, D. O. Ohlen, Y. Zhu, L. Yang and J. W. Merchant. 2000. Development of a global land cover characteristics database and IGBP DISCover from 1-km AVHRR Data. *International Journal of Remote Sensing* 21:1303–1330.
Mahowald, N. and C. Luo. 2003. A less dusty future? *Geophysical Research Letters* 30:doi:10.1029/2003GL017880.
New, M., D. Lister, M. Hulme and I. Makin. 2002. A high-resolution data set of surface climate over global land areas. *Climate Research* 21:1–25.
Ojima, D. S., B. O. M. Dirks, E. P. Glenn, C. E. Owensby and J. O. Scurlock. 1993. Assessment of C budget for grasslands and drylands of the world. *Water, Air and Soil Pollution* 70:95–109.
Ojima, D. S., X. Xiao, T. Chuluun and X. S. Zhang. 1998. Asian grassland biogeochemistry: Factors affecting past and future dynamics of Asian grasslands. Pp. 128–144 in *Asian change in the context of global climate change*, edited by J. N. Galloway and J. M. Melillo. Cambridge: Cambridge University Press.
Sheehy, D. P. 1992. A perspective on desertification in north China. *Ambio* 21:303–307.
Tucker, C. J., D. A. Slayback, J. E. Pinzon, S. O. Los, R. B. Myneni and M. G. Taylor. 2001. Higher northern latitude normalized difference vegetation index and growing season trends from 1982 to 1999. *International Journal of Biometeorology* 45:184–190.
Uno, I., G. R. Carmichael, D. G. Streets, Y. Tang, J. J. Yienger, S. Satake, Z. Wang, J. H. Woo, S. Guttikunda, M. Uematsu, K. Matsumoto, H. Tanimoto, K. Yoshioka and T. Iida. 2003. Regional chemical weather forecasting system CFORS: Model descriptions and analysis of surface observations at Japanese island stations during the ACE-Asia experiment. *Journal of Geophysical Research-Atmospheres* 108(D23):Art. No. 8668.
Wang, Q. and K. Otsubo. 2002. A GIS based study on grassland degradation and increase of dust storm in China. Pp. 110–115 in *Fundamental issues affecting sustainability of the Mongolian steppe*, edited by C. Togtohyn and D. Ojima. Ulaanbaatar, Mongolia: International Institute for the Study of Nomadic Civilizations.
Xiao, X., D. S. Ojima, W. J. Parton, Z. Chen and D. Chen. 1995. Sensitivity of Inner Mongolia grasslands to global climate change. *Journal of Biogeography* 22:643–648.
Xiao, X. M., J. Shu, Y. F. Wang, D. S. Ojima and C. D. Bonham. 1996. Temporal variation in aboveground biomass of *Leymus chinense* steppe from species to community levels in the Xilin River basin, Inner Mongolia, China. *Vegetatio* 123:1–12.

Zhang, X. 1992. Northern China. Pp. 39–54 in *Grasslands and grassland sciences in Northern China*. National Academy Press.

Zhang, X. Y., S. L. Gong, Z. X. Shen, F. M. Mei, X. X. Xi, L. C. Liu, Z. J. Zhou, D. Wang, Y. Q. Wang and Y. Cheng. 2003. Characterization of soil dust aerosol in China and its transport and distribution during 2001 ACE-Asia: 1. Network observations. *Journal of Geophysical Research-Atmospheres* 108(D9):Art. No. 4261.

Part IV

MARINE/COASTAL SYSTEMS

Chapter 15

TRANSPORT OF MATERIALS INDUCED BY HUMAN ACTIVITIES ALONG THE YELLOW AND YANGTZE RIVERS

DUNXIN HU, CHONGGUANG PANG, QINGYE WANG,
WEIJIN YAN and JIONGXIN XU

1. Introduction

Changes in sediment and nutrient transport from land to coastal seas may have serious impacts upon the coastal system globally. Such effects include reduction or even destruction of living resources, hindrance to marine activities, deterioration of amenities and damage to the coastal ecosystems of mangrove, coral reef, and benthic communities. Deforestation and spread of agriculture has increased sediment discharge during the last 2000 years. In the areas of southern Asia and Oceania, riverine sediment discharge to the sea is much larger than that previously estimated. Some of the increase may be due to accelerated deforestation. In other areas sediment discharge has decreased due to human activities such as water diversion, dam construction for irrigation, flood control and hydroelectric power generation, and water/soil conservation. The decrease of sediment delivery to the coastal zone induces coastal erosion, which is enhanced further by sea-level rise. Enhanced fertilization practices on land has increased riverine nutrient input in the last 40–50 years, which has resulted in eutrophication of water bodies and red-tide in the coastal zone.

Progress in the study on sediment and nutrient discharge to the coastal zone is reviewed in this chapter, using the Yellow and Yangtze Rivers as examples of the influence of human activities on sediment discharge and nutrient input, respectively.

2. Yellow River

The Yellow River (Huanghe) in China (Fig. 15.1) is a large and heavily sediment-laden river. According to the hydrometric records for the period 1919–1960, before the construction of large reservoirs, the annual mean input of sediment into the Bohai Sea was 12×10^8 t (Chien and Zhou 1965), occupying 19.4% of the world total, thus playing an important role in global sediment transport (Xu 2002).

Northern China suffered from flooding disasters of the Yellow River, which occurred over 1,500 times within the last 4,000 years. These floods are attributed mainly to heavy rainfall and variation (increase and decrease) of discharged sediment resulting from human activities. The impact of human activities on sediment discharge from the Yellow River is obvious (EGWCHYR 1982).

2.1. Increased sediment discharge

The Yellow River sediment discharge has changed remarkably in the last 5,000 years. The river was not muddy 2,500 years ago and its sediment discharge, based on Holocene sediment budgets (Milliman et al. 1987) and delta growth (Saito and Yang 1994), would have been only one-tenth of its present level.

In Chinese history, large-scale migrations occurred in two different periods. During the Han and Tang Dynasties, 700,000 and 60–70 million people moved to the loess plateau. About 600–700 years ago pastoral areas

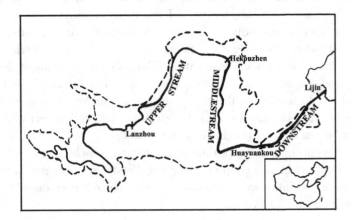

Fig. 15.1. The Yellow River basin.

and forests were converted to agricultural areas (Chen 1990), and the green area in the loess plateau was decreased from 53% to 3% (Wang and Aubrey 1987). Cultivation is thought to be the reason for this decrease (EGWCHYR 1982; Milliman et al. 1987). The rate of land formation can be used as an estimate of the increase in sediment discharge by the Yellow River. From 7,000 years ago to 1,128 AD, assuming that the trapping efficiency in the delta was about 90% (Bornhold et al. 1986), the Yellow River sediment flux to the Bohai Sea would have been about 1.1×10^8 t yr^{-1}. During the period 1128–1855 AD, when the Yellow River discharged into the Yellow Sea, approximately 250 km^3 of sediment was deposited along the Jiangsu coast, which equates to a mean sediment load of 4.3×10^8 t yr^{-1}, reflecting enhanced erosion from increased agricultural activity in the loess hills of northern China (Liu et al. 2002). After 1855, when the river changed course again to the Bay of Bohai, and until the mid-1980s, the annual riverine sediment load increased to 9.0×10^8 t yr^{-1} (Milliman and Meade 1983). These figures illustrate the rapid increase in sediment discharge from the Yellow River about 1,200 years ago due to the cultivation of the loess plateau.

2.2. Decreased water and sediment fluxes

During the last 40–50 years, water and sediment discharges by the Yellow River decreased rapidly (Ye 1994; Wang et al. 1997; Pang et al. 1999; Xu and Sun 2003). Figure 15.2 shows the water and sediment discharges at the upper, middle, and lower reaches of the Yellow River from 1950 to 2002. A clear decline in discharge since the 1970s occurred in the middle, and lower reaches, especially at Lijin station. The Lijin station is the most downstream sampling point on the main stream of the Yellow River and the water and sediment transports measured at this station are usually regarded as the water and sediment fluxes into the sea.

Between 1950 and 2000, the water flux into the Bohai Sea decreased from 480 to 46×10^8 m^3 yr^{-1}, and the sediment discharge decreased from 13 to 0.3×10^8 t yr^{-1}. The decline in water volume was so great that the river failed to flow downstream. Although there were some very dry years, such as 1875–1878, and 1922–1932, there was no record of the Yellow river failing to flow downstream during these periods. The first flow-cut-off of the Yellow River at Lijin station occurred in 1972, but it happened regularly after that and became more and more severe. The flow-cut-off of the Yellow river downstream occurred in 21 of the years from 1972 to 1998. The longest

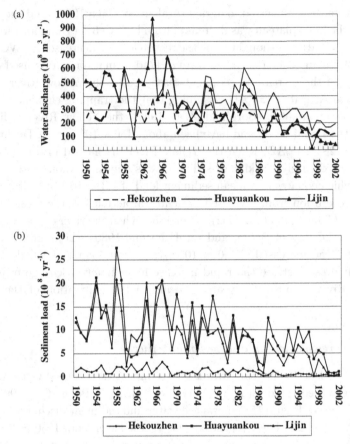

Fig. 15.2. (a) Water and (b) sediment discharge at upper, middle, and lower reaches of the Yellow River.

no-flow period was for 256 days in 1997 (Li et al. 1997). Since then the Yellow River had become a seasonal river. The maximum course length with no-flow was 683 km in 1995, which is 97% of the whole downstream course length (Pang et al. 1999). Since 2000, when a modulated water policy was applied, the Yellow River has not dried up. However, the water discharge is limited, being about 40 m^3 s^{-1} in spring.

2.2.1. *Water consumption*

Most of the drainage area of the Yellow River is located in a semiarid zone, and water resources are far from meeting water demand. In the past 30–40 years, water consumption from the river increased, especially in

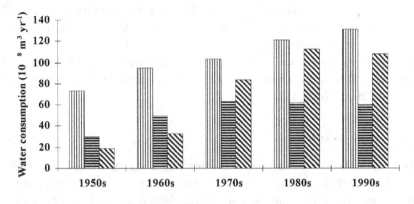

Fig. 15.3. Human consumption of water in the Yellow River basin in different decades (Pang et al. 1999).

the lower reaches of the Yellow River (Fig. 15.3). From 1950s to 1970s the volume of water consumed downstream increased from about $19 \times 10^8 \, m^3 \, yr^{-1}$ to about $84 \times 10^8 \, m^3 \, yr^{-1}$ which was greater than that used in any other period (Pang et al. 1999). In the 1990s, the quantity of water consumed amounted to more than 60% of the total "natural" runoff. Reduction in runoff by water consumption lowers the river's ability to transport sediment to the sea. Consequently, there is a negative correlation between the annual water and sediment fluxes and annual water consumption.

Pang et al. (1999) pointed out that flow-cut-off of the Yellow downstream did not happen even during the continuously dry years from 1922 to 1932. Therefore, the main factor responsible for the river flow-cut-off must be the rapid increase in consumption of water by human activities in the river basin.

2.2.2. *Water/Soil conservation*

Since 1960, water/soil conservation has been practiced in the drainage basin. By 1998, conservation had been applied to about $17.1 \times 10^4 \, km^2$ in the loess plateau (IRTCES 2001), which led to a decrease in water and sediment fluxes into the sea. Land terracing, and tree- and grass-planting also significantly reduced erosion on hillslopes. Check dams trapped eroded sediments in gullies and increased sediment storage. As a result of the above practices, transport of sediment into the Yellow River and then into the sea has been considerably reduced. However, the practice of water/soil conservation is relatively small compared to other human activities along the Yellow River. Increasing the total area of land terracing, and tree- and grass-planting by 10^2 km^2, would decrease the sediment and water flux to the sea by 0.19 million t yr^{-1}, and 0.02 billion m^3 yr^{-1}, respectively (Xu and Sun 2003; Xu 2003).

2.2.3. *Dam construction*

An investigation of sedimentation in reservoirs along the whole basin (1990–1992) found that there were eight reservoirs along the main stream of the Yellow River, with a total capacity of 41.3 billion m^3, and that 8.0 billion m^3 of sediment had been deposited in these reservoirs. Along the main tributaries, there are 483 reservoirs with a capacity larger than 10^5 m^3, and their total capacity is nearly 7.6 billion m^3. Up to 1997, the total cumulative volume of sediment within these reservoirs was 3.3 billion m^3 (Xu 2003). A large amount of sediment, together with water, had been trapped by these reservoirs, which led to a sharp reduction in sediment and water fluxes to the sea at the beginning of water storage of the reservoirs. Sediment trapped in reservoirs is very important, but not the crucial factor responsible for the continuous decline of sediment flux to the sea.

The Sanmenxia Reservoir is one of the most important reservoirs along the Yellow River. In terms of operation mode, the use of this reservoir can be divided into three stages. From 1960 to 1964, the reservoir was used for water storage, trapping almost all sediment in the reservoir. The total volume of sediment was 4.5 billion m^3. From 1964 to 1973, the reservoir was used for flood retention, and after flooding, sediment can be released through the dam. During this period, the total volume of sediment deposited was 1.2 billion m^3. After 1973, the reservoir was used under the mode of "storing clear water and releasing sediment", which significantly reduced sedimentation in the reservoir. During the 18-year period from 1973 to 1990, only 0.4 billion m^3 of sediment was trapped in the reservoir.

The Xiaolangdi Reservoir located between Sanmenxia and Huayuankou was built in 1999. The reservoir was designed to hold 12.7 billion m^3 water initially, and after sedimentation to hold 5.1 billion m^3 water. That is to say, the Xiaolangdi reservoir is expected to trap as much as 7.6 billion m^3 sediment from the Yellow River within 20–30 years. From 1999 to 2002, 0.9 billion m^3 of sediment has been trapped in this reservoir (IRTCES 2000, 2001, 2002).

2.2.4. Fluctuations of water and sediment discharges with climate

Spectral analyses were made with a long-time series (1950–2002) of water/sediment discharges at Lijin station. The most protruding peaks corresponded to the time of El Niño events suggesting that water and hence sediment discharges from the Yellow River are strongly correlated with El Niño events through precipitation, as suggested by Hu et al. (1998). Another peak in both water and sediment spectra, appears to be a quasi-biennial oscillation signal associated with the monsoon. It is suggested that climate change associated with El Niño and the monsoon can have a significant effect upon the water and sediment discharges of the Yellow River.

On the other hand, a tendency of climate change in the last 40–50 years in northern China is for the area covering the Yellow River basin to be continuously dry. For example, according to Xu (2003) and Xu and Sun (2003)'s regression equations, the annual precipitation in the area upstream of Huayuankou decreased by about 70 mm from the 1960s to 1990s, and sediment and water discharge to the sea would have decreased by more than 414.4 million t yr^{-1} and 12.3 billion m^3 yr^{-1}, respectively.

3. Yangtze River

The Yangtze River (Changjiang; Fig. 15.4) discharges almost the same amount of water to the sea as the Ganges River, but has a greater basin, a much larger drainage area, and contains more paddy fields with high

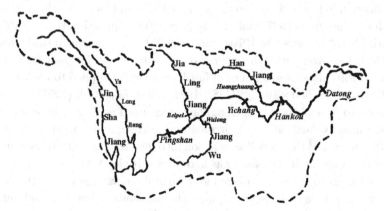

Fig. 15.4. The Yangtze River basin.

fertilizer inputs. Because of the extensive, high input agriculture in the basin, it provides a suitable example for demonstrating the impact of human activities on nutrient input to the coastal zone.

3.1. Anthropogenic impact on nitrogen flux

Riverine fluxes of nutrients from the Yangtze River have greatly increased in recent years (e.g. Zhang et al. 2003). The transport of dissolved inorganic nitrogen in the Yangtze to the estuary increased by 0.775 Tg per year between 1980 and 1989 (Duan et al. 2000), and Shen et al. (2003) observed that $2,849 \times 10^3$ Gg total nitrogen and $1,746 \times 10^3$ Gg dissolved inorganic nitrogen were discharged at the mouth of the Yangtze river in 1998. The reason for the increase in nitrogen concentration in the Yangtze River is not known and the topic has been controversial since the 1980s. Gu et al. (1981, 1982) suggested that the increase was due to the rapid increase in fertilizer nitrogen use along the basin of the Yangtze River. Edmond et al. (1983) deemed, however, that the high value of 0.9 mg L^{-1} measured in 1980 resulted from the biological fixation of nitrogen by a water fern. They suggested that eutrophication by fertilizer nitrogen was negligible, and Shen (1996) claimed that precipitation was responsible for the high concentration of inorganic nitrogen. Recently, Yan et al. (2003) pointed out that nitrogen fixation was often a dominant factor before 1978, while fertilizer dominated nitrogen input after 1983.

Four factors are considered in the attempt to identify the real cause for the increase of dissolved inorganic nitrogen, nitrogen fertilizer use, nitrogen fixation, population growth, and precipitation. Nitrogen fertilizer application in the basin began in the early 1950s and increased dramatically in the period up to 1997. Only 2.3 kg N ha^{-1} was applied in the late 1960s, 15.0 kg N ha^{-1} in the late 1970s, and 35 kg N ha^{-1} was applied to the basin in 1997. Fertilizer nitrogen application grew by about 11% annually from 1968 to 1997. The total amount of fertilizer nitrogen applied to the basin in 1997 was about 600×10^7 kg (CSSB 1981–1998). Zhu et al. (1997) reported that up to 50% of the nitrogen applied to cropped fields in China was lost by ammonia volatilization, denitrification, leaching and run-off. Shen et al. (2003) estimated that fertilizer and soil nitrogen lost through surface runoff from the Yangtze River basin was about 28×10^7 kg N yr^{-1}.

Nitrogen fixation in managed and natural ecosystems within the basin is another source of nitrogen inputs. The amount of nitrogen fixed varied only slightly during the period 1968–1997. The amount fixed in 1968 was

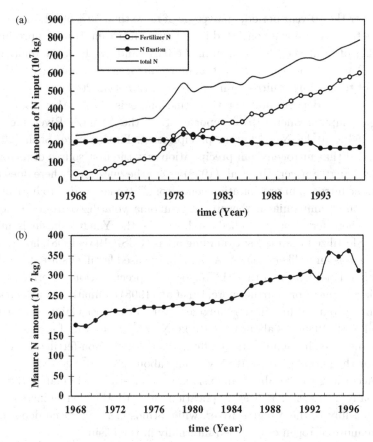

Fig. 15.5. (a) Nitrogen inputs to the Yangtze River basin. (b) Manure produced in the Yangtze River basin (Yan et al. 2003).

about 215×10^7 kg N, corresponding to 12.5 kg N ha^{-1} over the total area, and increased to a peak of 258×10^7 kg N in 1978. It then gradually decreased to 184×10^7 kg N in 1997 (Fig. 15.5; Yan et al. 2003).

This decrease should be attributed to the land-use changes during the period, particularly, the dramatic decrease in the cultivation of green manure (CSSB 1981–1998). Population growth leads to an increase of manure nitrogen. Manure nitrogen is not a new input to the basin, but rather it is stored and joins the internal nitrogen cycling in the basin. The amount of manure nitrogen voided in 1968 was 177×10^7 kg, equivalent to 70% of the total nitrogen input to the basin at that time, and increased to 311×10^7 kg, or 40% of the total nitrogen inputs in 1997 (Fig. 15.5).

Over the 30-year period, an average of more than 50% of total nitrogen inputs to the basin was converted into manure nitrogen. Humans produced only 2% of total manure nitrogen in the basin and livestock the remaining 98%. Poultry produced the least manure nitrogen, accounting for about 0.6% of total manure nitrogen in the basin (Yan et al. 2001).

Nitrogen deposition from the atmosphere is an another source of nitrogen inputs. Shen (1996) reported that the Yangtze River receives about 56×10^7 kg N yr^{-1} from precipitation. Furthermore, Shen (2003) suggested that nitrogen from precipitation was the first source of nitrogen for the Yangtze River. Hu et al. (1998) asked the question "where does the inorganic nitrogen in precipitation come from?" It may be due to horizontal advection of ammonia in the air, or ammonia volatilized from the local lands where fertilizer is applied, or both. As the Yangtze basin is more heavily fertilized than the surrounding areas (CSSB 1981–1998), horizontal advection is an unlikely source. As ammonia-based fertilizers are dominant in China, it is suggested that the nitrogen in precipitation is derived from fertilizer application. Furthermore, Hu et al. (1998) estimated that the mean fertilizer produced flux in the Yangtze River to the East China Sea for the period 1980–1985 was about 66×10^7 kg N yr^{-1} accounting for about 74% of the dissolved inorganic nitrogen flux, and the mean non-fertilizer affected flux for the period of 1963–1985 was only about 23×10^7 kg N yr^{-1}.

According to the above analysis and other studies, human activities, especially nitrogen fertilizer application lead to the sharp increase in nitrogen flux in the Yangtze River, while atmospheric nitrogen deposition and manure nitrogen only cycled internally in the basin.

3.2. *Variations in sediment and water fluxes*

Water discharge in the Yangtze River changed little from 1950 to 2002, while the sediment load dropped rapidly at Datong station by about 40% from 1984 to 2002 (Fig. 15.6). Shen et al. (2001) deemed that the decline in sediment load from 1984 resulted from dam construction and water/soil conservation. Pang and Wang (2004) and Huang et al. (2004) drew similar conclusions.

The sediment in the Yangtze River comes mainly from its upperstream. The major sediment yield is from the Jinshajiang basin between the Yalongjiang and Pingshan station and the main tributary of the Jialingjiang. The mean sediment load originating from the Jinshajiang and

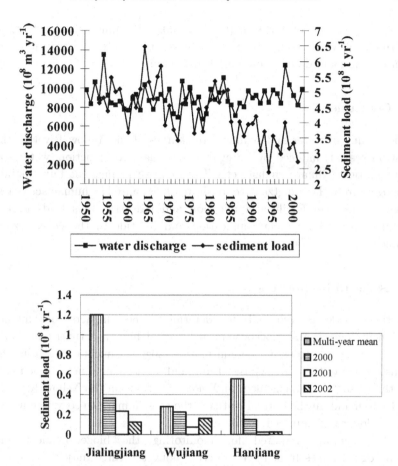

Fig. 15.6. Water and sediment discharges at Datong station (Shen et al. 2001) and sediment discharge of main tributaries in the Yangtze River.

Jialingjiang is about 3.75×10^8 t yr^{-1}, which accounts for about 75% of the sediment load at Yichang station (CWRC 2002).

The sediment yield of Jialingjiang dramatically decreased from a multi-year mean (1956–2002) of 1.20×10^8 t yr^{-1} to an average mean in 2000–2002 of about 0.25×10^8 t yr^{-1}. This is clearly due to water/soil conservation along the Jialingjiang basin (Zhang and Wen 2002). In contrast, the multi-year mean of sediment yield of Wujiang is small, less than 0.3×10^8 t yr^{-1} with little decrease for the period 2000–2002. Even though the multi-year mean of sediment yield in the Hanjiang is less than 0.6×10^8 t yr^{-1}, the individual annual mean decreased markedly. This is attributed mainly to dam or reservoir construction (CWRC 2002). In summary, the decrease of sediment yield along the Yangtze River is mainly due to water/soil conservation and dam or reservoir construction.

4. Conclusions

The decline of water and sediment discharges of the Yellow River in the last 50 years is induced by water diversion, dam construction, water/soil conservation, climate change, etc. The increase in nutrient flux (mainly nitrogen) from the Yangtze River to the ocean is caused by human activities, such as nitrogen fertilizer application and industry. This leads to the frequent occurrence of eutrophication and red-tide in the coastal zone influenced by the Yangtze River.

5. Issues to be Addressed

Further work is required to determine the impacts on fluvial, sedimentological, and biogeochemical fluxes of human activities on land, the consequences to estuarine and marine biogeochemical cycles from the construction of the Three Gorges Dam, and what changes need to be made to the regulation management at Xiaolangdi Dam on the Yellow River in order to maintain the normal status of the coastal marine ecosystem where the Yellow river empties into the sea.

Systems are required for monitoring the biogeochemical and hydrological fluxes from river to ocean along the three major rivers, the Yellow, Yangtze, and Pearl, with special emphasis on organic carbon and nitrogen measurements, and for simulating the transport of material from land to the ocean.

Literature Cited

Bornhold, B. D., Z. S. Yang, G. H. Keller, D. B. Prior, W. J. Wiseman, Q. Wang, L. D. Wright, W. D. Xu and Z. Y. Zhuang. 1986. Sedimentary framework of the modern Huanghe Delta. *Geo-Marine Letters* 6:77–83.

CWRC (Changjiang Water Resources Commission). 2002. *Bulletin of sediment in the Yangtze River, China.* Beijing: National Ministry of Water Resources (in Chinese).

Chen, B. S. 1990. *On the population of the Loess Plateau of China.* Beijing: China Economy Press (in Chinese).

Chien, N. and W. H. Zhou. 1965. *Channel processes in the lower Yellow River.* Beijing: Science Press (in Chinese).

CSSB (China State Statistical Bureau). 1981–1998. *China statistical yearbook.* Beijing: China Statistical Publishing House (in Chinese).

Duan, S., S. Zhang and H. Huang. 2000. Transport of dissolved inorganic nitrogen from the major rivers to estuaries in China. *Nutrient Cycling in Agroecosystems* 57:13–22.

EGWCHYR (Editorial Group of the Water Conservancy History of the Yellow River). 1982. *Summary of the water conservancy history of the Yellow River.* Beijing: China Water and Electricity Press (in Chinese).

Edmond, J. M., A. Spivak, B. C. Grant, M. X. Hu and Z. X. Chen. 1983. Chemical dynamics of the Changjiang Estuary. Pp. 229–239 in *Proceedings of International Symposium on Sedimentation on the Continental Shelf with Special Reference to the East China Sea.* Beijing: China Ocean Press.

Gu, H. K., X. X. Xiong, M. X. Liu and Y. Li. 1981. Nitrogenous geochemistry near the Yangtze mouth. I. Nitrates near the Yangtze Estuary. *Journal of Shandong College of Oceanography* 11:37–46 (in Chinese).

Gu, H. K., X. X. Xiong, X. N. Ma, Y. Li, Q. R. Wei, M. X. Liu, W. T. Shen, G. F. Ren, H. X. Diao and G. J. Li. 1982. The nitrogen cycle near the estuary of the Yangtze River. *Collected Oceanic Works* 5:48–66.

Hu, D., Y. Saito and S. Kempe. 1998. Sediment and nutrient transport to the coastal zone. Pp. 245–270 in *Asian change in the context of global climate change*, edited by J. N. Galloway and J. M. Melillo. London: Cambridge University Press.

Huang, H. J., F. Li and X. R. Zhang. 2004. A primary comparison of water and sediment flux between the Changjiang River and the Yellow River. *Studia Marina Sinica* 46:79–90 (in Chinese).

IRTCES (International Research and Training Center on Erosion and Sedimentation) 2000–2002. *Bulletin of rivers and sediment in China.* Beijing: National Ministry of Water Resources (in Chinese).

Li, D., X. H. Yang and Y. Q. Tian. 1997. The situation and characteristics of flow-cut-off of the Yellow River in its downstream. *The Yellow River* 19:10–12 (in Chinese).

Liu, J. P., J. D. Milliman and S. Gao. 2002. The Shandong mud wedge and post-glacial sediment accumulation in the Yellow Sea. *Geo-Marine Letters* 21:212–218.

Milliman, J. D. and R. H. Meade. 1983. World-wide delivery of river sediment to the oceans. *Journal of Geology* 91:1–21.

Milliman, J. D., Y. S. Qin, M. E. Ren and Y. Saito. 1987. Man's influence on the erosion and transport of sediment by Asian rivers: the Yellow River example. *Journal of Geology* 95:751–762.

Pang, C. G. and F. Wang. 2004. The distributing features and temporal variability of suspended matter concentration in the East China Sea. *Studia Marina Sinica* 46:22–31 (in Chinese).

Pang, C. G., Z. S. Yang and W. Liu. 1999. Characteristics of the Yellow River flow-cut-off in its downstream and cause analysis. *Chinese Journal of Natural Disasters* 8:132–142 (in Chinese).
Saito, Y. and Z. S. Yang. 1994. The Huanghe River: Its water discharge, sediment discharge, and sediment budget. *Journal of Sediment Society Japan* 40:7–17 (in Japanese).
Shen, H. T., H. L. Wu, X. C. Liu, J. X. Wu, Q. H. Huang and R. B. Fu. 2001. *Material flux of the Changjiang Estuary*. Beijing: China Ocean Press (in Chinese).
Shen, Z. L. 1996. A preliminary discussion on the control of inorganic nitrogen in the Changjiang Estuary. *Marine Sciences* 20:61–62 (in Chinese).
Shen, Z. L., Q. Liu, S. M. Zhang, H. Miao and P. Zhang. 2003. A nitrogen budget of the Changjiang River catchment. *Ambio* 32:65–69.
Shen, Z. L. 2003. Is precipitation the dominant controlling factor of high inorganic nitrogen content in the Changjiang River and its mouth? *Chinese Journal of Oceanology and Limnology* 21:368–376.
Wang, L. Y., P. Lin and J. Z. Wang. 1997. Causes analysis of flow-cut-off of the Yellow River in its down stream. *The Yellow River* 19:13–17 (in Chinese).
Wang, Y. and D. G. Aubrey. 1987. The characteristics of the China coastline. *Continental Shelf Research* 7:329–349.
Xu, J. X. and J. Sun. 2003. Influence of precipitation and human activities on water fluxes from the Yellow River into the sea in the past 50 years. *Advances in Water Science* 14:690–695 (in Chinese).
Xu, J. X. 2002. Sediment flux into the sea as influenced by different source areas in the drainage basin: Example of the Yellow River, China. *Hydrological Sciences* 47:187–202.
Xu, J. X. 2003. Sediment flux to the sea as influenced by changing human activities and precipitation: example of the Yellow River. China. *Environmental Management* 31:328–341.
Yan, W. J., S. Zhang and J. H. Wang. 2001. Nitrogen biogeochemical cycling in the Changjiang drainage basin and its effect on Changjiang River dissolved inorganic nitrogen: Temporal trend for the period 1968–1997. *Acta Geographica Sinica* 56:506–514 (in Chinese).
Yan, W. J., S. Zhang, P. Sun and S. P. Seitzinger. 2003. How do nitrogen inputs to the Changjiang basin impact the Changjiang River nitrate: A temporal analysis for 1968–1997. *Global Biogeochemical Cycles* 17:1091–1100.
Ye, Q. C. 1994. *Research on environmental changes of the Yellow River basin and laws of water and sediment transportation*. Jinan: Shandong Science and Technology Press (in Chinese).
Zhang, S., H. B. Ji, W. J. Yan and S. W. Duan. 2003. Composition and flux of nutrients transport to the Changjiang Estuary. *Journal of Geographical Sciences* 13:3–12.
Zhang, X. B. and A. B. Wen. 2002. Variations of sediment in upper stream of Yangtze River and its tributary. *Journal of Hydraulic Engineering* 2002:438–445 (in Chinese).
Zhu, Z. L., Q. X. Wen and J. R. Freney. 1997. *Nitrogen in Soils of China*. Dordrecht, The Netherlands: Kluwer Academic Publishers.

Chapter 16

IMPACTS OF GLOBAL WARMING ON LIVING RESOURCES IN OCEAN AND COASTAL ECOSYSTEMS

QISHENG TANG and LING TONG

1. Introduction

Both natural and anthropogenic factors affect the living resources in ocean and coastal ecosystems. Recent decades have seen more frequent and stronger climate variations and changes compared to the relatively stable climate from 1840 to 1923 (Gedalof and Smith 2001). Understanding the functioning of marine ecosystems and how they respond to global change is essential in order to effectively manage global marine living resources, such as fisheries. Fish constitute the bulk of the living marine resources which have sustained human communities over centuries, but which are now increasingly under threat. However the food webs, composed of small microscopic organisms like plankton, which sustain the larger visible components in the marine ecosystems, are less well known.

El Niño events are the primary drivers of inter-annual climate variability which affect many fish resources. The management responsible for the protection of fisheries has often failed to allow for these variations. Ocean warming will have direct consequences for species distribution and spawning habitats and indirect consequences for food web stability in marine ecosystems. Understanding the environment and its impact on ecosystem variations may lead to sustainable management and development of living resources of ocean and coastal systems.

2. Induced Changes in Marine Ecosystems

Long-term variations in the physical parameters of the ocean and atmosphere play an important role in influencing the dynamics of the ocean ecosystem. For example, sea temperature variation not only affects directly the metabolic rates of the organisms, but also influences other oceanic conditions, such as local currents. In turn, these results in the exposure or submergence of inter-tidal organisms and impacts on the movements of planktonic larvae. Variations of oceanic temperature can also influence the substrate structure, photosynthetic light intensity, water-column stratification and nutrient cycling, and therefore the productivity of living marine resources. The atmosphere also has strong impact on the ocean ecosystem via the air–sea interaction.

Climate variations and changes impacting on the ocean ecosystem can be marked by the accumulated anomalies of the air temperature, monthly mean sea surface temperature and sea surface salinity. Ecosystem investigations were conducted in the Bohai Sea in 1959, 1982–1983, and 1992–1993. Biological data from these investigations show that species diversity and biomass of phytoplankton, zooplankton, benthos, and fish resources, changed in response to the changes in air temperature, and sea surface temperature and salinity. In 1992–1993, the total biomass had decreased by 30% and the total biomasses of phytoplankton and zooplankton had decreased by 50% compared with 1982–1983 (Jin and Tang 1998). All of the 1992–1993 indexes of species diversity, biomass, fish resources, and recruitment of living marine resources had decreased compared with those in 1982–1983 indicating that air temperature, and sea surface temperature and salinity may have affected the ecosystem significantly. The increase in air temperature and sea surface temperature in the Bohai Sea region was consistent with the recent rise in air temperature in northern China and the increase of sea surface temperature in the Yellow Sea and the East China Sea during the period investigated.

3. Abundance of Living Resources

Global warming and climate change perturb the biomass yields of populations in marine and coastal ecosystems. Natural factors may have an important impact on long-term changes in dominant fish species. The alteration in the relative abundance of species belonging to various ecotypes is closely related to climate change. It seems to be accepted that small pelagic fish such as the Pacific herring (*Clupea pallasi*) and prawn (*Penaeus orientails*) species are more sensitive to climate change.

The fauna of the Yellow Sea belong to the sub-East Asia region of the North Pacific temperate zone in respect of the zoogeography of the world.

The biotic communities of the Yellow Sea ecosystem are complex in species composition, spatial distribution, and community structure. The main characteristic of biotic communities is the remarkable seasonal variation. Compared with other shelf regions in the northwest pacific, the Yellow Sea ecosystem has a relatively low primary production (about $60\,\mathrm{g\,C\,m^{-2}\,yr^{-1}}$), and the phytoplankton production is estimated to be $0.52\ 10^9\,\mathrm{t\,km^{-2}\,yr^{-1}}$. The phytoplankton biomass in the Yellow Sea ecosystem has been relatively stable over the past 30 years, but the annual zooplankton biomass declined by about 50% from 1959 to 1968. There was a similar trend in the East China Sea (Tang 1989; Chen et al. 1990).

It is believed that extreme overexploitation is one of the factors causing the decline in abundance of the most important commercial demersal species in the Yellow Sea, such as the small yellow croaker (*Pseudosciaena polyactis*), hairtail (*Trichiurus jaumela*), large yellow croaker (*Pseudosciaena crocea*), flatfish (*Pleuronectidae*), and Pacific cod (*Gadus macrocephalus*). The fluctuation in abundance of some demersal species may be affected by both natural and anthropogenic factors.

The population dynamics of the fleshy prawn (*Penaeus orientalis*) clearly show the influence of both natural and anthropogenic factors. This commercially important crustacean is widely distributed in the Bohai Sea and the Yellow Sea. Most individuals have one-year life span. Spawning occurs in late spring and the prawns grow rapidly and reach commercial size in September. When water temperature begins to drop significantly in late autumn, it migrates out of the Bohai Sea to the deep region of the Yellow Sea for over-wintering. The biomass yield of the prawn varies from year to year with the annual catch ranging from 7,000 to 50,000 t during 1953–1990. We found that both the environment and spawning-stock size were related to fluctuations in recruitment. In 1976 and 1979 the spawning stock of prawn was limited, but the 1979 class produced three times the recruitment compared with the 1976 class because of the favorable environment in 1979. A favorable environment and a large spawning stock result in a larger recruitment. The major environmental factors affecting recruitment are river runoff, rainfall, irradiation, and salinity (Tang et al. 1989). It has also been reported that there is a correlation between El Niño and the prawn biomass; the prawn catch is a reduced in the El Niño years.

The living resources in the Yellow Sea are multi-species in nature. Warm temperature species are the major component of the fish biomass, accounting for about 60% of the total biomass. Warm-water species and cold temperature species account for only 15% and 25%, respectively. Demersal and semi-demersal species account for 58% of the fish and pelagic species for about 42%. The diversity and abundance of the community in the Yellow Sea are comparatively lower than those in the East China Sea and the South China Sea. About 20 major species contribute 92% of the total biomass. Dramatic shifts of species dominance in the ecosystem occurred from the 1950s through the 1980s. The dominant species in the 1950s and early 1960s were small yellow croaker and hairtail, while Pacific herring and chub mackerel became dominant in the 1970s. Some small-size, fast-growing, short-lifespan and low-value species such as Japanese anchovy (*Engraulis japonicus*) and half-fin anchovy (*Setipinna tatyi*) increased markedly in abundance in the 1980s and have taken a prominent position in the ecosystem resources.

The pelagic species of Pacific herring (*Clupea harengus pallasi*), chub mackerel (*Scomber japonnicus*), Spanish mackerel (*Somberomorus maculates*), and silver pomfret (*Stromaleoides argentues*) are the major large size pelagic species in the Yellow Sea. The annual catch from 1953 to 1988 fluctuated greatly from 30,000 to 300,000 t (Tang 1995). The reasons for the fluctuations in the abundance of these pelagic species are complex. Both abundance and catch have tended to increase steadily since the species began to be utilized fully in the 1960s. The changes are particularly large for Pacific herring and chub mackerel stocks.

The fishing of Pacific herring in the Yellow Sea has a long history and is full of drama. In the last century its yield experienced two peaks: one around 1900 and the other around 1938. These peaks were followed by periods of little or no catch. In 1967 a large number of 1-year old herring began to appear in the bottom trawling catch because of recovery of the stock. The catch then increased rapidly to a third peak of 200,000 t in 1972 because the very strong abundance of 1970 class reached a maximum historic record of 26.8×10^8 (age 1). The catch had declined to <1000 t in 1989–1990. Herring in the Yellow Sea had been characterized by strong fluctuation of recruitment which affects the fishable stock directly (Tang 1981, 1987). Environmental effects on herring recruitment in the Yellow Sea were investigated by cluster analysis based on the data of recruitment and environmental conditions of the spawning ground from 1968 to 1977. Results of the analysis indicate that rainfall, wind, and irradiation are the major factors affecting the recruitment. The long-term changes in abundance may be correlated with the 36-year cycle of dryness–wetness in eastern China, which is apparently affected by the El Niño Southern

Oscillation (Li 1992). Trends in the catch of chub mackerel are similar to those of Pacific herring.

The major resource populations in 1958-1969 were small yellow croaker, flatfish, Pacific cod, hairtail, Skates (*Raja* and *Dasystic*), searobin and angler (*Lophins litulon*) which accounted for 71% of the total biomass. Eleven percent of them were planktophagic, 46% benthophagic, and 43% inchthyophagic. In 1985-1986 the major resources were Japanese anchovy, half-fin anchovy, Japanese squid (*Loligo japonica*), seasnail (*Liparis tanakae*) flatfish, small yellow croaker, and scaled sardine (*Harengula zunasi*) which accounted for 67% of the total biomass. Fifty-nine percent were planktophagic, 26% benthophagic, and 16% inchthyophagic. The study indicated that some commercially important demersal species of large size and higher trophic level were replaced by less valuable pelagic species of smaller size and low trophic level.

There are two types of species shifts in the ecosystem resources. One is systematic replacement and the other is ecological replacement. Systematic replacement occurs when a dominant species declines in abundance or is depleted by overexploitation and a competitive species uses the surplus food and vacant space. Ecological replacement occurs when a big change in the environmental conditions induces a shift in dominant species. The ecological replacement always happens in the pelagic resource. Figure 16.1 gives the relative abundance of various ecotypes comprising the annual catch in the Yellow Sea, and long-term changes in environmental conditions. It shows that during the warm years (1980s) the warm and temperate species such as half-fin fish trend to increase in abundance while during the cold years (1970s) the boreal species such as Pacific herring trend to increase. The natural environmental variations have an important effect on long-term changes in dominant species of various ecotypes.

4. Productivity of Small Pelagic Fish

Small pelagic fish are abundant in Chinese coastal and offshore waters. The mean annual catch of pelagic fish (4.15 million tons) accounted for 39% of the total marine catch, ranging from 30% to 48% in the last decade. The highest annual catch was 7.18 million tons in 1998. Species with an annual production of more than 100 thousand tons are Japanese anchovy, Japanese scad, Spanish mackerel, chub mackerel, silver pomfret, Japanese pilchard, and Chinese herring. The pelagic fish catch increased in the last decade due

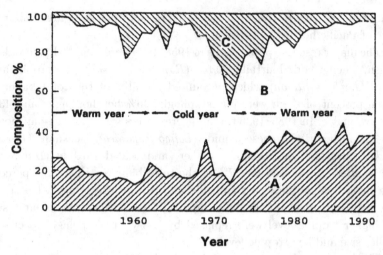

Fig. 16.1. Relative abundance of various ecotypes comprising the annual catch in the Yellow Sea, and long-term changes in environmental conditions. (A) Warm-water species; (B) temperate-water species; (C) Boreal species.

to the fast development of fisheries of Japanese anchovy, Japanese scad, and Spanish mackerel.

The bulk of Chinese pelagic catch is composed of ~20 major species, most of which are local stocks (Table 16.1). The warm-water species and warm-temperate species include mainly Japanese anchovy (*Engraulis japonicus*), Japanese scad (*Decapterus maruadsi*), Spanish mackerel (*Scomberomorus niphonius*), chub mackerel (*Scomber japonicus*), Japanese pilchard (*Sardinnops melanosticta*), silver pomfret (*Pampus argenteus*), Chinese herring (*Ilisha elongata*), Japanese jack mackerel (*Trachurus japonicus*), half-fin anchovy (*Setipinna taty*), scaled sardine (*Harengula zunasi*), and spotted sardine (*Clupanodon punctatus*). Only a few pelagic fishes belong to cold water species, such as Pacific herring (*Clupea harengus pallasi*) in the Yellow Sea. Japanese anchovy is the most abundant pelagic species and is distributed widely in the northwestern Pacific. In Chinese waters, anchovy stocks inhabit the Yellow, Bohai, East China, and the northern South China Seas. In the Yellow and East China Seas anchovy stocks are located mainly in areas with water depth of 40–80 m and water temperatures of 7°–14°C in winter. The main areas are the central and southern Yellow Sea and the northern part of the East China Sea. The biomass of anchovy in these two seas fluctuated between 2.5 and 4.3 million tons, with an average of 3.0 million tons, between 1986 and 1995.

Table 16.1. The annual catch of important commercial small pelagic fish in China during the last decade (10^3 tons).

Species	1990	1991	1992	1993	1994	1995	1996	1997	1998	1999	2000
Japanese anchovy	54	113	193	557	439	489	671	1,202	1,373	1,097	1,143
Japanese scad	381	420	392	261	431	515	608	506	532	503	502
Spanish mackerel	209	201	147	146	203	227	284	340	517	566	497
Chub mackerel	197	243	243	273	336	372	374	409	385	403	350
Silver pomfret	83	95	73	117	138	209	220	243	303	338	339
Japanese pilchard	42	63	53	47	69	58	93	125	121	147	154
Pacific herring	4	3	2	1	1	2	2	16	21	18	15
Chinese herring	24	31	30	29	33	47	51	78	84	110	108
Pelagic catch	2,105	2,415	2,619	2,837	3,438	4,227	3,802	5,381	7,181	5,868	5,758

The biomass was at a maximum of 4.3 million tons in 1992–1993 (Iversen and Zhu 1993). So far, the anchovy stock is mainly caught at spawning time in spring and feeding time in summer near the coast of the three provinces, Liaoning, Hebei, and Shandong, in the north of China. The over-wintering ground for anchovy is located in the offshore area and the stock there is under-exploited.

The causes of fluctuations in abundance of pelagic species are more complicated than those for demersal species. Overexploitation is one reason for the decline, but natural and ecologic factors affect abundance also. In the long history of exploitation, the yield of Pacific herring in the Yellow Sea varies greatly, but Spanish mackerel and silver pomfret stock appear to be relatively steady. The abundance and catch of Spanish mackerel and silver pomfret have tended to increase steadily since the 1960s. There is no strong relationship between spawning stock and recruitment of Pacific herring. Rainfall, wind, and irradiation appear to be the main factors affecting recruitment.

Even though the small pelagic fish are abundant, the utilization of most stock in Chinese waters has been fully exploited. However, these fish are characterized by fast recruitment and renewability since they have shorter life spans and a larger recruitment stock than the demersal species. As small pelagic fish are the food of large carnivorous fish they are an important link in the food web of the Yellow and East China Seas. The ecological links of small pelagic fish with other components of the marine community, and their relationship with environmental conditions need to be studied further. These studies will be the basis for the sustainable utilization of the marine resource.

5. Ecosystem Productivity

Indices of productivity and community structure, including primary production, zooplankton biomass, and fish productivity indicate that there has been large variation in ecosystem productivity in the Bohai Sea during the past four decades, and this might be a distinct character of coastal ecosystems.

The chlorophyll a concentration, primary production and phytoplankton abundance in the Bohai Sea were at a relatively high level before 1982. The yearly average phytoplankton abundance showed an declining trend from 1959 to 1998 with a drop from 222×10^4 cell m^{-3} in 1982 to 99×10^4 cell m^{-3} in 1992. Phytoplankton abundance in 1992 and in 1998 was about 38% of that in 1959 and 1982 (Tang et al. 2003). The pronounced change in phytoplankton abundance happened when both the chlorophyll a concentration and primary production were relatively low. In the same period, the number of phytoplankton species also decreased. Summer was the peak season for primary production and phytoplankton abundance in both 1982 and 1992.

In the Bohai Sea, the annual mean zooplankton biomass, which is roughly an estimation of secondary production of the ecosystem, decreased by about 40% from the late 1950s (107.3 mg m^{-3}) to the early 1990s (64.0 mg m^{-3}). However, remarkably high values were observed in the spring and summer of 1998, which was equal to three to six times the means at the same season from 1959 to 1982. The possible causes for the extremely high value observed in 1998 were as follows. First, the abundance of the dominant species, particularly *Calanus sincus* and *Centropages mcmurrichi* increased greatly in 1998 (Table 16.2). Second, some species, which were seldom found in previous surveys, were abundant in 1998. For instance, Macruran larvae were the most abundant species in Laizhou Bay, reaching 1898 m^{-3}.

The seasonal variation of fish productivity in the Bohai Sea is remarkable as most of the species emigrate to the over-wintering ground

Table 16.2. Dominant species abundance (number m^{-3}) of zooplankton in the Bohai Sea.

Year	Season	Sagitta crassa	Calanus sinicus	Labidocera euchaeta	Centropages mcmurrichi
1959	Spring	24.3	78.2	5.1	13.7
	Summer	79.3	1.8	42.3	0.0
	Autumn	32.2	3.4	48.8	0.2
	Mean	45.3	27.8	32.1	4.0
1982–1983	Spring	5.1	33.6	3.9	24.2
	Summer	56.3	9.5	85.4	0.6
	Autumn	49.1	0.9	50.0	0.0
	Mean	36.8	14.7	46.4	8.3
1992–1993	Spring	13.0	35.5	4.3	2.4
	Summer	72.8	6.5	19.6	0.0
	Autumn	22.4	2.1	23.4	0.0
	Mean	36.1	14.7	15.8	0.8
1998	Spring	55.6	285.8	22.5	185.8
	Summer	96.7	96.7	72.2	0.0
	Autumn	23.5	23.5	45.7	0.0
	Mean	58.6	135.3	46.8	61.9

Table 16.3. Biodiversity indices of fisheries in the Bohai Sea in autumn (October).

	Year		
	1982	1992	1998
Species number	77	74	52
Richness	4.90	4.84	3.77
Diversity	3.09	2.82	2.31
Evenness	0.71	0.65	0.58

in the Yellow Sea in late autumn and come back to the Bohai Sea in spring for spawning and nursery. Normally, the peak biomass in the Bohai Sea occurs in autumn and the lowest in winter. The bottom trawl surveys data of catch per hour of hauling in different years clearly indicated that the production had decreased since the 1950s.

The community structure of the fish resource tended to become simpler during the past decades. The species richness, diversity, and evenness decreased from the beginning of 1980s to late 1990s (Table 16.3). The shifts of dominant species were observed by the surveys in 1982. Large-sized demersal species were replaced by small pelagic species such as half-fin anchovy (*Setipinna taty*), anchovy (*Engraulis japonicus*), and Gizzard-shad (*Clupanodon punctatus*). This replacement has continued and as a result, the average trophic level of the species in the ecosystem decreased from 4.3 in 1959, to 3.3 in 1998–1999.

6. Issues to be Addressed

Some ocean areas such as coastal water, mesopelagic zones, upwelling regions, and high-latitude areas, which are likely to be particularly sensitive to long-term changes in climate, have been the subject of intensive study. The continental margins are a critical boundary because they are most directly relevant to human development. For example, input of nutrients, sediments, and pollutants are large in these areas and fishing pressure has been intense. There has, however, been little consideration of the role of marine food webs in continental margins that are heavily impacted by fishing, and their relationship to biogeochemical cycles in this region. The mesopelagic zone is an important ocean region for decomposition of organic matter and the recycling of nutrients. It is also an important region for pelagic food webs.

Future research on the ocean should identify elements and processes relevant to global change and develop the capability to observe them on appropriate scales to develop a predictive capability for the response of the ocean system to natural and anthropogenic perturbations. This would also enable the assessment and prediction of scenarios and options in order to enable society to make choices about sustainable futures.

The living resources in marine ecosystems will be the foundation of food security in human development. However, changes in the outputs of the marine ecosystems in recent years are cause for concern; primary production has decreased, biodiversity has been reduced, eutrophication has been aggravated, and high value fish resources have decayed in recent years. It is crucial that the cause of these changes be determined so that the healthy and sustainable development of the marine ecosystem is ensured. Studies on the East China Sea and the Yellow Sea ecosystem dynamics by Chinese scientists are focusing on the coupling of physical and biological processes in the region. The goals are to identify key processes of ecosystem dynamics, improve predictive and modeling capabilities, understand the ecosystem responses to climate change and provide a rational system for the management of living marine resources. Multiprinciple and comprehensive studies will be conducted, aimed at the following key scientific problems: (a) energy flow and energy conversion of key species, (b) recruitment of zooplankton population, (c) renovation of bio-elements, (d) ecological effects of key physical process, (e) coupling of pelagic and benthic systems, and (f) the contribution of the microbial food loop.

Literature Cited

Chen, G., X. Gu, H. Hao and G. Yang. 1990. *Marine Fisheries Environment of China*. Hangzhou: Zhejing Science and Technological Press (in Chinese).

Gedalof, Z. and D. J. Smith. 2001. Inter-decadal climate variability and regime-scale shifts in Pacific North America. *Geophysical Research Letters* 28:1515–1518.

Iversen, S.A. and D. Zhu. 1993. Stock size, distribution and biology of anchovy in the Yellow Sea and East China Sea. *Fisheries Research* 16:147–163.

Jin, X. and Q. Tang. 1998. The structure, distribution and variation of fishery resources in the Bohai Sea. *Journal of Fishery Sciences of China* 5:18–24 (in Chinese).

Li, Y. 1992. A preliminary analysis on the relationship between Southern Oscillation and dryness–wetness in the eastern China during the last 400 years. *Marine Science* 82:37–40 (in Chinese).

Tang, Q. 1981. A preliminary study on the cause of fluctuation in year class size of pacific herring in the Yellow Sea. *Transactions of Oceanology and Limnology* 2:37–45.

Tang, Q. 1987. Estimation of fishing mortality and abundance of pacific herring in the Yellow Sea by cohort analysis (VPA). *Acta Oceanologica Sinica* 6:132–141.

Tang, Q. 1989. Changes in the biomass of the Yellow Sea ecosystem. Pp. 7–35 in *Biomass and Geography of Large Marine Ecosystems*, edited by K. Sherman and L. M. Alexander. American Association for the Advancement of Science.

Tang, Q. 1995. The effects of climate change on resources populations in the Yellow Sea ecosystem. *Canadian Special Publication of Fisheries and Aquatic Sciences* 121:97–105.

Tang, Q., J. Deng and J. Zhu. 1989. A family of Ricker SRR curve of the prawn under different environment condition and its enhancement potential in the Bohai Sea. *Canadian Special Publication of Fisheries and Aquatic Sciences* 108:335–339.

Tang, Q., X. Jin, J. Wang, Y. Cui, Z. Zhuang and X. Meng. 2003. Decadal-scale variations of ecosystem productivity and control mechanisms in the Bohai Sea. *Fisheries Oceanography* 12:223–233.

Chapter 17

SEA-LEVEL CHANGES AND VULNERABILITY OF THE COASTAL ZONE

NOBUO MIMURA and HIROMUNE YOKOKI

1. Introduction

Global warming induces environmental changes which bring about changed external forces acting on the coastal zones, such as increase in seawater temperature, sea-level rise, changes in precipitation, exacerbated extreme events including typhoons, changes in ocean currents, and waves. Sea-level rise has attracted special attention, because it will seriously affect low-lying coastal lands, particularly deltas formed by large rivers, and small islands. These land forms occur widely in the Asian and Pacific region. As high population growth and economic development is expected in this region, the impact of sea-level rise and climate change are potential concerns for the safety and sustainable development of the region. In this chapter, we review present knowledge and status of research on sea-level change and its impact on the coastal zones.

2. Past and Future Sea-Level Change

The Intergovernmental Panel on Climate Change summarized the present status of the studies on the historical changes in mean sea-level and its future prediction in the Third Assessment Report published in 2001 (IPCC 2001a). On average, the global mean sea-level rise 10–20 cm during the past 100 years. For the future prediction of global warming, the Panel developed four basic scenarios for the future socioeconomic development (IPCC 2000). Based on these scenarios, the Panel indicated that the global mean temperature will increase $1.4°$–$5.8°C$ by 2100, which would result in a sea-level rise of 9–88 cm. Among various factors, thermal expansion of the ocean contributes about 30% of the rise, while 20% is due to melting of glaciers and icecaps on land, and the contribution of other factors is small or unclear (IPCC 2001a).

In China, analysis of tide gauge records showed that the mean sea-level along the coast over the last 30–40 years has increased by 1.5–2.0 mm yr^{-1} (ESD-CAS 1994). Li et al. (in press) indicated that sea-level changes relative to the land surface is much more profound in China (tens to hundreds of mm yr^{-1} for past decades). For example, the ground subsidence in Shanghai ranged from 2.5 to 110 mm yr^{-1} for the last 70 years. In addition, the river surface rises as a delta advances. The water level in the Yellow River delta has risen 2–3 m since 1920s due to delta expansion (ESD-CAS 1994). Therefore, it is essential that both mean sea-level rise induced by global warming and local changes relative to the land surface be included for impact assessments.

Cho (2003) reported an analysis of long-term sea-level trends along the Korean coasts using the records from 23 tide-gauge stations. After a correction for post-glacial rebound of 2.31 ± 2.22 mm yr^{-1}, the regional variation in the long-term sea-level trend along the Korean coasts was determined to be 0.57, 3.13, and 2.64 mm yr^{-1} for the east, south, and west coasts, respectively. Satellite altimeter data were also used to estimate the sea-level trends for the period 1992–2001. The changes for the east and south coasts were 4.6 and 4.8 mm yr^{-1}, respectively, which are larger than the global average of 3.1 mm yr^{-1} (Nerem 1999).

Numerous studies have also been made on the long-term trends for Japan. Uda et al. (1992) analyzed about 40 tide-gauge data for the period 1955–1989 and found that the average rate of sea-level rise was 2.2 mm yr^{-1}. Recently, Nakano et al. (2002) analyzed the data from 44 tide-gauge stations and found that the mean sea-level rise along the southwest Japanese coast has increased rapidly since 1985 (Fig. 17.1). This sea-level rise is correlated primarily with the recent increase in seawater temperature. Along the northern part of the Japanese coast the sea-level rise during the period 1980–2000 ranged from 2.7 to 15.3 mm yr^{-1}.

The global and Korean trends derived from the satellite altimeter, Topex/Poseidon, are larger than those obtained using the tide-gauge data. As the satellite seems to have advantages in terms of accuracy, further studies are needed to determine the reasons for the discrepancy. As the Chinese studies pointed out, real sea-level change is a combination of global

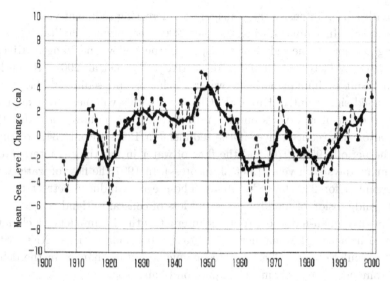

Fig. 17.1. Long-term variation of mean sea-level around Japan.

mean value and local changes relative to the ground movements. The local changes in some areas near mega cities and areas with rapid development are often larger than the global mean value.

3. Impacts on Environment

Sea-level rise results in a chain of events in the coastal environment, such as inundation of low-lying areas, exacerbated coastal erosion, higher storm surges and waves, and intrusion of seawater into aquifers and rivers. These changes, in turn, introduce a variety of impacts on the coastal systems. The exposed coastal systems consist of the natural environment, and man-made systems. In this section, we will see how the impacts are being studied according to the two groups.

3.1. Deltas

Typical deltas in East Asia include the Yellow River, Yangtze River, Pearl River, and Amur (Heilongjiang) River (Li et al. in press). As the rivers flow down relatively steep lands and the monsoon leads to high sediment yields, they transport large quantities of sediment to the sea. Human activities often intensify the transport by increasing sediment yields in the river basin. In China, during the past 2000 years, when civilization was being developed, the sediment transport in the Yellow and Yangtze Rivers increased by factors of 10 and 2, respectively (Saito et al. 2001). Recently, as a result of human activity, water discharge and sediment load have decreased significantly (Hu et al., Chapter 15, this volume). For example, construction of reservoirs on the Yellow River and removal of water for irrigation has reduced river-borne sediment in 1990s to only 34% of that in the period 1950–1970, and the flow of water in the river is interrupted for many days each year (Li et al. in press). In 1997 the flow of water was interrupted for 226 days (Ye 1998). These events induced retreat of the delta and erosion of muddy coasts and loss of coastal wetland.

Such human-induced problems interact with the impacts of climate change in a complex manner. As deltas accommodate mega cities, with large populations, and high socioeconomic activities changes in the delta environment have a significant impact on human society.

3.2. Coral reefs

Although the growth rates of individual corals vary from 10 to 100 mm yr^{-1}, the consolidation of a reef is much slower (1–10 mm yr^{-1}). After studying the responses of coral reefs to the post-glacial sea-level change, a range was identified (Woodroffe 2000). Kayanne (1992) showed that the maximum rate of upward reef growth was 4 mm yr^{-1} in the past 10,000 years. The predicted sea-level rise this century ranges from 9–88 mm yr^{-1}, which spans the threshold rate, indicating that sea-level rise will be a critical factor for survival of the reefs.

Global warming will also result in increased water temperature which is likely to increase the incidence of coral bleaching. When the water temperature continues to exceed the upper limit favorable for the growth of coral reefs (18°–28°C), corals discharge symbiotic algae which results in coral bleaching. The largest bleaching event took place when extraordinary warm water covered the ocean surface after the El-Niño event in 1997–1998 (Wilkinson 2000).

In the tropical and subtropical regions, human activities such as fishing, coral mining, tourism, and water pollution will continue to increase and have a bigger effect on coral growth. These pressures will reduce the resilience of coral reefs to global warming and climate change.

3.3. Mangroves

Mangrove forests are coastal ecosystems growing in tropical and subtropical areas. They have high biological productivity, and provide a buffer zone between the land and the sea, and act as a breakwater against waves and storm surges. As mangroves grow in the upper half of the inter-tidal zone, they are sensitive to sea-level rise and increasing water temperature. Sea-level rise may result in migration of the forests and changes in tree species.

The upward growth of mangrove forests depends on the rates of accumulation of sediments. Miyagi et al. (1995) and Fujimoto et al. (1999) indicated that mangrove forests with high productivities could adjust to a sea-level rise of $5\,\mathrm{mm\,yr^{-1}}$ and migrate landward. However, if there are barriers along the coast such as seawalls, they will not be able to move into the land and the area of mangrove forests will shrink (IPCC 2001b). The present northern limit of mangroves is Kagoshima, Japan, around 31.5°N, but with increasing water temperature, the limit may move further north. As with coral reefs, mangrove forests face degradation from activities such as formation of shrimp ponds. These human pressures weaken the resilience of mangroves to global warming and sea-level rise.

3.4. Sandy beaches and mud flats

Many countries suffer from beach erosion caused by lack of sediment discharge from rivers, interference of alongshore sediment transport by port and coastal structures, and sand mining. Sea-level rise can be another cause of beach erosion. As beaches tend to transform their profiles according to the increase in sea-level and incident waves, this self-adjustment induces offshore sediment transport from the foreshore, i.e. beach erosion.

Mimura and Kawaguchi (1996) estimated the area of erosion of Japanese sandy beaches using the Bruun rule model (Bruun 1992). The lengths of shoreline retreat and the areas of eroded sandy beach were estimated for all the sandy beaches in Japan. The study suggested that 57%, 82% and 90% of the existing Japanese sandy beaches would be eroded due to sea-level rises of 30, 65 and 100 cm, respectively. As beach erosion is severe even today because of the lack of sediment supply and alongshore sediment transport, erosion due to sea-level rise would make the situation much worse.

Mud flats will be another victim of sea-level rise. As the bottom sediment is more easily moved and the slope is very gentle, even a small sea-level rise would inundate vast areas and cause serious erosion. However, there is no quantitative estimate of these impacts.

3.5. Man-made systems

Various human activities are concentrated on the coasts, including cities, residential areas, ports and fishery harbors, roads and rail roads, natural disaster prevention facilities, industries and tourism activities. These will be affected by sea-level rise and climate change. In many countries, infrastructures such as port facilities, breakwaters, seawalls, dykes, sewage outfall, water gates, and drainage systems have accumulated on the coastal zone. In the face of sea-level rise, many of the structures will need to be raised or reinforced to maintain the present level of stability and function.

In Japan, Kitajima et al. (1993) estimated the costs of protecting port facilities and coastal structures against a 1 m rise in sea-level. Total costs for protection in 1992 were estimated at 115 billion US$, of which about 78 billion US$ were necessary for raising the port facilities, and another 36 billion US$ were for the coastal protection facilities. Protection costs for the 2,900 fishery harbors and 9,000 km of coastal protection structures would almost double these estimates.

4. Socioeconomic Vulnerability in China

The Chinese coastline including mainland and island coast is 32,000 km long. The area of coastal provinces and cities is 1.6 millions km^2 with 527.6 million people (41.9% of the total population). Gross domestic product of the coastal provinces is more than 1.0 trillion US$, about 72.5% of China's total (Li et al. in press).

About 70% of the sandy and muddy coasts suffer from erosion (Xia 1993); 1400 km^2 land in Jiangsu Province has been lost since 1855 and 202 km^2 was lost during 1985–1996 in the Diaokao abandoned sub-delta of the modern Yellow River delta (Ren 1986). Based on 20 years of observations on the 33 km coast of southwestern Shangdong Peninsula, Zhuang et al. (2000) concluded that 40% of the loss can be attributed to the reduction of river-borne sediment, 50% to beach sand mining and 10% may be due to sea-level rise.

Sea-level rise will exacerbate coastal flooding due to storm surges. In order to protect coastal lowlands from storm surge and strong wave attack, sea walls have been constructed and the protected coastline is about

Table 17.1. Flooded areas (km^2) affected by sea-level rises of 30, 65, and 100 cm in the major vulnerable areas of China (Du et al. 1997).

Main vulnerable area	Protection	30 cm	65 cm	100 cm
Zhu River delta	None	5,546	5,967	6,543
	Existing	1,152	3,457	6,520
Yangtze delta and Jiangsu coastal plain	None	54,547	58,663	61,288
	Existing	898	22,242	52,091
Yellow River delta and North China plain	None	21,255	23,106	25,428
	Existing	21,010	23,100	25,428
Total	None	81,348	87,736	93,259
	Existing	23,060	48,799	84,039

12,000 km long (Du et al. 1997). The target protection level in China varies from 1 storm per 1,000 years to 1 storm per 10 years though most of them are for 1 in 50 to 1 in a 100 year storm surges (Li 1993). High storm surges and heavy rainfall generated by typhoons during the flood season cause an average economic loss of 15 million US$ per year and 17 million people flooded in 1990-1998.

Du et al. (1997) presents a national study on the vulnerability of coastal zones. For sea-level rise of 30, 65, and 100 cm, the economic losses, wetland and land losses, protection/adaptation costs are all in the low vulnerability class, while people affected are in the medium-high class according to the classification used as part of the Intergovernmental Panel on Climate Change common methodology (Nicholls and Mimura 1998). Table 17.1 shows that flooded areas for a 30 cm sea-level rise are very different for protected and non-protected scenarios in the Yangtze and Zhu River deltas because these areas are protected by high-level dykes, while they are close together for the Yellow River delta which has only low-level dykes. The flooded areas estimated for non-protected and protected scenarios are almost the same for a sea-level rise of 100 cm, indicating the current level of coastal protection is insufficient for high sea-level rise. These analyses suggest that the Chinese coastal zone has a lower vulnerability to sea-level rise compared with the previous reports (Han et al. 1995).

5. Socioeconomic Vulnerability in Korea

Coastal zones are very important also for Korea for they provide a range of goods and services including fishery resources, sea transportation, and the natural environment. Korea has suffered from storm surges, high waves, and tsunamis.

A recent event was Typhoon 0314 in 2003. Typhoon 0314 passed from south to north over Korea causing serious damages. One hundred and thirty-two human lives were lost, and 366 were wounded, and asset damages amounted to 4.4 billion US$. Many port facilities were lost and seriously damaged in Pusan, Masan, and Sogido. City areas close to the coasts were flooded by storm surges, which drowned people in basements of buildings.

Recently, Cho (2003) reported on a vulnerability assessment for Korea, which used inundation area and inundated people as indices and sea-level rise scenarios with tide level and storm surges.

For the 1 m sea-level rise scenario with high tide and storm surge, the maximum inundation area is 2,643 km^2 (~1.2% of total area of Korea) and the population affected is 1.255 million (~2.6% of the total population). This result suggests that the coastal zone of the Korean Peninsula is quite vulnerable to future sea-level rise. The west coast is more vulnerable than the east and south coasts (Fig. 17.2). On the western coast of the Korean Peninsula, North Korea appeared to be more vulnerable than South Korea. As the present study does not consider such factors as duration of the typhoon, and existence of coastal dyke and seawalls the results represent the worst possible scenario.

6. Socioeconomic Vulnerability in Japan

Japan's coastline is over 34,000 km long and the majority of the population and economic activities are concentrated in the coastal zones. Although coastal municipalities occupy only about 32% of the total area of 370,000 km^2, they account for 46% of the population, 47% of the industrial output, and 77% of the total expenditure for retail business or market goods. A notable feature is the artificial coast which occupies 29% of the total coastline, while the natural beaches cover about 57%, and semi-natural beaches 14%. Marine transportation and fisheries are highly developed, with 1,094 commercial and industrial ports and 2,950 fishing ports (Kojima 2000). Land reclamation and artificial islands have been constructed to create land for industrial factories, power plant facilities, farming ground, and other development.

In a macroscopic analysis, Matsui et al. (1992) calculated the coastal vulnerability, and area, population, and amount of assets are at risk by sea-level rise and storm surges. Three water levels (mean, high, and high

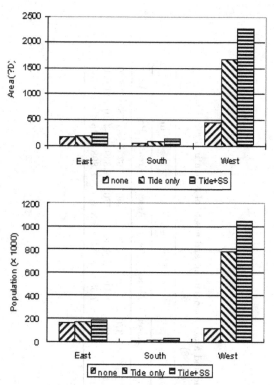

Fig. 17.2. Inundation area (top) and population (bottom) in the east, south, and west coast of Korea vulnerable to 1 m sea-level rise (Cho 2003).

tide caused by storm surge or tsunami) were used to evaluate the effect of inundation and flooding. Table 17.2 shows the comparisons for the present and the sea-level rise conditions.

Even under the present situation, 861 km² of land is already below high water level, where about 2 million people live and 540 billion US$ worth of assets exist. These areas are concentrated in large cities like Tokyo, Osaka, and Nagoya, which are Japan's economic centers. If a 1 m sea-level rise occurs, the area at risk will expand to 2,339 km², 2.7 times the present amount, and the population at risk will increase to 4.1 million.

7. Vulnerability of the Asia–Pacific Region

A region-wide assessment was performed for the impacts of sea-level rise and storm surges using global datasets on climatic, environmental, and societal

Table 17.2. Impacts of sea-level rise on the inundated areas, population, and assets of Japan in 1992. Unit: area (km^2), population (million persons), assets (billion US$).

Water level	Present			0.3 m rise			0.5 m rise			1.0 m rise		
	Area	Pop.	Asset	Area	Pop.	Asset	Area	Pop.	Asset	Area	Pop.	Asset
Mean	364	1.02	340	411	1.14	370	521	1.40	440	679	1.78	530
High	861	2.00	540	1,192	2.52	680	1,412	2.86	770	2,339	4.10	1,090
High tide	6,268	11.7	2,880	6,662	12.3	3,020	7,583	13.5	3,330	8,898	15.4	3,780

information to draw a picture of the future threats (Mimura, 2000). The target area of this study is the whole Asia and Pacific region, which covers 30°E to 165°S, and 90°N to 60°S. Land area and population within this area are about 6.5 million km^2 and 3.8 billion people as of 1994, which is estimated to increase to 7.6 billion in 2100.

Even today, the areas below high tide level and storm surge level, i.e. inundated and flooded areas, are 311,000 km^2 or 0.48% of the total area and 611,000 km^2 or 0.94%, respectively. They increase to 618,000 and 858,000 km^2 (0.98% and 1.32%) with a 1 m sea-level rise. The flooded area increased by sea-level rise amounts to 247,000 km^2. Regarding the affected people today, about 47 million people or 1.21% of the total population lives in the area below high tide level, while 270 million people or 5.33% live below storm surge level. This shows that the Asia and Pacific region are already vulnerable to flooding by storm surge. If the mean sea-level rises 1 m and the population growth by 2100 is taken into account, the population affected become about 200 and 450 million people. The increase in population in the flooded area reaches 249 million.

The areas which may be affected seriously are distributed in the deltas of Yangtze River, Mekong River, and Ganges and Brahmaputra Rivers, and the southern part of Papua New Guinea. Given the high concentration of population and socioeconomic activities in the coastal areas, countries in East Asia are among those with high future risks.

8. Future Research Needs

This chapter discusses the vulnerability of East Asia to sea-level rise and climate change based on existing studies. The profile and characteristics of vulnerability are quite diverse, in some cases also very serious, reflecting the variety of natural and man-made systems in the region. These results also lead us to a consideration of response strategies. Vulnerability is an overall concept for coping with the adverse effects of sea-level rise and climate change. It consists of several components, such as susceptibility and resilience to environmental changes and natural disasters, and capacity of the system to adapt (IPCC 2001b). In the face of climate change, each country needs to take action in the form of engineering measures, and institutional and planning arrangements, to reduce the adverse effects. The degree and range of the possible impacts identified in this review is quite large. On the other hand, the adaptive capacity of the coastal systems and society is limited, particularly in the developing countries, and the impacts may be overwhelming. In order to obtain a clearer vision on coping with sea-level rise and climate change, we need further studies to fill gaps in the present knowledge of prediction of future global change, vulnerability in both qualitative and quantitative manners, adaptive capacity of the coastal systems, and response strategies. Concrete themes of future research needs are as follows:

1. Monitoring to detect the changes in mean sea-level, typhoon characteristics, and other elements of climate change.
2. Development of climate models and inter-comparison of models. The reliability and spatial resolution of present climate models seems to be insufficient to provide more accurate vulnerability assessments. We need closer collaboration with the modelers so that more suitable models can be developed.
3. Integrated studies of vulnerability which include both impacts of climate change and other human activities. In many areas, human intervention has more significant effects on the coastal environment. Degradation leads to decreased capacity of the natural systems to return to the original state after impact. Based on such studies, we can identify the most vulnerable systems and places in the region to develop response strategies.
4. Response strategies and options. The actions to reduce the impacts and cope with them are called adaptation. Adaptation is necessary for the safety of the future society and an important component of sustainable development in the region.

Literature Cited

Bruun, P. 1992. Sea-level rise as a cause of shore erosion. *Journal of Waterways, and Harbors Division* 88:117–130.
Cho, K. 2003. Vulnerability of Korean coast to the sea-level rise due to 21st global warming. Pp. 139–146 in *Proceedings of international symposium on diagnosis, treatment and regeneration for sustainable urban systems.* Ibaraki, Japan: Ibaraki University.
Du, B. L., S. Z. Tian and C. H. Lu. 1997. *The impacts of sea-level rise on the major vulnerable areas of the China's Coastal zones and relevant strategies.* Beijing: China Ocean Press (in Chinese with English abstract).
ESD-CAS (Earth Science Division, Chinese Academy of Sciences). 1994. *Impact of sea-level rise on the deltaic regions of China and its mitigation.* Beijing: Science Press (in Chinese).

Fujimoto, K., T. Miyagi, T. Murufushi, Y. Mochida, M. Umitsu, H. Adachi and P. Pramojanee. 1999. Mangrove habitat dynamics and Holocene sea-level changes in the Southwestern coast of Thailand. *Tropics* 8:239–255.
Han, M. K., J. J. Hou and L. Wu. 1995. Potential impacts of sea-level rise on China's coastal environment and cities: A national assessment. *Journal of Coastal Research* 14:79–95.
IPCC (Intergovernmental Panel on Climate Change). 2000. *Special report on emission scenarios.* Cambridge, UK: Cambridge University Press.
IPCC (Intergovernmental Panel on Climate Change). 2001a. *Climate change 2001: The scientific basis,* edited by J. T. Houghton, Y. Ding, D. J. Griggs, M. Noguer, P. J. van der Linden, X. Dai, K. Maskell and C. A. Johnson. Cambridge, UK: Cambridge University Press.
IPCC (Intergovernmental Panel on Climate Change). 2001b. *Climate change 2001: Impacts, adaptation and vulnerability.* Cambridge, UK: Cambridge University Press.
Kayanne, H. 1992. Deposition of calcium carbonate into Holocene reefs and its relation to sea-level rise and atmospheric CO_2. Pp. 50–55 in *Proceedings of the 7th International Coral Reef Symposium.*
Kitajima, S., T. Ito, N. Mimura, Y. Tsutsui and K. Izumi. 1993. Impacts of sea-level rise and cost estimate of countermeasures in Japan. Pp. 115–123 in *Vulnerability assessment to sea-level rise and coastal zone management.* Proceedings of the Intergovernmental Panel on Climate Change Eastern Hemisphere Workshop, edited by R. Mclean and N. Mimura.
Kojima, H. 2000. Vulnerability and adaptation to sea-level rise in Japan. Pp. 81–88 in *Coastal impacts of climate change and adaptation.* Proceedings of the Asia Pacific Network for Global Change Research/Synthesis and Upscaling of Sea-Level Rise Vulnerability/Land–Ocean Interactions in the Coastal Zone Joint Conference, 14–16 November 2000, Kobe. Asia Pacific Network for Global Change Research and Ibaraki University.
Li, C. X. 1993. Impact of relative sea-level rise on the coastal lowland of China. *Advance of Earth Sciences* 8:26–30 (in Chinese with an English abstract).
Li, C., D. Fan, B. Deng and V. Korotaev. The coasts of China and issues of sea-level rise. *Journal of Coastal Research* Special Issue. In press.
Matsui, T., H. Tatcishi, M. Isobe, A. Watanabe, N. Mimura and A. Shibazaki. 1992. Vulnerability assessment of sea coastal zone in Japan under the sea-level rise. *Proceedings of Coastal Engineering* 39:1031–1035 (in Japanese).
Mimura, N. 2000. Distribution of vulnerability and adaptation in the Asia and Pacific Region. Pp. 21–25 in *Proceedings APN/SURVAS/LOICZ Joint Conference on the Coastal Impacts of Climate Change and Adaptation in the Asia–Pacific Region,* APN and Ibaraki University.
Mimura, N. and E. Kawaguchi. 1996. Responses of coastal topography to sea-level rise. *Proceedings of Coastal Engineering* 96:1349–1360.
Miyagi, T., T. Kikuchi and K. Fujimoto. 1995. Late Holocene sea-level changes and mangrove peat accumulation/habitat dynamics in the western Pacific area. Pp. 19–26 in *Rapid sea-level rise and mangrove habitat,* edited by T. Kikuchi. Institute of Basin Ecosystem Studies, Gifu University.

Nakano, S., S. Tadokoro, K. Uno and M. Fujiki. 2002. Long-term variation in the mean sea-level and effects of global warming. *Proceedings of Coastal Engineering* 49:1351–1355 (in Japanese).
Nerem, R. S. 1999. Measuring very low frequency sea-level variations using satellite altimeter data. *Global and Planetary Change* 20:157–171.
Nicholls, R. J. and N. Mimura. 1998. Regional issues raised by sea-level rise and their policy implications. *Climate Research* 11:5–18.
Ren, M. E. 1986. *Investigation of coastal zones and tidal flat sources in Jiangsu.* Province. Beijing: China Ocean Press.
Saito, Y., Z. Yang and K. Hori. 2001. The Huanghe (Yellow River) and Changjiang (Yangtze River) deltas: A review on their characteristics, evolution and sediment discharge during the Holocene. *Geomorphology* 41:219–231.
Uda, T., H. Ito and Y. Ohtani. 1992. Mean sea-level change along the Japanese coasts since 1955. *Proceedings of Coastal Engineering* 39:1021–1025 (in Japanese).
Wilkinson, C. 2000. *Status of coral reefs in the world.* Townsville, Queensland: Australian Institute of Marine Science.
Woodroffe, C. 2000. Biogeomorphological response of tropical coasts to environmental change. Pp. 15–18 in *Coastal impacts of climate change and adaptation.* Proceedings of the Asia Pacific Network for Global Change Research/Synthesis and Upscaling of Sea-Level Rise Vulnerability/Land-Ocean Interactions in the Coastal Zone Joint Conference, 14–16 November 2000, Kobe. Asia Pacific Network for Global Change Research and Ibaraki University.
Xia, D. X. 1993. Overview of Chinese coastal erosion. *Acta Geographica Sinica* 48:468–475.
Ye, Q. C. 1998. Flow interruptions and their environmental impact on the Yellow River delta. *Acta Geographica Sinica* 53:385–392 (in Chinese with an English abstract).
Zhuang, Z. Y., P. Yin, J. Z. Wu and L. H. Zhuang. 2000. Coastal erosion and its influence on southern Shangdong sandy coast. *Marine Geology and Quaternary Geology* 20:15–21 (in Chinese with an English abstract).

Part V

DRIVING FORCES

Chapter 18

CHANGES IN POPULATION NUMBER, COMPOSITION, DISTRIBUTION, AND CONSUMPTION IN EAST ASIA

XIZHE PENG and XUEHUI HAN

1. Introduction

The population of East Asia is dominated by two of the world largest populated countries, China and Japan. The former passed the 1.3 billion population mark on 6 January 2005, and the latter recorded a population of 127 million in 2003. Over the last half a century, East Asia witnessed a very profound demographic transition. Both mortality and fertility have declined in every country and area of the region, more rapid and more dramatic than in any other part of the world. Between 1950 and 2000, life expectancy in East Asia rose from 43 to around 75 years, and fertility level dropped from more than five to less than two children per woman. As a result, population growth speeded up to alarming levels before it started to slow down in recent decades.

The huge population base with widely diversified demographic dynamics in East Asia, together with unprecedented economic development and changes in consumption and life styles, has inevitably caused great pressure on resources and environment, not only in East Asia but in the whole world as well (EWC 2002).

2. Demographic Transition

2.1. Fertility and mortality

The driving forces for the shift to a slower population growth regime in most of the East Asian countries come from two competing factors. The first is the decline in death rates, particularly among infants. Due mainly to the very high infant mortality rate, population growth in East Asia used to be modest in the past few centuries even though women were, on average, giving birth to six or more children during their reproductive span. As death rates declined, East Asia's population growth accelerated to close to 2% per year around 1950s. Japan was the first country in this region that started demographic transition and reached a below replacement fertility (2.1 births per woman through her reproductive period) by the early 1960s. It was followed by a dramatic fertility decline in Hong Kong, Macao, Taiwan, South Korea, and China. With the fertility level in China below the replacement level in the early 1990s, all countries in East Asia, except Mongolia, where the total fertility rate is 2.3 births per woman, have completed demographic transition.

To an extent, lower birth and death rates are by-products of development. Higher incomes contribute to infant mortality decline by raising nutritional standards and to lower birth rates by raising the market value of women's time and, hence, the opportunity costs of childbearing. Other features of social and economic development, particularly the educational attainment of women, may play an even more important role. It can be seen from Table 18.1 that a high female literacy rate, higher percentage of the population living in urban areas, lower infant mortality rates, and higher life expectancies at birth are, in general, major factors resulting in lower fertility levels. However, it has also become increasingly clear in recent years that birth and death rates can decline rapidly in countries that have not yet achieved significant levels of economic development such as in China.

The commitments by Governments to provide effective family planning and reproductive programs are another determinant of rapid fertility decline in East Asia. The process of fertility transition in China is worth noting. China's total fertility rate has dropped from 5.8 in the early 1970s to 2.8 in the early 1980s and to 1.8 in the late 1990s. However, there are remarked regional and rural–urban diversities in fertility level that are caused mainly by differences in socioeconomic development and decentralized family planning programs in China.

2.2. Future population growth

The future East Asian population will be largely determined by population dynamics in China. The medium-growth scenario projected by the United Nations (UN 2001) anticipates that in the next half century the population

Table 18.1. Major demographic and economic indicators of East Asian countries and regions.

Countries or areas	Population (millions)*	Total fertility rate*	Life expectancy at birth (Years)#		Infant mortality rate (per 1,000)*	Percentage urban#	Female adult literacy rate (%)#
			Male	Female			
China	1,298.85	1.88	69	73	30.00	32	77
Hong Kong	6.86	0.96	76	82	3.20	100	90
Macao	4.54	0.93	75	80	10.00	99	90
Japan	127.33	1.33	77	84	3.00	79	—
North Korea	22.70	2.07	70	76	42.00	60	—
South Korea	48.60	1.45	69	77	5.00	82	96
Mongolia	2.75	2.43	65	68	58.00	64	99

Sources: *Data for 2002 (World Bank 2004); # Data for 2000 (ESCAP 2000).

of China will reach to 1.5 billion, a net increase of 200 million. About two thirds of the growth is anticipated to occur between 2000 and 2025, and China's population will eventually stop growing before the middle of the 21st Century. Population decline is projected for Japan and Korea over the coming decades. If the current demographic trend continues, the population of Japan will fall from 127 million at present to 110 million by 2050 and 90 million by 2100, but will recover somewhat by 2300. The population of South Korea is expected to grow to 53.0 million in 2030, and then decline to 51.3 million in 2050.

2.3. Population aging

The number of persons in East Asia aged 65 and over is expected to increase by 3.4 times over the first half of the 21st century. In addition to size, the proportion of the elderly is anticipated to rise quickly in the near future. In 2000, Japan's population was by far the most aged in Asia: 23.2% of its population were aged 60 and over compared to 7.7% in the 1950s.

Other East Asian countries and regions followed Japan's path in an even more rapid manner. It took Japan 28 years, from 1967 to 1995, to increase its proportion of persons aged 60 and over from 10% to 20%. The comparable demographic shift in the Republic of Korea and China will be achieved in 22 years and 27 years (Ogawa 2002). The median age for East Asia, which is the most aged sub-region in Asia, is projected to almost double from 23.5 to 44.3 years over the period 1950–2050.

While the major cause of rapid aging in East Asia is fertility decline, improvement in longevity is another salient factor. In 2000, the expectation of life at birth in Japan was 77 years for males and 84 years for females, among the highest in the world. The same indices for China are 69 and 73, for Hong Kong 76 and 82, for Macao 75 and 80, and in South Korea 69 and 77 (Table 18.1). However, marked regional variations exist in China. The life expectancy at birth for people living in big Chinese cities such as Shanghai has already reached a level similar to that of Japan, but the figure is much lower in some inland rural areas.

Among the elderly, females account for a much larger proportion. This could be called the feminization of the aged in East Asia. Old women are likely to be widows. Most of the elderly, particularly in China, live in rural areas with their children. A substantial proportion of the elderly are engaged in economic activities.

Table 18.2. Number of adults and children with HIV and Aids, 2003 (UNAIDS 2004).

Regions	Estimates	Range
East Asia	900,000	450,000–1,500,000
China	840,000	430,000–1,500,000
Japan	12,000	5,700–19,000
South Korea	8,300	2,700–16,000

2.4. HIV and Aids

East Asia is not the region hit most seriously by the AIDS epidemic. However, according to a recent estimation from UNAIDS (2004), in 2003, East Asia was home to nearly 1 million men, women, and children living with HIV (Table 18.2). In China, about 840,000 people were reported as HIV positive or AIDS patients in 2003. There were sharp rises in HIV among people with identifiable risky behaviors such as drug injectors, male, transvestite, and female sex workers and their clients. In addition, some people caught the AIDS virus through blood contamination.

Given the very large population numbers in East Asia, if HIV continues to spread widely among those with risk behaviors and their immediate sex partners, several million new infections will result. Prevention efforts that seek to limit exposure to HIV and reduce infection levels have been taken by governments of East Asian countries and regions. However, some of these efforts do not seem to be succeeding, and more societal and government input are urgently required.

3. Migration and Urbanization

Another path taking by the population is migration. Urbanization is the migration of population from rural areas to urban areas or the rural areas transform into urban areas by some means. Since the migrating population move from one place to another, the culture, economics, religion, and society maybe completely different at the destination. Therefore, the migration will induce great changes in the targeting population.

East Asia is relatively under-urbanized given its level of economic development. On an average, in 2000, 40% of the population lived in urban areas that are mainly determined by the low urbanization level in China. The urbanization of Japan, South Korea, and Mongolia is much more advanced than that of China. The low level of urbanization in East Asia

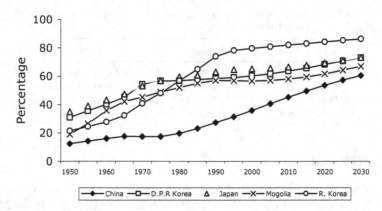

Fig. 18.1. Trend of urbanization in East Asia (UN 2003).

compares with the average of 76.8% for Latin American countries, 49.3% for North African countries, and 34.0% for Sub-Saharan African countries.

The potential for urban growth in China is substantial (Fig. 18.1). In 2001, there were 662 cities, and 20,358 towns in China with a total urban population of 480 million. Furthermore, China has launched its ambitious urbanization program aiming at transferring more than 10 million rural residents annually into urban areas over the next 20 years.

East Asia has witnessed unprecedented large migration flows in the past decade. It has been reported that in China 140 million people, most of them farmers, are on the move. An overwhelmingly majority of them are temporary migrants. The major destinations are cities along the east coast and southern China. This large scale population movement has, to a substantial extent, changed the population distribution in East Asia and will exert great impact on socioeconomic development.

With the process of urbanization, large numbers of people are increasingly concentrated in small geographic regions. In 2000, of the 19 megacities of the world with population exceeding 10 million three, Tokyo, Shanghai, and Beijing, were in East Asia. In the coming 15 years, four additional cities, including Tianjin, are expected to join the megacity group in the region.

4. Demographic Impacts on Environments and Resources

Quoting an Asian Development Bank study (ADB 1997), the UN reported recently that "In Asia, rivers typically contain four times the world average

Changes in Population Number, Composition, Distribution, and Consumption 273

of suspended solids, and 3 times the world average of bacteria from human wastes. The reported median fecal coliform count is 50 times higher than the World Health Organization guidelines" (UN 2001). Only one of every three Asians has access to safe drinking water (EWC 2002).

4.1. Ecological footprints

The environmental impact (I) inflicted by population size (P), affluence, indicated by consumption per capita (A), and the technologies used to supply each unit of consumption (T) can be described by the equation $I = PAT$, that was introduced by Ehrlich and Holdren (1971). The equation is a useful framework for identifying the drivers of environmental impacts and for estimating potential changes in impacts due to changes in any of the drivers. However, as an accounting equation, it does not lend itself to straightforward hypothesis testing. More recently, ecological footprints have been used to examine the issue. The Ecological Footprint is a resource management tool that measures how much land and water area a human population requires to produce the resources it consumes and to absorb its wastes, taking into account prevailing technology. According to Loh and Wackernagel (2004), most East Asian countries and regions are facing serious ecological deficits. The total ecological footprints in 2001, measured by global hectares per person for China, North Korea, South Korea, Japan, and Mongolia were 1.5, 1.5, 3.4, 4.3, and 1.9, respectively, and the corresponding ecological deficits are 0.8, 0.8, 2.8, 3.6, and −9.9, respectively. Only in Mongolia, was there is a negative ecological deficit, indicating an ecological reserve. This is caused mainly by its small population and large land territory. The increase in ecological footprints is the result of population growth and consumption. In East Asia, consumption has increased by an average of 6.1% per year, which reflects an increasing standard of living.

China has one of the fastest growing ecological footprints of any country on the Earth. The footprint increased from 0.8 in 1961 to 1.5 in 2001. In 2001, China's total ecological footprint ranked 75 in the world, a figure well below the world average. However, due to its huge population size, its ecological deficit is higher than the world average. There are also remarkable regional and rural-urban diversities. The residents of wealthy urban centers extend their eco-footprints deeply into the backward rural areas. Certain enterprises and higher income groups contribute disproportionately to these footprints. The ecological footprint of a low-income household is much less than that of a wealthier one. In 2004, the average income of urban residents in China was 3.23 times that of the rural population, which results in different consumption patterns. It is reported that consumption in six major Chinese cities including Beijing and Shanghai accounted for one fifth of China's total annual consumption in 2004. Along with the process of urbanization, the ecological footprint in China is expected to grow even faster in the near future.

4.2. *Arable land*

China supports approximately 20% of the world's population but has only 7% of the world's arable land, with the result that per capita arable land (0.086 hectare) is well below the world average.

Not only are the opportunities for expanding cultivated areas limited in most East Asian countries, but as a result of rapid development, population growth, and urbanization process, East Asia counties are actually experiencing a decline in cultivated areas. South Korea's arable land has decreased from 2.03 million ha in 1961 to 1.68 million ha in 2002 (FAO 2001). In Mongolia, arable land has decreased from 1.3 million ha in 1990 to 0.9 million ha at present due mainly to desertification caused by climate change and development. While cultivated land in China has declined, land for residential, industrial, and mining uses grew by 1.30% in 2002. Future population growth will accentuate the problem of declining arable land. Partially as a result of such a trend, grain production in China has declined from a recent peak of 512 million tonnes in 1998 to 457 tonnes in 2002. Agricultural output per hectare in China and Mongolia is well below the high levels sustained in Japan and South Korea. While developments in biotechnology hold promise of improved crop varieties in the future, output to produce sufficient food to feed the ever increasing population forces increasingly intensive cultivation. Consequently, there is a danger of depleting the region's scarce water and land resources. Both institutional and technological innovations will be necessary to manage those resources.

4.3. *Water*

Water scarcity has become a salient problem in East Asia. According to the Population Institute of Canada (2005), per capita water availability in 1995 was 9,988, 4,374, 2,295, and 1,472, cubic meters in Mongolia, Japan, China, and South Korea, respectively, and is likely to decline to 6,071, 4,508, 1,891, and 1,258 cubic meters in 2025 as a result of population dynamics. Water availability for the Chinese population is about one quarter of the world average. In addition, water resources in China are distributed unevenly. Most of the surface water (77%) comes from the seven major river systems and 90% of the population conduct socioeconomic activities in the basins of these rivers. Chronic or recurrent shortages of freshwater have been reported, particularly in northern China. According to the government report, China consumed 549.7 billion cubic meters of water in 2002. Urban residents used 5.8%, rural farmers 5.4%, industries 20.8%, and agriculture and other primary productions used 68%. In 2002, China's urban residents used 219 liters per capita per day, while rural farmers consumed an average of 94 liters per capita per day. The urbanization process, together with population growth, will certainly increase the demand for water, which will result in the already severe water shortage in China's northern cities becoming even worse.

4.4. General energy consumption

During the past few decades, the East Asian countries have experienced many changes in their energy consumption patterns both in quantitative and qualitative terms. This can be explained by two factors. The first is the natural increase caused by population growth and demographic changes such as age groups and household size. Another factor is the increase in economic activity and development. The population of the region is expected to reach about 1,646 million in 2025 and substantial economic growth is expected. Therefore, energy consumption will continue to increase in the next few decades and there will be an increasing demand on external sources of energy.

China is the second largest energy consumer, after the United States, surpassing Japan for the first time in 2003. At present, coal makes up 65% of China's primary energy consumption. While coal's share of Chinese energy consumption is projected to fall, coal consumption will still increase in absolute terms. The demand for oil is increasing rapidly with a current total oil demand of 5.56 million barrels per day and the demand is projected to reach 12.8 million barrels per day by 2025.

Energy consumption in South Korea is typical of fuel switching patterns. Coal, which accounted for the largest share in 1986, only accounted for 1.8% of household energy consumption in 1998. Other energy sources such as bio-fuels and low quality coal accounted for 18.5% of residential energy consumption in the 1980s but was less than 0.1% in 1998.

Household energy consumption is expected to increase throughout the East Asian region together with economic growth and rising per capita income. The Asia Pacific Energy Research Centre projects that energy consumption of the residential and commercial sectors will grow by 40% from 2002 to 2010 for the whole of Asia, reaching 1,082 million tonnes of oil equivalent per annum in 2010.

In the developing areas of the region, households still rely largely on non-commercial energy. They rely heavily on bio-fuels, such as wood-fuel, crop and livestock residues, and charcoal, which are generally not included in official energy data. The contribution of fuel wood to total final energy consumption varies widely (1.5% to 85.3%) but its share in household energy consumption has remained above 50% in many countries.

The choice of bio-fuel is often based on accessibility and insufficient income to use commercial energy. In many countries, urbanization and increased family income has resulted in a shift from bio-fuels to other forms of energy. There is an indication in a number of developed economies that with higher incomes the residential fuel mix is shifting away from bio-fuel to oil and coal and finally to electricity and gas.

During the past few decades, electricity consumption (Table 18.3) has grown faster than the use of any other fuel, and the growth rate of per capita household electricity consumption in this region has outnumbered the growth rate of per capita income. During the period 1988–1998, the average annual growth rate of electricity consumption in the South Korea was 10.9%. China is expected to be the largest electricity consumer in East Asia by 2020. Residential electricity consumption in China has tended to grow faster than national electricity consumption.

Due to rapid dissemination of home appliances, electricity use has increased more than fourfold during the period 1980–1998. Electricity

Table 18.3. Electric power consumption in East Asia (kWh per capita; World Bank 2004).

	1998	1999	2000	2001
China	721.74	757.74	827.05	893.40
Hong Kong	5,325.34	5,267.44	5,446.66	5,541.12
Japan	7,286.50	7,446.73	7,406.02	7,236.87
North Korea	778.49	832.30	886.83	913.56
South Korea	4,179.79	4,595.21	4,967.98	5,288.38

currently accounts for 37% of household energy consumption in China. However, coal remains the main fuel used to generate electricity so households still indirectly consume a large amount of coal. The use of liquefied petroleum gas, town gas, and natural gas has risen about 14.5%.

Research shows that small households in South Korea use more energy per person than large families. A person in a two member family consumes 4 million tonnes of energy annually, whereas a person in a household with more than six members consumes only 1.3 million tonnes of energy per year. In China, household energy consumption increased in both urban and rural areas by 21% and 48%, respectively. Trends toward smaller families and continued urbanization will influence future household energy consumption needs in East Asia.

Among the four conventional fuels (liquefied petroleum gas, natural gas, diesel, and kerosene) most households used liquefied petroleum gas and kerosene in 1995. The household consumption amounted to 503,000 tones of liquefied petroleum gas, and 776 cubic meters of kerosene. When compared with 1989 figures, these show increases of 56.7% for liquefied petroleum gas and 56.4% for kerosene. Consumption of wood-fuel dropped by 20.5% over the same period.

Current levels of per capita commercial energy consumption are still comparatively low in most developing areas of the region. However, the projected rapid expansion of energy consumption calls for pro-active energy efficiency promotion policies, including demand side management programs. In addition to the energy planning and development authorities, utilities have an important role in demand side management implementation. Demand side management programs recognize the increasing importance of information and awareness creation campaigns aimed at creating preference among consumers for more energy efficient products and consumption patterns.

Literature Cited

ADB (Asian Development Bank). 1997. *Second water utilities data book: Asian and Pacific region.* Manila: Asian Development Bank.
EWC (East-West Center). 2002. *The future of population in Asia,* Honolulu: East-West Center.
Ehrlich, P. and J. Holdren. 1971. The impact of population growth. *Science* 171:1212–1217.
ESCAP (Economic and Social Commission for Asia and the Pacific). 2000. *ESCAP population data sheet.* Bangkok: Economic and Social Commission for Asia and the Pacific.

FAO (Food and Agriculture Organization). 2001. *FAOSTAT database collections.* http://www.apps.fao.org. Rome, Italy: FAO.

Loh, J. and M. Wackernagel. 2004. *Living planet report 2004.* Switzerland: World Wildlife Fund. http://www.footprintnetwork.org

Ogawa, N. 2002. Ageing trends and policy responses in the ESCAP region. *Background paper for fifth Asian and Pacific population conference,* 11–14 December 2002. Bangkok: ESCAP.

Population Institute of Canada. 2005. *Water availability over time — country by country.* http://www.populationinstitute.ca/tables_and_charts/water_availability_ables/JAP-NEW.htm

UN (United Nations). 2001. *World population monitoring 2001: Population, environment and development.* New York: United Nations.

UN (United Nations). 2003. *World population prospects: The 2002 revision, highlights.* New York: United Nations.

UNAIDS. 2004. *2004 report on the global HIV/AIDS epidemic: 4th global report.* New York: United Nations.

World Bank. 2004. *World development indicators 2004.* Washington DC: World Bank.

Chapter 19

APPLICATION OF CLEAN TECHNOLOGIES AND NEW POLICIES TO REDUCE EMISSIONS AND THE IMPACTS OF ENVIRONMENTAL CHANGE ON HUMAN HEALTH

KEJUN JIANG and GANG WEN

1. Role of Clean Technologies and New Energy

Several studies stress the importance of clean technology and new sources of energy for achieving low pollutant emissions in the future (IPCC 2001). Clean technology refers to very low or no pollutant emissions, and new energy refers to renewable sources of energy such as, solar, wind, hydrogen, and advanced nuclear energy. Many clean technologies are already available and in use in the developed countries, and they have potential for application in developing countries. From a long-term view point, some advanced technologies that are in development will have strong impact for a clean future (IPCC 2001).

Many studies focus on energy as a major source of air pollution and greenhouse gas emission in East Asia (Kainuma et al. 2003). In Japan, because the energy supply relies mainly on imported sources, many new technologies have been developed and efficiency is high. In China, highly efficient and clean utilization of coal and other fossil energy sources are emphasized in a sustainable energy development strategy. Renewable and new energy development is also a common trend for technology development both in China and other East Asian countries.

For developing countries like China, high efficiency and clean technology will play important roles in reaching a low greenhouse gas emission pathway. Figure 19.1 shows an example of advanced technology and improved energy efficiency in the steel making industry.

In some countries improvements in energy efficiency have allowed increases in production without increasing energy use, and as a consequence

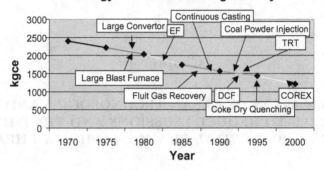

Fig. 19.1. Technology progress leading to improved energy efficiency in the steel making industry (China Statistical Year Book 2002). EF, electric furnace; DCF, direct current furnace; TRT, top pressure recovery turbine; COREX, one technology for smelting reduction processes in steel making process; kgce, kg coal equivalent.

industrial sector emissions have stabilized or declined. Thus, opportunities exist to use advanced technologies to reduce emissions in the major industries in developing countries (IPCC 1996, 2001; Jiang et al. 1998). China is a large country with a rapidly developing economy where technology is important for meeting energy demands and environment standards. Technological progress is a driving force for greenhouse gas emission reduction in East Asian countries, and it is targeted at the demands for energy and the environment, in the short and long terms. Technological strategies could well be combined with both energy and environment policies. Detailed studies of technologies to reduce CO_2 emissions at the sector level show how technological progress is desired by sectors for economic reasons rather than those of climate change (see Table 19.1; Hu et al. 1996, Jiang et al. 1998).

Many of these technologies already appear in sector development plans made by governments or enterprises. We need to raise the demand for these technologies by including climate change as a factor. The long-term scenario study for China (Jiang et al. 1999; Nakicenovic et al. 2000; Kainuma et al. 2003) suggested that the following key technologies be included so that the emission of greenhouse gases will be reduced; (i) renewable energy production (solar energy, etc.), (ii) advanced nuclear power generation, (iii) fuel cells, (iv) advanced clean coal technologies, (v) advanced gas turbines, (vi) unconventional natural gas and crude oil production technologies, (vii) syn-fuel production technology, and (viii) carbon

Table 19.1. Technologies contributing to greenhouse gas emission reduction in short- and medium-term.

Sector	Technologies
Steel industry	Large scale equipment (coke oven, blast furnace, basic oxygen furnace, etc.), equipment for coke dry quenching, continuous casting machine, top pressure recovery turbine, continuous rolling machine, coke oven gas equipment, gas recovery equipment, direct current electric arc furnace
Chemical industry	Large scale equipment for chemical production, waste heat recovery system, ion membrane technology
Paper making	Facilities for residual heat utilization, co-generation system, black liquor recovery system, continuous distillation system
Textile	Shuttle less loom, high speed printing and dyeing, co-generation system
Non-ferrous metal	Reverberator furnace, waste heat recovery system, system for lead and zinc production
Building materials	Dry process rotary kiln with pre-calciner, electric power generator with residual heat, Hoffman kiln, tunnel kiln
Machinery	High speed cutting, electric-hydraulic hammer, heat preservation furnace
Residential	Cooking by gas, centralized space heating system, energy saving electric appliances, highly efficient lighting
Service	Centralized space heating system, centralized cooling–heating system, co-generation system, energy saving electric appliances
Transport	Diesel truck, low energy use car, electric car, natural gas car, hybrid vehicle, electric railway locomotives
Common use technology	High efficiency boiler, fluid bed combustion boiler, high efficiency electric motor, speed adjustable motor, energy saving lighting
Power generation	Super critical unit, natural gas combined cycle, pressured fluid bed combustion boiler, wind turbine, integrated gasification combined cycle, smaller scale hydropower, biomass based power generation, nuclear power

sequestration and capture. With the exception of unconventional energy technologies, these are on the list for government action.

While development of these technologies is a common requirement around the world, it is expected that most of them will be developed in other countries because of the lack of investment on technology research and development in China. There is an argument for the development of some of these technologies to be supported by further investment in China. For example, integrated gasification combined cycle and clean coal technologies have a large potential market in China, but it is not known how much development of these technologies will occur in a country with little coal use. If China could be a leader in the development of these technologies, benefits would be obtained on both environmental and economic grounds. In this case, a policy for technology development could be devised based on the additional driver of climate change.

2. Policies

In order to promote clean technologies and new energy development, various policies could be adopted. Many countries have government supported technology research and development programs. In China the "863" and "973" programs organized by the Ministry of Science and Technology focus on new technology development, and recently the National Development and Reform Commission announced a development program for key energy technologies.

Beside a technology development strategy, energy and environment related policies could be developed to reach the mitigation target of the UNFCC (UN 1992). Experience from other countries is also useful for the development of relative policies. By thinking simultaneously about the policy framework for economic development, energy, and environment, the following countermeasures could be considered.

2.1. *Environment strategy covering climate change*

Domestic environmental issues are well recognized in some developed regions in China, and concrete policy frameworks have been published by many local governments. However, these policies do not include consideration of climate change issues. Most of these policies focus on short-term action to deal with various local environmental problems, many of which could be addressed within several years. Environment policies could include the climate change issue as medium- and long-term targets. Pilot action should be encouraged for local governments and capable enterprise for various purposes.

2.2. Developed regional trade system

Some countries have started to introduce domestic environmental trading systems, such as sulfur dioxide and carbon dioxide emission trading. This could also be incorporated in the policies adopted by China. Global Environment Facility is working in China on a renewable energy portfolio system. Experience from this regime and other countries could be a good basis for a possible trading system on carbon dioxide emission.

2.3. Energy tax system

The energy tax system is commonly used in OECD countries and China is preparing to introduce such a system. In China a tax will soon be levied on vehicle fuel to replace the road construction fee. This is a trend for future reform of the energy price system. Many experts in China believe that a carbon tax system will not be introduced in the near future, but it is useful to consider the possibility of a mixed tax system which includes a response to the climate change issue.

2.4. Regional and international collaboration on technology research and development

Because of the close economy and social relationship among East Asian countries, it is important for these countries to work together on technology research and development. Japan is taking the lead for technology research and development in the world, and Korea and China have their own national plans for technology research and development. These countries should work more closely together when there is a good basis for joint development.

To get a better understanding of the potential impacts of different policies on emission mitigation in China, many were simulated by energy and environmental modeling studies. Table 19.2 presents the popular policies that could be applied in China based on the experience of developed countries (Detlef et al. 2003).

3. Eco-City Development

In Japan, an environment path is described as "pollution first then recovery". When rapid economic growth started in the 1950s, four famous environmental disasters occurred in Japanese cities. After that, many regulations and laws were enacted to improve the environment in the cities. After more than 30 years improvement, there is almost no

Table 19.2. Overview of policy options, instruments and applicability.

Policy options	Possible policy instruments	Applicability		Implementation explored in modeling experiment
		A1b-C	B2-C	
1. Incentives for energy-efficiency investments	• Taxes/subsidies • Low/zero-interest loans • Information campaigns • Appliance labels/standards • Investment in public transport systems • Voluntary agreements with industry	+/− + ++ + ++	++ ++ +++ ++ ++	Reducing the gap in final energy intensity between Western Europe and China in 2050 by another 30% beyond the baseline
2. Energy taxation inducing a series of responses	• Tax on gasoline/kerosene, as part of "greening tax" policy	+	+++	Adding an energy tax — equal for all fuel types — for industry and transport equal to the current Western European tax level for oil and gas
3. Influencing market penetration of secondary energy carriers	• Taxes/subsidies, e.g., on natural gas or biofuels • Emission standards	+/− +	+ +++	Reducing the use of coal in the building sector to zero
4. High-efficiency, gas-fired combined-cycle technologies in central electric power generation	• Technology and emission standards • Institutional reforms • R&D projects. • Investments	++	++	In 2050, 15–20% of all electricity is generated by gas-fired combined cycle

(Continued)

Table 19.2. (Continued)

Policy options	Possible policy instruments	Applicability		Implementation explored in modeling experiment
		A1b-C	B2-C	
5. Advanced clean coal options including Integrated Gasification Combined Cycle		++	++	All new coal power plants from 2010 onwards are highly efficient
6. Reduce transmission losses		++	++	Losses in distribution and transmission of electricity are reduced in 2050 to the level of OECD countries (8%)
7. Increasing the share of nuclear power generation	• Technology and emission standards • Portfolio standards/ renewable energy obligation • Institutional reforms • R&D projects • Investments	++	+	Use of nuclear power is increased from 10% (A1b-C) and 7% (B2-C) to 20% of all electricity generated
8. Increasing the share of renewables such as solar and wind in electric power generation		++	++	The use of new renewables in electric power generation is increased from 7% (both A1b-C and B2-C) to 20% of all electricity generated. In 2020, the required share is 10%
9. Increasing the share of hydropower generation		++	+	Use of hydropower is increased from 68% to 90% of maximum implementable potential of 378 GW

(Continued)

Table 19.2. (Continued)

Policy options	Possible policy instruments	Applicability A1b-C	Applicability B2-C	Implementation explored in modeling experiment
10. Accelerating the penetration of biomass-derived fuels	• R&D projects • Tax exemption/subsidies to farmers • Low/zero-interest loans • Portfolio standards/renewable energy obligation	+	+++	Overrule market dynamics with expansion targets; 10% market share in oil/gas market in 2020 and 20% market share in 2050
11. Carbon taxation inducing a series of responses	• Carbon tax on fuel use in all sectors	++	++	Implementation of a US$30 carbon tax
	Not explored in model experiments			
Small-scale decentralized electricity generation: • Cogeneration • Solar, wind, mini-hydro, biomass	• R&D projects • Low/zero-interest loans • High payback rates to grid for distributed generators • Institutional reforms	−	++	
Accelerating the availability of natural gas by opening up import, construct infrastructure, etc.	• Bi/multilateral co-operation • Natural gas exploration and exploitation	+++	+	
Carbon removal and storage for large coal-fired power stations	• Emission standards	+	−	

Note: A1b-C and B2-C are two scenarios for China based on IPCC SRES scenario A1B and B2. A reference scenario was also given for the study.

environmental pollution in Japanese cities, and more concern is given to global environmental issues.

In many large cities in China, air pollution is a serious problem. Sulfur dioxide, nitrogen oxides, carbon monoxide, and aerosol particles are major pollutants. In Beijing, prior to the mid-1980s, the main pollutant was sulfur dioxide from the burning of coal. Since then air pollution, mainly by nitrogen oxides from the transportation sector, has become worse, and in 1997 Beijing was classed as one of the most polluted cities in the world (WHO 1997). In recent years the annual concentration of nitrogen oxides in Beijing was greater than that in all of the northern cities in China, and in 1998 the concentration was three times the national standard. In addition, the concentrations of sulfur dioxide and total suspended particles were still greater than the national standards.

It is well recognized that there are environmental problems in the cities of China, and many have set up targets for sustainable development, and made their own plans for reaching those targets. Sustainable development in a city covers four aspects: (1) the city's ecological landscape, which includes park land per capita, the proportion of land devoted to open space, oxygen equivalent per capita, and oxygen supply capacity; (2) environment quality including clean air, safe reliable water supplies, food security, sanitation, and low noise level; (3) coordinating ability in the city including resource conservation, renewable energy, meeting human needs, efficient transportation, and industry awareness; and (4) the ability of the city to cope with impacts of population, city space and economy, and to provide protection against disasters (Shenzhen Declaration 2003).

Because of serious air pollution in Beijing, the local government announced a group of policies in 1998 to control air pollution. Twenty eight countermeasures were adopted including prohibiting the use of coal in the central area of the city, increasing the use of natural gas, moving industry outside Beijing, and introducing emission standards for vehicle design. In 1999 more policies were announced, and by the end of 2003 there was a significant improvement in air quality.

In addition to these controls on air pollution, reforestation projects have been instituted by the central and Beijing governments to control sandstorms originating in nearby provinces as a result of deforestation, and overgrazing and over-cultivating land.

Starting from 2001, the State Environment Protection Administration of China started the Model City for Environment Protection Project, and certification indicators were published. The criteria that cities must meet are, a prosperous society, an economy with a high growth rate, a balanced ecological movement, proper use of natural resources, good environmental quality, a clean and beautiful urban landscape, a developed infrastructure, comfortable and convenient lifestyle, and a healthy population (China Internet Information Center 2002). A total of 27 indicators were identified covering social, economic, and environmental quality indices, as well as a number of different indicators of urban environmental construction and management level. In 2003 a revised and more detailed list of indicators was published. Hangzhou and Ningbo were the first to be classified as model cities, and by 2003 there were 32 cities or districts which fell into this category.

In 1999, the State Environment Protection Administration launched a program in which Model Districts will reach the standards set by the International Standards Organization for organizations in the global marketplace. The purpose of this program (designated ISO14000) is to simultaneously address environmental and developmental needs, and to improve the environment in the area. It establishes a group of districts in which all enterprises reach the ISO14000 standard. The program was conducted in 46 key environment protection cities.

4. Policies to Reduce Emission

In order to explore policies which will enable China to reach a low emission trajectory, a set of emission scenarios was developed using the Integrated Policy Assessment model for China (IPAC) in the Energy Research Institute. A set of scenario storylines for China was formulated by defining several key-driving factors such as gross domestic product, population growth, and energy efficiency improvement. The future development patterns used were mainly taken from different trends in China. The B2 scenario taken from the IPCC Special Report on Emission Scenarios (SRES) was adopted for the global scenario (Nakicenovic et al. 2000; IPCC 2001). In order to reflect the possible development trend for China, six storylines were developed; three economic development patterns (high, medium, and low domestic product growth rate), and two development paths for population growth (low and high). The government is focusing on the medium growth rate and the high population path. By combining these key factors, six storylines were formulated. A brief definition of these storylines is given in Table 19.3.

Table 19.3. Description of scenarios.

Scenario	Code	Description
Traditional development scenario	S1	Future energy and environment development follows the experience of industrialized countries during their initial stage of industrialization. Large space for energy intensive industry because of relying on raw material production and low innovation of knowledge which makes slow technology progress and high energy demand
Conventional scenario	S2	Economy development and energy industry follows the experience of China in last several decades. Industry will continue to remain dominant for next several decades. Energy supply relies mainly on domestic resource
Energy policy intervention scenario	S3	Energy industry is promoted by government with planning, emphasis on clean energy and improvement of energy efficiency. International energy market is regarded as one of the important sources for clean energy. Energy policies from government could be well implemented
Environment driven scenario	S4	Based on the understanding of domestic environment problems, more environment policies will be introduced beside existing energy and environment policies. Energy supply and use will satisfy the requirements of domestic environment. Clean energy and clean production is a mainstream of the society
Tiger development scenario	S5	A higher economic growth is assumed. Conventional development pattern is the same as that in scenario S2, when higher technology progress is assumed because of financial ability for technology R&D
Gray development scenario	S6	A lower economic growth is assumed. Conventional development pattern is the same as that in scenario S2, when lower technology progress is assumed because of financial ability for technology R&D

The simulation studies suggest that there will be a significant increase in primary energy demand, which ranges from 2.3 billion tons coal equivalent (tce) to 3.7 billion tce in 2020, and 2.7 billion tce to 4.4 billion tce in 2030. The results also suggest that CO_2 emission will increase quickly with economic development in China. Compared with the base year (1990) CO_2 emissions are projected to be 1.7, 2.5, and 3.8 times larger in 2000, 2010, and 2030 for the medium growth rate (which may be regarded as a possible development rate), and 1.6, 2.1, and 3.3 times for the lowest growth rate. Simulated results on the distribution of CO_2 emissions in 2010 in China suggest that the coast region will account for most of the emissions. The simulations also give the potential for energy saving and CO_2 emission reduction. There could be significant potential, ranged by 8% to 46% in 2030. We emphasize the need for policy implementation in the areas of economic development, energy, technology and environment to reach a clean development pattern. This may be achieved by following scenarios S2 to S4.

5. Impact on Human Health

Several studies examining the relationship between air pollution and human health have been conducted in China. One study which examined the relationship between the air pollution and mortality in Beijing showed that the risk of mortality increase by 11% with each doubling of sulfur dioxide concentration, and by 4% with each doubling of total suspended particles. Pulmonary heart disease also increased significantly with higher pollution levels (Xu et al. 1994; Xu et al. 1996). Respiratory diseases, hospitalization, and visits to the doctor are often a more sensitive measure of the impact of air pollution on human health than mortality. Problems occurred even when levels were below the World Health Organization's recommended guidelines (Xu et al. 1995).

Health problems apart from respiratory diseases are also associated with air pollution. Between 65% and 100% of children in Shanghai have blood lead levels greater than 10 μg per deciliter and those in industrialized areas had levels averaging between 21 and 67 μg per deciliters. These statistics are of great concern because lead poisoning in children leads to neurological damage (World Resources 1998). Zhao Yihong reported that the economic loss in China in 1995 due to health problems caused by air pollution was ~57 billion Yuan. These health problems can be avoided by the development of clean technologies, new energy sources, and eco-city policies which will mitigate pollutant emission in China.

Literature Cited

China Internet Information Center. 2002. *Urban environment in China.* http://us.tom.com/English/1871.htm

China Environment Year Book. 1996. Beijing: China Environment Yearbook Press (in Chinese).

China Statistical Year Book. 2002. Beijing: China Statistics Press.

Detlef van Vauren, L. Yun, B. de Vries, J. Kejun, C. Graveland and Z. Fengqi. 2003. Energy and emission scenarios for China in the 21st century — Exploration of baseline development and mitigation options, *Energy Policy* 4.

Hu, X., K. Jiang and J. Liu. 1996. Application of AIM/emission model in P.R. China and preliminary analysis on simulated results. AIM interim paper, IP-96-02, Tsukuba, Japan.
IPCC (Intergovernmental Panel on Climate Change). 1996. *Climate change 1995: The science of climate change. Contribution of working group 1 to the second assessment report of the Intergovernmental Panel on Climate Change*, edited by J. T. Houghton, L. G. Meira Filho, B. A. Callander, N. Harris, A. Kattenberg and K. Maskell. Cambridge: Cambridge University Press.
IPCC (Intergovernmental Panel on Climate Change). 2001. *Climate change 2001: The scientific basis. Contribution of working group I to the third assessment report of the Intergovernmental Panel on Climate Change*, edited by J. T. Houghton, Y. Ding, D. J. Griggs, M. Nogure, P. J. van der Linden and X. Dai, Cambridge, U.K.:Cambridge University Press.
Jiang, K., T. Morita, T. Masui and Y. Matsuoka. 1999. Long-term emission scenarios for China, *Environment Economics and Policy Studies* 2:267–287.
Jiang, K., X. Hu, Y. Matsuoka and T. Morita. 1998. Energy technology changes and CO_2 emission scenarios in China. *Environment Economics and Policy Studies* 1:141–160.
Kainuma, M., Y. Matsuoka and T. Morita. 2003. *Climate policy assessment.* Tokyo: Springer.
Nakicenovic, N., J. Alcamo, G. Davis, B. de Vries, J. Fenhann, S. Gaffin, K. Gregory, A. Grübler, T. Y. Jung, T. Kram, E. L. La Rovere, L. Michaelis, S. Mori, T. Morita, W. Pepper, H. Pitcher, L. Price, K. Riahi, A. Roehrl, H.-H. Rogner, A. Sankovski, M. Schlesinger, P. Shukla, S. Smith, R. Swart, S. van Rooijen, N. Victor and D. D. Zhou. 2000. *Special report on emissions scenarios for the Intergovernmental Panel on Climate Change.* Cambridge: Cambridge University Press.
Shenzhen Declaration. 2003. The Shenzhen Declaration on EcoCity Development. DOC from internet conference on ecocity development (February–June 2003). www.UrbanEcologyAustralia-Ecocities.htm
UN (United Nations). 1992. *United Nations framework convention on climate change.* Geneva: United Nations.
WHO (World Health Organization). 1997. *Health and environment in sustainable development.* Fact sheet 170. Geneva: World Health Organization.
World Resources. 1998. Health impacts associated with air pollution. In *Environmental change and human health.* A joint publication by the World Resources Institute, the United Nations Environment Programme, the United Nations Development Programme and The World Bank.
Xu, X. P. et al. 1994. Air pollution and daily mortality in residential areas of Beijing, China. *Archives of Environmental Health* 49:216–222.
Xu, X. P., B. L. Li and H. Y. Huang. 1995. Air pollution and unscheduled hospital outpatient and emergency room visits. *Archives of Environmental Health* 103:286–289.
Xu, Z. Y. et al. 1996. The effect of air pollution on mortality in Shenyang City. *Journal of Public Health in China* 15:61.

Chapter 20

TRENDS AND PROACTIVE RISK MANAGEMENT OF CLIMATE-RELATED DISASTERS

PEI JUN SHI, NORIO OKADA, JOANNE LINNEROOTH-BAYER and YI GE

1. Introduction

Weather-related natural disasters in the form of droughts, windstorms, floods, avalanches, and other extremes continue to pose a serious threat to East Asia. Recent examples include a devastating typhoon in Japan in 1991, the 1996 and 1998 floods in the Yangtze and Song Nen river basins in China (Zhang and Wang 1997; Li and Wang 1998), and severe flooding in Korea in 1998. These disasters not only resulted in huge economic losses and casualties, but also severely disrupted the lives of the people in these countries. Moreover, the losses from weather-related disasters are increasing, a trend that will be likely worsened by climate warming. To date, increasing losses can be explained mainly by people and capital locating to vulnerable regions (Miletti 1999). However, the Third Assessment Report of the Intergovernmental Panel on Climate Change predicts that over the next half century climate change will increase the intensity and frequency of extreme weather (McCarthy et al. 2001). There is scientific agreement that even in the (unlikely) event that deep reductions in greenhouse gas emissions were embarked upon now, greenhouse gases already in the atmosphere would give rise to significant climate change impacts. This adds to the urgency of pre-disaster or proactive "risk management" for reducing the burdens of weather disasters.

2. Disaster Damage Trends

Since the late 1960s, the losses caused by natural disasters in East Asia have been increasing. Figure 20.1 shows direct disaster losses in East Asia

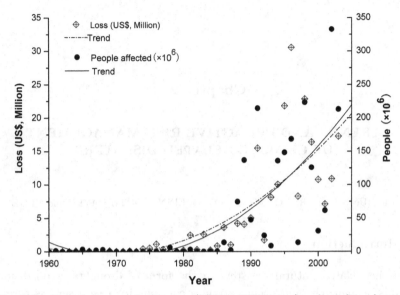

Fig. 20.1. Reported economic damages (USD at 2000 prices) and number of people affected by natural disasters in East Asia (OFDA/CRED 2004).

(normalized for exchange rate fluctuations) from 1960 to 2003. It is striking that the direct economic losses rose in this period from a decadal average of USD 80,914 in 1960s to USD 14,433,097 in the 1990s, or near 178-fold, whereas Gross National Product in this period (excluding North Korea and Mongolia) increased only eight-fold. As shown in Fig. 20.1, the number of persons affected is also accelerating, from over 1.03 million people in the period 1960–1969, to over 112.6 million people in the period 1990–1999 — over a 109-fold increase.

3. Disaster Losses and Climate Change

The increase in the natural disaster losses in East Asia is alarming, and the same trend also exists in other regions of the world. Weather-related disasters have taken the lives of well over one million people worldwide in the past two decades, and the economic losses from these and other natural disasters have increased almost ninefold from the decade of the 1960s to the 1990s. Because of high concentrations of capital, it is not surprising that losses are higher in the developed world; yet, in terms of Gross National Product, poor countries and regions suffer disproportionately higher losses,

and 90% of fatalities from natural disasters occur in developing countries (Linnerooth-Bayer and Amendola 2000; UNDP 2002).
Annual damages caused by extremes in weather cause 70% of total reported disaster damages throughout the world (UNDP 2002). Floods cause the greatest number of deaths, or about 40%. Moreover, it is estimated that half of the global population (3 billion) live in coastal regions or along main rivers, so the exposure degree is very high to floods and ocean disasters. Escalating economic losses have been attributed mainly to changing land-use practices and concentration of capital in high-risk areas. Still, there is some evidence that this trend is also resulting from a changing climate (Jiang and Shi 2003), and climate change may be contributing to weather disasters in Asia. For instance, the comparative increase in the numbers from climate-related and geophysical disasters in East Asia from 1960 to 2003 is shown in Fig. 20.2. Whereas the numbers (5-year averages) of geophysical disasters have remained relatively stable, increasing only slightly after 1985, the numbers of climate-related disasters increased rapidly since the 1980s.

Table 20.1 shows the detailed economic losses caused by natural disasters in East Asian countries between 1980 and 2003. In these countries, the proportion of weather-related disaster losses exceeds 50%, and even reached 100% in South Korea, North Korea, and Mongolia.

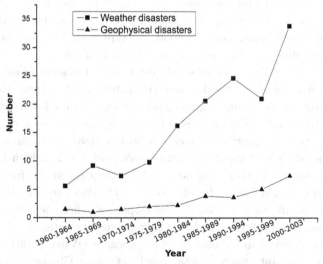

Fig. 20.2. Climate related and geophysical disasters in East Asia from 1960 to 2003. (OFDA/CRED 2004).

Table 20.1. Economic losses caused by natural disasters in East Asian countries. (OFDA/CRED 2004).

1980–2003 (USD)	China	Japan	Korea DPR	Korea Rep	Mongolia
Drought	1,818,714	0	0	0	0
Extreme temperature	3,000,000	0	0	6,000	0
Famine	0	0	0	0	0
Flood	104,822,508	19,20,300	16,709,400	3,061,800	270
Slides	63,190	210,000	0	0	0
Wave/surge	0	0	0	0	0
Wild fires	110,000	0	0	0	1,822,800
Wind/storm	23,906,157	11,148,700	6,144,276	7,570,981	885,000
Earthquake	16,361,819	3,288,500	0	0	0
Epidemic	0	0	0	0	0
Insect infestation	0	0	0	0	0
Volcano	0	0	0	0	0
Ratio*	89.10%	80.15%	100%	100%	100%

This evidence shows that in the past two to three decades, the frequency and damage of weather-related disasters in East Asia are increasing at a rate higher than that of geophysical disasters (excluding recent years). This suggests that climate change may be playing a role. However, other factors may also be causing this differential increase. More persons and capital may be locating in weather-threatened versus other disaster prone regions (for example, there is a large migration in China to the coast, which is at higher risk of windstorms, coastal flood, and typhoon). Moreover, additional phenomena play a role in the increasing losses from weather.

The most significant may be deforestation and other land-use changes that are implicated in flood losses (Bronstert 2003). Given the many contributing factors to increasing disaster losses, it may be instructive to examine the trend in the hazards, for example, the extreme precipitation. In the Northern Hemisphere, precipitation has increased by an estimated 0.5–1% per decade in the 20th century over most mid and high latitudes, and it is likely that there has been a 2–4% increase in the frequency of heavy precipitation events. In addition, the El Niño Southern Oscillation phenomenon, affecting weather in the tropics, has become more frequent and intense. In some regions of Asia, the frequency and intensity of droughts have been observed to increase in recent decades (Watson 2001). For the 21st century, the Intergovernmental Panel on Climate Change expects, with high to medium likelihood, that in many regions of Asia there will be worsening heat waves, more intense precipitation events, including increased Asian summer monsoons, and increased intensity of storms (Houghton et al. 2001).

4. Proactive Risk Management

With the increasing losses from weather-related disasters and a recognition that disaster prevention is less costly in many cases than disaster relief and reconstruction, the disaster community is placing great emphasis on pre-disaster or pro-active risk management, and especially on non-structural prevention strategies combined with new forms of solidarity for sharing the disaster burdens (Freeman et al. 2003). To date, the policy emphasis has been on structural engineering measures to reduce the *hazard*, for example, levees and dams. However, structural mitigation projects, especially for the flood hazard, have the potential to provide short-term protection at the cost of long-term problems. In many countries, flood control systems have exacerbated rather than reduced the extent of flooding. Sediment deposit in river channels has raised the height of river channels and strained dike systems. Now when floods occur, they tend to be of greater depth and more damaging than in the past (Benson 1997a,b), sometimes affecting settlements downstream from the protective measures. Furthermore, structural mitigation projects have the potential to provide people with a false sense of security. The damages from the 1993 flooding of the Mississippi River in the United States were magnified because of misplaced confidence in structural mitigation measures that had encouraged development in high-risk areas (Linnerooth-Bayer et al. 2000).

During the past decades, there has been a continuous evolution in the common understanding and practice of disaster management (ISDR 2002). Currently, the international disaster community as well as some progressive national governments are taking a more holistic approach, which focuses on risk and vulnerability (Freeman et al. 2003). Disaster reduction strategies thus include improved vulnerability and risk assessment, as well as changes in institutional capacities and operational abilities. The assessment of the vulnerability of critical facilities, social and economic infrastructure, the use of effective early warning systems, and the application of many different types of scientific, technical, and other skilled abilities are essential features of a disaster reduction strategy. In addition, a new concept is emerging for proactive disaster financing, which will assure victims and governments of sufficient capital after a disaster to finance reconstruction.

Especially in developing countries, where disasters have the potential to set back economic growth for the long term, timely reconstruction of critical private and public infrastructure are important. The core of current disaster risk management is recognition that public policies must take account of the social, economic, and political situation, and that there are multiple perspectives to the issues at stake.

4.1. Integrated disaster risk management

In recent years, many countries in Asia, including China and Japan, have updated existing acts and regulations related to disaster management, and are increasingly reorienting their institutions away from emergency response to disaster prevention and financing. The current state of administrative and legal arrangements for disaster risk management in China and Japan is reviewed in Table 20.2 (ISDR 2002). Integrated disaster risk management involves, of course, more than national legislation and disaster reduction plans. It will also require "mainstreaming" disaster risk reduction into development activities and other forms of legislation.

Many countries have adopted legislation dealing with a broad range of issues including protection of water resources or biodiversity conservation, and often disaster risk reduction is "mainstreamed" into this environmental legislation. Some of the features of integrated environmental/disaster reduction policies include water policies (water pricing and hydropower regulation) which may (or may not) have a beneficial impact on disaster reduction. Environmental and other policies can be designed to promote sustainable use of water and allow adjustments depending on seasonal forecasts to avoid floods. Mainstreamed water policies can guarantee the protection of wetlands and floodplains for storing water. Furthermore, flood and drought risk management are increasingly looked at in the context of water resources, and therefore depend on effective international water management. Similarly, policies of fuel wood and development of alternative sources of energy will reduce deforestation and contribute to flood, avalanche and landslide control. Programs and conventions for protecting biodiversity, preventing climate change, and desertification can also reduce vulnerability through enhanced natural resource management (ISDR 2002).

Despite extensive and comprehensive disaster risk management legislation, without the commitment of local communities and individual households/businesses, the best intended disaster management will fall

Table 20.2. The conditions of administrative and legal arrangements for disaster risk management. (ISDR 2002).

	Focal point for disaster management	National action plans	State and provincial disaster reduction plans
China	China International Reduction Disaster Decade Commission (CIRDDC, 1989–2000) China National Committee for International Disaster Reduction (CNCIDR, 2000–present)	The National Natural Disaster Reduction Plan of the People's Republic of China; Laws of People's Republic of China on Protecting Against and Mitigating Earthquake Disaster; Laws of People's Republic of China on Water and Soil Conservation; Flood Control Law of the People's Republic of China; Forest Law of the People's Republic of China; Meteorology Law of the People's Republic of China	
Japan	Cabinet Office	Disaster Countermeasure Basic Act (basic plan for disaster reduction)	Operational plans for disaster reduction, local plans for disaster reduction

short. Moreover, the market is essential for complementing, and in some cases counter-balancing, the top-down management plans of the central authorities. In what follows, we briefly review the disaster risk management landscape for China and Japan.

4.2. Financial disaster risk management

Risk management policies should include mitigation and prevention efforts as well as consideration of risk-financing instruments. The victims of economic damages from natural disasters (households, businesses, agriculture, and the public sector) do not always absorb the full losses. The financial burdens can be transferred to others through the assistance of four main institutions; (1) the government or taxpayers, (2) insurance and other risk-transfer arrangements, (3) family and neighbors, and (4) domestic and international donors and lenders. Viewed globally, national and state governments take on the largest share of post-event assistance. We refer to this route for redistributing the burdens as *collective loss sharing*. In developing countries, however, informal arrangements involving relatives and neighbors appear to be the most important form of assistance. Domestic and international humanitarian assistance and loans are generally small compared to the global disaster bill, but in well-publicized disasters donor assistance can play a large role. The market also plays an increasingly important role in distributing catastrophe losses. Insurance, reinsurance, and capital market-based securities transfer losses from the immediate victims to a wide and increasingly global web of premium payers and investors. A *risk transfer instrument* is a pre-disaster arrangement in which the purchaser incurs a relatively small cost in order to avoid the risk of a large loss after a disaster occurs. The important distinction between risk transfer and collective loss sharing is that the former is purchased by the persons or community at risk before the disaster, whereas the latter post-disaster assistance is provided by the state and thus (usually) funded by taxpayers. (Linnerooth-Bayer et al. 2003).

Risk transfer and financing can be an important pre-disaster or pro-active disaster risk management measure at both the scale of the individual household as well as the government. Public policies and private actions for disaster risk reduction and reconstruction require access to financial resources. Not surprisingly, however, there are little insurance or other forms of risk transfer in developing countries. Catastrophe insurance coverage is correlated with per-capita income and thus very limited in low-income countries, where it is purchased mainly by international firms and the small middle class. Only at a threshold of per-capita income about $3,000 does insurance become affordable on a larger scale.

For the most part, flood and other types of insurance are not available to the poor, and there is little institutional structure in many low-income countries to provide the necessary legal and regulatory institutional support for a viable insurance market. One should not, however, interpret these figures as meaning that more insurance in low-income countries is desirable. Only under special conditions, particularly if there is no less costly ways of recovering from a disaster (e.g., by family contributions or, in the case of the government, by relying on its many post-disaster financing sources) it is advisable for governments or private individuals in low-income countries to purchase insurance or other risk transfer instruments. Recognizing that neither private insurance nor public assistance can stand on its own, some countries, including the United States, France, Norway, New Zealand, and Japan, have legislated national insurance programs that combine private and public responsibility. The Japanese earthquake insurance system, for example, is backed by the government, thus sharing the risks between the private insurers and taxpayers (Linnerooth-Bayer et al. 2000). Financial risk management must also be a part of the government's planning process. Three sources of government risk can be distinguished, relief to victims, reconstruction of private homes and other assets, and reconstruction of public infrastructure. Usually governments raise funds after the disaster occurs (*ex post-financing*), but there is increasing interest in arranging pre-disaster (*ex ante-financing*). Many governments set aside funds into a catastrophe reserve in order to meet their obligations after a disaster. Rather than holding large sums in reserve, governments can make pre-disaster arrangements for receiving credit after a disaster at a pre-determined interest rate, or purchase insurance from commercial reinsurers. Recently, attention has been given to novel risk-transfer or hedging instruments, including catastrophe bonds. A catastrophe bond is an instrument offered by the insurer or industry at risk, whereby the investor receives an above-market return when catastrophes do not occur, but shares the insurer's losses by sacrificing interest or principal when catastrophes do occur. With these bonds or other capital market instruments, insurers can pay to transfer catastrophe risk to investors and, therefore, directly to the global capital markets. There are clear advantages to public risk transfer, but there are also costs. For this reason it is important to examine the costs and benefits of alternative financial arrangements (Hochrainer et al. 2004).

5. China

The aim of the Chinese government is to build disaster resilient communities by promoting increased awareness of the importance of disaster reduction as an integral component of sustainable development. Institutionally, in October 2000 the government established the Chinese National Committee for International Disaster Reduction. This inter-ministerial coordinating institution led by a State Councilor responsible for designing a national disaster reduction framework, developing guiding policies, and coordinating relevant departments in supervising disaster reduction works undertaken by local governments.

Moreover, China is mainstreaming disaster risk management into other areas of legislation. After the Yangtze River floods in 1998, the Chinese government banned logging in the upper watershed and increased reforestation efforts and prohibited additional land reclamation projects through several relevant legislations. The China National Wetlands Conservation Action Plan finalized in 2000 is an example of a specific environmental legislation supporting disaster reduction. The key element of this process is the progressive implementation of the *National Disaster Reduction Plan of the People's Republic of China* running from 1998 to 2010. Significantly, it was formulated on the basis of the overall national development policies reflected in the "Ninth Five Year Plan for National Economic and Social Development", and the "2010 Prospective Target Outline" for national accomplishments (Gao and Gao 1995; CIRDDC 1998).

In order to implement the National Disaster Reduction Plan, the Chinese National Committee for International Disaster Reduction begins at the provincial level, and then moves to the local. Several provinces have issued mid-term plans on disaster reduction in their areas, such as, the Provinces of Guangdong, Jiangxi, Yunnan, and Shanxi, and the government of Heilongjiang Province is working to initiate a local program strategy. This plan emphasizes non-structural measures, such as relocating population out of high-risk areas and other changes in land use, as key to implementation of the National Disaster Reduction Plan. Since the devastating floods in 1998, the government has carried out an ambitious plan. For example, on the Yangtze River, extensive prevention measures have been taken, such as reinforcing necessary levees, but also removing levees in many areas to return reclaimed land to the flood plain, building new towns for resettlement and reforesting land now in cultivation. Statistical data shows that the benefit of disaster reduction during 1998–2000 is of the order of 700 billion Yuan. (*Chang Jiang Water Resources Commission* 2003).

China is also improving its capacity to respond to disasters by a more expedite damage analyses. Based on a new management system and its national network, disaster information can be sent to the central government within 24 h of a disaster occurring (CNCIDR 2003). To quickly respond to emergencies, China has built a national network for storing relief food. Finally, in 2002, the development of "Chinese Environment and Disaster Monitoring and Forecasting Satellites" has been approved by the State Council of China. Three satellites will be launched in the first phase (2005–2006) (CNCIDR 2003).

Despite extensive attempts to promote non-structural mitigation measures, the main mitigation measures in China have been structural.

As reported by an International Decade for Natural Disaster Reduction document, during the past half century China has constructed 247 km of checking dams, 84,000 reservoirs, 12,000 km of barrages, and more than 49,000 irrigation and drainage pumping stations (CPCU 2003). Recently, China has invested heavily in the huge Three Gorges and Xiaolangdi projects. Climate change may require extensive investments to improve current levees which are designed only for floods under 10–20 years return period. Moreover, the quality of reservoirs is considered to be unsatisfactory, and 37% of reservoirs in 13 provinces are considered to be unsafe (Peng 1997).

Regarding the current legislation, there is a general sense that new laws are needed to promote disaster prevention, so developing a comprehensive mitigation law is on the agenda (Peng 1997). Of course, it is not only top-down legislation and implementation that will prevent human and economic disaster losses in China, the emerging market players, and local initiatives, must also be part of a holistic program. The notion of insurance, and combining insurance with individual initiatives for loss reduction, is only beginning in China. This does not mean that there is no insurance for floods and other types of disasters. Flood insurance is offered by the Public People's Insurance Company of China, and commercial companies may be entering the market. Very recently, the insurance market has been opened for foreign companies, and Swiss Re, among others, is now licensed to offer reinsurance to public and private insurance companies operating in the Chinese market. Of course, insurance will be unaffordable to the majority of Chinese citizens in high-risk areas, which underlines the importance of combining national solidarity with more individual responsibility if it is affordable. A challenge that is being addressed in a project for the Dongting Lake area is developing a public–private insurance system for China that will provide victims with more protection and at the same time encourage individual loss-reduction measures.[1]

Finally, local initiatives are crucial for any disaster reduction plan. In China, the population has traditionally relied heavily on the provincial and national governments for protection and also for providing funds for reconstruction. This may be changing, and in the future it is likely that there will be more individual and community activities and interest on the part of non-Government Organizations.

---

[1] This project is being carried out by Beijing Normal University and the International Institute for Applied Systems Analysis (IIASA).

6. Japan

The Disaster Countermeasures Basic Act was passed in 1961 as a result of a typhoon that hit the Bay of Ise in 1959 and caused more than 5,000 deaths. The act aimed to remedy inadequacies in the old disaster-reduction framework and promote comprehensive, systematic efforts by the government to reduce disasters. It had five main thrusts; (1) to clarify disaster-reduction responsibilities and implement programs to prepare for, provide emergency response to, and recover from disaster, (2) to promote comprehensive administrative efforts toward disaster-reduction, (3) to promote systematic administrative efforts toward disaster-reduction, (4) to provide public financial resources to cope with disaster, and (5) to set out procedures for proclaiming disaster emergencies. There are other laws besides the very general Disaster Countermeasures Basic Act that impinge on disaster-reduction efforts, including the flood control laws, the Disaster Relief Law, and the Large-scale Earthquake Countermeasures Act.

Moreover, the disaster-reduction plans focus on:

(a) *Basic Plan for Disaster Prevention.* Created by the Central Disaster Prevention Council, this plan defines basic guidelines for the establishment of disaster-reduction organizations and systems, promotion of disaster-prevention programs, timely and appropriate recovery and reconstruction from disaster, and pursuit of scientific and technical research into disaster-reduction.
(b) *Operational Plans for Disaster Prevention.* Formulated by designated government agencies and designated public institutions in accordance with the Basic Plan for Disaster Prevention, these plans define measures to be taken to reduce the disasters for which the agency or institution is responsible or to which it may be subject.
(c) *Local Plans for Disaster Prevention.* Formulated by the prefectural and municipal Disaster Prevention Councils or heads of municipalities in accordance with the Basic Plan for Disaster Prevention and local conditions, these plans contain specific measures to be taken by local disaster-reduction institutions (ADRC 1998, 1999, 2002).

7. Conclusions

While the role that climate change is playing in disaster losses today cannot be definitively assessed, the Intergovernmental Panel on Climate

Change and others point to some evidence that climate change is increasing the frequency and intensity of weather-related disasters. Experts also predict that climate change will increase the intensity and frequency of weather-related disasters in the next decade. We have argued that disaster authorities should focus more on pre-disaster or proactive risk management, including loss reduction and loss sharing.

Literature Cited

ADRC (Asian Disaster Reduction Centre). 1998. Disaster management in Japan. Cabinet Office Government of Japan Country Report — Japan [A]. http://www.adrc.or.jp/country_report.php

ADRC (Asian Disaster Reduction Centre). 1999. Disaster management in Japan. Cabinet Office Government of Japan Country Report — Japan [A]. http://www.adrc.or.jp/country_report.php

ADRC (Asian Disaster Reduction Centre). 2002. Disaster management in Japan. Cabinet Office Government of Japan Country Report — Japan [A]. http://www.adrc.or.jp/country_report.php

Benson, C. 1997a. The economic impact of natural disasters in Fiji. London: Overseas Development Institute.

Benson, C. 1997b. The economic impact of natural disasters in Vietnam. London: Overseas Development Institute.

Bronstert, A. 2003. Floods and climate change: Interactions and impacts, in Special issue on flood risks in Europe, edited by J. Linnerooth-Bayer and A. Amendola. Risk Analysis 23:537–639.

Chang Jiang Water Resources Commission. 2003. A great success of the construction of flood-controlling system in Yangtze River. Natural Disaster Reduction in China 2:53 (in Chinese).

CIRDDC (China International Reduction Disaster Decade Commission). 1998. The national disaster reduction plan of the People's Republic of China (1998–2010). Natural Disaster Reduction in China 8:1–8 (in Chinese).

CNCIDR (China National Committee for International Disaster Reduction). 2003. The review of disaster reduction works of the member units in China International Disaster Reduction Commission (1998–2002). Natural Disaster Reduction in China 4:34–38 (in Chinese).

CPCU 2003. Managing flood losses: An international review of mitigation and financing techniques, Part I. CPCU (Chartered Property and Casualty Underwriter) Journal 75–93.

Freeman, P., L. Martin, J. Linnerooth-Bayer, K. Warner, G. Phlug and R. Mechler. 2003. Disaster risk management: National systems for the comprehensive management of disaster risk; financial strategies for natural disaster reconstruction. Washington, DC: Inter-American Development Bank.

Gao, W. and Q. Gao. 1995. Progress in disaster reduction of China and its prospect in the "ninth five". Journal of Natural Disasters 4:1–8 (in Chinese).

Hochrainer, S., R. Mechler and G. Pflug. 2004. Financial natural disaster risk management for developing countries. Proceedings of the 13th annual conference of the European Association of Environmental and Resource Economists, Budapest.

Houghton, J. T., Y. Ding, D. J. Griggs, M. Noguer, P. J. van der Linden and D. Xiaosu. *Climate change 2001: The scientific basis. Contribution of working group I to the third assessment report of the Intergovernmental Panel on Climate Change.* Cambridge: Cambridge University Press.

ISDR (International Strategy for Disaster Reduction). 2002. *Living with risk: A global review of disaster reduction initiatives.* Preliminary version prepared as an interagency effort coordinated by the ISDR Secretariat, Geneva, Switzerland.

Jiang, T. and Y. Shi. 2003. Global climatic warming, the Yangtze floods and potential loss. *Advances in Earth Science* 18:277–284 (in Chinese).

Li, J. and A. Wang. 1998. The analyses of flooding disaster in the Chang Jiang River valley in 1998. *Climate and Environmental Research* 3:390–397 (in Chinese).

Linnerooth-Bayer, J. and A. Amendola. 2000. Global change, natural disasters and loss sharing: Issues of efficiency and equity. *The Geneva Papers on Risk and Insurance* 25:203-219.

Linnerooth-Bayer, J., M. J. Mace and R. Verheyen. 2003. Insurance-related actions and risk assessment in the context of the UN FCCC. Background paper for UNFCCC workshop on insurance-related actions and risk assessment in the framework of the UNFCCC. http://unfccc.int/sessions/workshop

Linnerooth-Bayer, J., S. Quiano, R. Loefstedt and S. Elahi. 2000. *The uninsured losses of natural catastrophic events; seven case studies of earthquake and flood disasters.* Summary Report. Laxenburg, Austria: IIASA.

McCarthy, J. J., O. F. Canziani, N. A. Leary, D. J. Dokken and K. S. White. 2001. *Climate change 2001: Impacts, adaptation and vulnerability. Contribution of working group II to the third assessment report of the Intergovernmental Panel on Climate Change (IPCC).* Cambridge: Cambridge University Press.

Miletti, D. 1999. *Disasters by design.* Washington, DC: Joseph Henry Press.

OFDA/CRED (Office of Foreign Disaster Assistance/Centre for Research on the Epidemiology of Disasters). 2004. International Disaster Database. http://www.cred.be/emdat

Peng K. 1997. Research on the countermeasures of fighting and defending against natural calamities in economic development in China. *Chongqing Environmental Science* 19:21–26 (in Chinese).

UNDP (UNDP Expert Group Meeting). 2002. A climate risk management approach to disaster reduction and adaptation to climate change. http://www.undp.org/bcpr/disred/documents/wedo/icrm/riskadaptationintegrated

Watson, R. T. 2001. *Climate change 2001: Synthesis report. Contribution of working groups I, II, and III to the third assessment report of the Intergovernmental Panel on Climate Change.* Cambridge: Cambridge University Press.

Zhang, J. B. and K. Peng. 2001. The significant effects of non-engineering measures on disaster prevention and control. *Journal of Chongqing Institute of Technology* 15:1–8.

Zhang, Y. and G. Wang. 1997. Analysis on flood in mid-lower reaches of Yangtze River in 1996. *Express Water Resources and Hydropower Information* 18:28–30 (in Chinese).

Zhong. G. and H. Wang. 2000. Analyzing 1997–1998 El Niño and the relative weather phenomenon. *Marine Forecasts* 17:34–40 (in Chinese).

Zhou, W. G. and P. Shi. 2001. The sustainable development and disaster reduction public policy of flooded area in China. *Journal of Beijing Normal University (Social Science)* 2:110–117.

Chapter 21

ASSESSMENT OF VULNERABILITY AND ADAPTATION TO CLIMATE CHANGE IN WESTERN CHINA

YONGYUAN YIN, PENG GONG and YIHUI DING

1. Introduction

Policy makers and scientists are increasingly recognizing that some of the greatest challenges arising from the interactions between human development and the global environment entail complex system responses to multiple and interacting stresses originating in both the social and environmental realms. Since conventional climate impact assessment selects a particular environmental stress of concern (global climate change) and seeks to identify its most important consequences for a variety of economic or ecosystem properties, impact assessment has been relatively unhelpful in addressing such challenges. In this respect, vulnerability assessment is more desirable to offer such guidance.

The main goal of vulnerability assessment is to develop effective methods to measure vulnerability and to assess the environmental risks in dealing with climate stresses. This chapter presents a brief review of existing methods for measuring climate vulnerability. This is followed by a description of an ongoing research project aiming at developing an integrated assessment approach for identifying regional vulnerability to climate change, and for prioritizing adaptation options to deal with it. The chapter presents a conceptual research framework which integrates climate change scenarios, socioeconomic scenarios, current climate vulnerability identification, sustainability indicator specification, adaptation option evaluation, and multi-stakeholder participation.

2. Vulnerability Assessment

Vulnerability can be expressed as a statistical measure of the extent or duration of a system failure under a scenario of climate stress (i.e., an unsatisfactory value that is outside the coping range). Statistics or observed data above the upper threshold or below the lower limit are considered as unsatisfactory or vulnerable. It should be noted that these ranges may change over time. The extent of a system failure is the amount an observed value is above or below the threshold of the coping range. For example, the vulnerability of a water supply can be measured to show its success or failure to supply a certain amount of water for a municipality. If we use river flow, F, as an indicator to measure vulnerability, the water system vulnerability can be calculated by

$$EV_f = \text{Max}[0, LF_t - F_t, F_t - UF_t],$$

where EV_f is the water system's maximum-extent vulnerability based on river flow indicator; LF_t and UF_t are the lower and upper critical thresholds of the coping range, respectively; and F_t is the observed river flow data. Vulnerability can also be measured as the maximum duration of failure, which is calculated by Maximum Duration − Vulnerability (p) of DV_f = Maximum duration (number of time periods) of a continuous series of failure events for indicator F, occurring with probability p or that may be exceeded with probability $1 - p$.

In resource vulnerability assessment, vulnerability indicators are commonly used as decision criteria or standards by which the degree of resource vulnerability class can be identified. For example, an environmental vulnerability index approach, developed by the South Pacific Applied Geoscience Commission, uses 47 indicators with assigned weights to calculate vulnerability composite indices for several countries (Kaly et al. 1999; Kaly and Pratt 2000). The purpose of this approach is to evaluate the significance of environmental vulnerability of a nation facing alternative stresses or hazards.

The USAID Famine Early Warning System study adopted various methods to assess vulnerability in poor southern Africa countries (USAID 1997). While the study focuses on food security, availability, sources of income, and assesses food trends in areas of vulnerability in the country, climate variables and stresses are major concerns. Some methods used focus more on adaptive capacity. For example, the income estimation approach aims at estimating income at an administrative level to see if sufficient income can be obtained to purchase enough food. This method attempts to identify geographic locations which are vulnerable to food insecurity and famine. The assessment results in classifications of slightly, moderately (have enough stocks to face a shock and weather it), highly or extremely vulnerable (experienced past shocks, liquidated savings or assets, and thus are more vulnerable to future shocks) locations. Household income was used as the key factor for analysis, and thus was more relevant to adaptive capacity estimation. Another similar method called domestic resource capacity approach, is based on communities' ability to either collectively or individually allocate resources to reduce disaster risk (e.g., access to land and ownership, family labor availability, livestock resources).

A vulnerability analysis mapping project initiated in Mozambique for vulnerability assessment applied classification methods to generate flood risk maps, drought risk maps, food system maps, land use maps, market access maps, and health and nutritional profiles maps.

3. Integrated Assessment Project in Western China

The integrated assessment project combines computer modeling and non-model-based methods including a series of training workshops, survey, expert judgement, community engagement, multi-stakeholder consultation, ecological simulation modeling, geographical information system, remote sensing, and multi-criteria decision-making. The project addresses the following questions:

1. How vulnerable is Western China to current and future climate change in some key sectors?
2. What can the vulnerabilities of these key sectors teach us about future vulnerability? and
3. What are the desirable adaptation options to deal effectively with future climate changes?

The integrated assessment framework facilitates the participation of regional stakeholders in the whole process. Training workshops have been undertaken by as many local scientists as possible. Figure 21.1 shows the general approach of the study.

3.1. Climate change and socioeconomic scenarios

The project begins with a careful study of present-day climate conditions, impacts, and stresses to provide a baseline for assessing societal vulnerabilities to climate change. Three types of scenarios have

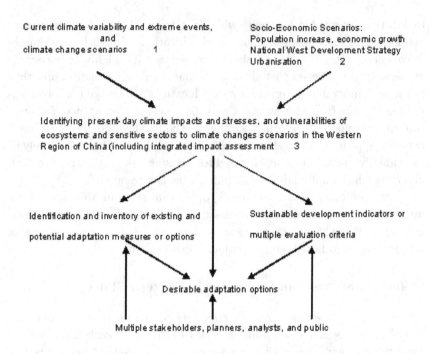

Fig. 21.1. Flow-chart showing the research structure of the project.

been specified: climate change, future socioeconomic conditions, and adaptation options. The climate change scenario represents possible future climatic conditions with various assumptions. Changes in population, income, technology, consumption rates, and China's new Western Region Development Strategy have been taken into account in developing the socioeconomic scenarios.

3.2. Identifying present-day climate impacts and vulnerability

Component 3 of the project (Fig. 21.1) can be divided into two parts. While part one examines present-day climate impacts of various key sectors in the region, part two identifies societal vulnerabilities to future climate change scenarios. Results of part one establish a base that can be used to measure progress toward reducing vulnerability to future climate change. Once these measurements are identified for each economic, ecological, and social vulnerability indicator, they can be applied to project vulnerabilities of the sensitive sectors to future climate change scenarios.

3.3. Multi-criteria evaluation

There is a need for new research approaches and tools that can evaluate alternative adaptation strategies or policies, which many impact assessment methods cannot do. IPCC (2001) suggests a list of high priorities for narrowing gaps in vulnerability and adaptation research. Among them is the integration of scientific information on impacts, vulnerability, and adaptation in decision-making processes, risk management, and sustainable development initiatives. In this respect, component 4 of the framework focuses on methodology development to link impact assessment with sustainability evaluation assisted by multi-criteria policy analysis and multi-stakeholder consultation.

4. Application to Heihe Basin

Western China includes predominantly arid and semi-arid areas in the north is dominated by mountains in the south, and has extremely fragile ecological conditions, few financial resources, poor infrastructure, low levels of education, and restricted access to technology and markets. The region suffers from climate variations and may experience severer impacts of climate change on food production, water resources, and ecosystem health in the future. Moreover, the region's adaptive capacity is lower than that in the coastal region of China. People in the Western region are facing substantial and multiple stresses, including rapidly growing demands for food and water, poverty, degradation of land and water quality, and other issues that may be amplified by climate change.

4.1. Heihe Basin

Heihe Basin is the second largest inland river basin in the arid region of Northwest China (latitude 35.4°–43.5°N, longitude 96.45°–102.8°E; Fig. 21.2). It includes parts of two provinces (Qinghai and Gansu) and

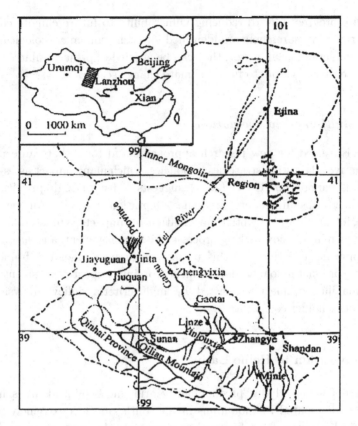

Fig. 21.2. Map of the Heihe Basin (Zhang et al. 2004).

the Inner Mongolia Autonomous Region. With an area of 128,000 square kilometers, the Basin is composed of diverse ecosystems including mountain, oasis, forest, grassland, and desert but provides limited agricultural land and water resources.

4.2. Scenarios

A set of scenarios representing current climate variation and future change have been identified. Results of seven general circulation models used in the IPCC scientific assessment report (IPCC 2001) and results taking into account increasing greenhouse gas and sulfate aerosol concentrations have been used to project climate change over Western China during the 21st century. The results suggest that temperature will increase continuously with increasing greenhouse gas concentration, and the increase will be larger than that in Eastern China. While the precipitation projection has an increasing trend for the next 100 years, especially in Western China and northwestern China, the increase will not be as large as that in southwestern China. The following tasks were carried out by the project:

- To project climate change under two scenarios using a global coupled model.
- To conduct dynamic downscaling simulation of various episodes in the 21st century with the regional climate model with boundary conditions provided by the global model.
- To apply simulation results of the global and regional climate models into a hydrological model, to simulate the future hydrological cycle over Western China; and
- To provide projected climate change simulations to impacts and adoption studies.

Various methods were employed to set future population increases and economic growth scenarios. Chinese government development strategies and plans were reviewed to collect additional data for the design of the scenarios, and workshops and community consultations with multi-stakeholders in the region were held to identify regional concerns related to the scenarios.

4.3. Sensitivity analysis

The purpose of sensitivity analysis is to identify the climate variables which are important in determining system vulnerability. In addition, sensitivity analysis can indicate those key sectors, systems, or areas which are sensitive to certain climate variables.

Since relationships between climate variables and system activities are based on historical statistics or experience, this information can be obtained by consultation with experts or stakeholders in workshops or surveys. In the consultation process, stakeholders may also identify current adaptation options and critical thresholds or criteria they use in management. Climate sensitivity analysis through stakeholder consultation were used in this project because of its success in the Hunter Valley, Australia (Hennessy and Jones 1999). Table 21.1 lists some climate variables and activities which are sensitive to climate change in Western China.

Table 21.1. Potential sensitivity matrix showing the climate variables with the greatest forcing and activities with the broadest sensitivity in Western China (Hennessy and Jones 1999).

	Climate and related variables (forcing)	Activities (sensitivity)
High	Rainfall — variability	Water supply
	Drought	Water management, cropping, grazing
	Evaporation	
	Soil moisture	Water supply
	Stream flow	Cropping, salinity
		Water supply
Moderate	Temperature — min	Cropping
	Wind	Soil erosion, sandstorm
	Irrigation	Cropping, salinity, soil erosion
Low	Hail	Cropping, properties
	CO_2	Crop yield, carbon sequestration

4.4. Vulnerability and adaptive capacity assessment

4.4.1. Climate vulnerability indicators

Vulnerability indicators are criteria by which the effects of climate change or/and the efficiency of alternative adaptation options can be measured. Some indicators of resource vulnerabilities are presented in Table 21.2.

It is apparent that economic return is one of the most important indicators for measuring adaptive capacity, and that improvement in economic return will enhance adaptive capacity.

There has been an increasing concern about China's ability to feed itself. The provision of adequate food on a continual basis is a major indicator of regional sustainability. Agricultural production can be considered as a security indicator of a country's ability to achieve higher levels of self-sufficiency, or it may be used to represent a vulnerability indicator to check whether the resource base can provide sufficient food.

Many of the industrial and housing developments occur in productive farmlands, forestry lands, and areas that are perceived to be of natural, scenic, or scientific importance. Slowing the conversion of farmland to urban and industrial uses is seen to be critical for regional sustainability in China, and thus another indicator to reflect this concern is required.

It is now generally realized that an environmental concern should be incorporated in decision-making in an effort to achieve sustainable development (WCED 1987). There are a large number of parameters that can be used as indicators of ecological vulnerability. In Western

Table 21.2. Regional vulnerability and adaptive capacity indicators used in the case study.

Sectors	Indicator	
Water resources	Vulnerability	Water demand, water storage stress, water stress, hydropower
	Exposure	Water supply climate variables, Palmer drought severity index, low flow event
	Adaptive capacity	Economic return, industry productivity, regulated annual supply, institutional frameworks
Agriculture	Vulnerability	Population growth, agricultural production, arable land loss, food consumption
	Exposure	Cold snap, heat stress days, monsoon pattern, accumulated degree days, water shortage, Palmer drought severity index
	Adaptive capacity	Farm income, agricultural product price, technology
Ecosystems	Vulnerability	Soil erosion, desertification, sandstorms, population growth rate, population density
	Exposure	Water supply, high wind days, sandstorms, Palmer drought severity index, heat stress days, cold snap days
	Adaptive capacity	Forest area protection, emission reduction of CO_2, ecological protection

China, environmental concern is reflected in the indicators of soil erosion, desertification, greenhouse gas emission, sandstorms, and shortage of water.

There is increasing concern about the implications of climate change for water management (Gleick 1990). Access to water in the Heihe Basin has already led to disputes, confrontation, and in many cases violent clashes. If climate change increases the periods of drought, water shortages, and conflicts over water use are likely to increase. Dealing with potential water use conflicts with changing climate is therefore considered as an important indicator.

4.4.2. *Critical thresholds*

Critical thresholds of indicators set the boundaries beyond which systems feel significant effects of climate change. Thus, critical thresholds divide the coping zone and the vulnerability zone of a system. For example, drought hazards are based on rainfall amount, or aridity index, and if conditions remain below the threshold levels for a sustained period, drought hazards are declared. The threshold levels (drought index) can be used to measure the frequency of droughts over time. It is clear that identifying critical thresholds is extremely important in adaptive capacity evaluation.

In vulnerability and adaptive capacity measurement, many of the indicators can be expressed in numerical terms, particularly climate and physical variables. It is also recognized, however, that many indicators cannot be quantified, and many of the threshold levels can only be qualitatively described. Yohe and Tol (2001) point out that relationships between adaptive capacity and its determinants are difficult to quantify. Functional representations of these relations are only useful when they can offer insights into the complexity. With this respect, stakeholders, policy makers, and analysts jointly identify critical threshold levels commonly used in indicator measurement (Jones and Page 2001).

4.4.3. Setting priorities

Given the fact that not all the indicators can be considered equally important in determining system vulnerability, a choice must be made to prioritize the different indicators in a multi-criteria decision-making system. Since indicator priority identification is a difficult and complex process, the analytic hierarchy process method developed by Saaty (1980), was used to assist indicator prioritization in this study. This method provides a means by which indicators can be compared and evaluated in an orderly manner.

The analytic hierarchy process requires decision-makers to provide judgements on the relative importance of the indicators. The result of the analysis is a prioritized ranking indicating the overall preference for each of the indicators. Another feature of the process is that it provides a flexible framework for public participation in decision-making or problem solving. Yin and Cohen (1994) presented a systematic approach, assisted by the analytic hierarchy process, to identify and prioritize regional sustainability indicators relating to climate change. In the results, a stakeholder's preference is expressed by the relative importance of indicators on an ordinal scale. That is, indicators are ranked as "most important," "next most important," and so on.

4.4.4. Fuzzy set method

Fuzzy pattern recognition and classification methods have been applied to a multitude of environmental problems involving uncertainty. For example,

Yin et al. (1999) adopted a fuzzy model for selecting desirable adaptation options in dealing with water level fluctuations in the Great Lakes Basin. Numerous applications of fuzzy set theory in environmental analysis have indicated that fuzzy recognition method is an effective technology in dealing with environmental risk.

For each critical factor (indicator), a fuzzy criterion of vulnerability level is provided. Assuming n vulnerability levels exist, the criteria for the indicators are represented by vectors with n elements, respectively:

$$C_i = [c_{i1}, c_{i2}, \ldots, c_{in}], \quad i = 1, 2, 3, \ldots, m,$$

where n is the nth vulnerability level, and m is the indicator determining vulnerability.

A fuzzy pattern recognition model can be applied to calculate the aggregated ratings of vulnerability ranks for all the land units in the region. The problem under consideration is how to assign different land units into proper categories of overall vulnerability level on the basis of the given data and criteria, and thus partition the whole region into several sub-regions with unique vulnerability patterns. For illustration purpose, the two fuzzy sets, U, of classification criteria and V of water vulnerability levels are specified as follows:

$U = \{$(temperature), (rainfall), (low flow event frequency), (low flow event duration), (causality and/or injury), (damage to ecosystem), (water use conflicts)$\}$,

$V = \{$(extremely vulnerable), (high risk), (moderate risk), (low risk), (acceptable)$\}$.

The next step of the analysis is to define grades of the fuzzy criterion set for each of the criteria. The grades of the fuzzy criterion set corresponding to each criterion are derived from some threshold information. For each criterion, the score corresponding to the highest vulnerability level is assigned as the grade of the fuzzy membership. After assigning fuzzy membership grades, the following step is to identify the corresponding vulnerability class for each land unit (Yin and Huang 1990).

4.4.5. *Adaptive capacity classification*

The fuzzy recognition model can also be used to assign adaptive capacity classes to all the land units under study. Obviously, the sets, U, of classification criteria and V of adaptive capacity levels of water system should be modified accordingly:

$U = \{$(economic return), (industry productivity), (technology advancement), (regulated annual supply), (institutional frameworks), (water storage capacity)$\}$,
$V = \{$(extremely adaptive), (high adaptive), (moderate adaptive), (low adaptive), (acceptable)$\}$.

Once these vulnerability measurements are identified for economic, ecological, and social vulnerability indicators, they can be applied to project potential vulnerabilities of the resource systems to future climate change scenarios. A set of existing adaptation options to deal with current vulnerabilities to climate variations will be identified for adaptation option evaluation.

4.5. Adaptation policy evaluation

One task of the study is to determine whether alternative adaptation options or policies can lead to a reduction in damages or allow one to take advantage of opportunities associated with climate change. The adaptation evaluation studies usually adopt two approaches. The first examines the effectiveness of alternative short term or autonomous adaptation options by using impact assessment modeling (Carter et al. 1994). Impact assessment models are run with different climate change scenarios with or without certain adaptation options. Another approach deals with anticipatory or planned adaptation strategies and government policies. Thus, the tools used for the second approach are related to policy evaluation or analysis (Stratus Consulting Inc. 1999). Results generated from this step can be used for public consultation and adaptation evaluation processes.

4.6. Evaluation of adaptation options

Multi-criteria options evaluation of adaptation measures is one of the major activities of the study. To select desirable measures among alternatives, multi-stakeholder consultation and multi-criteria options evaluation were used to relate vulnerability information to decision-making which requires subjective judgement and interpretation. In this study, alternative options were evaluated by relating their various impacts to a number of relevant indicators. The results of various impacts generated in the impact assessment were used as references for ranking the performance of each adaptation option against each sustainability indicator. These indicators were used as multi-criteria by which the strengths and weaknesses of the various adaptation options can be evaluated. Two multi-criteria decision making (MCDM) techniques, one designed by Yohe and Tol (2002) and the analytic hierarchy process (AHP) method discussed above, were adopted in the case study to assess the potential contributions of various adaptation options in improving adaptive capacities of resource systems. While the modified Yohe and Tol method used adaptive capacity determinants for evaluating alternative adaptation options, the AHP was applied to identify the priorities of evaluation criteria and to rank desirability of alternative adaptation measures (Yin 2001; Yin et al. 2007a).

5. Conclusions

This chapter presents an integrated assessment approach which was developed for identifying the patterns of climate vulnerability by using quantitative and qualitative information. The method provides a viable means for the synthetic analysis of climate vulnerability and adaptive capacity based on a multitude of critical indicators. The integrated assessment can provide holistic analysis of climate change and adaptation policy to improve the scientific understanding of interactions among indicator setting, vulnerability assessment, adaptation policy evaluation, and sustainability.

The integrated assessment framework integrates simulation models, fuzzy set classification, multi-criteria decision-making, multi-stakeholder involvement, and climate change scenarios. Moreover, alternative adaptation options can be evaluated against multiple sustainability indicators, and linkages established between vulnerability assessment and decision-making, and between climate change and system sustainability. For more information on detailed application of the IA in the case study and the project results, readers are referred to Yin et al. (2007a and 2007b).

Literature Cited

Carter, T. R., M. L. Parry, H. Harasawa and S. Nishioka. 1994. *IPCC technical guidelines for assessing climate change impacts and adaptations.* London: Department of Geography, University College.
Gleick, P. H. 1990. Vulnerabilities of water systems. Pp. 223–240 in *Climate change and U.S. water resources,* edited by P. Waggoner. New York: John Wiley and Sons, Inc.

Hennessy, K. J. and R. N. Jones. 1999. *Climate change impacts in the hunter valley: Stakeholder workshop report.* Melbourne: CSIRO Atmospheric Research.
IPCC (Intergovernmental Panel on Climate Change). 2001. *Climate change 2001: Impacts, adaptation, and vulnerability. Summary for policymakers.* A Report of Working Group II. Geneva, Switzerland: IPCC.
Jones, R. N. and C. M. Page. 2001. Assessing the risk of climate change on the water resources of the Macquarie river catchment. Pp. 673–678 in *Integrating models for natural resources management across disciplines, issues and scales*, Vol. 2, edited by F. Ghassemi, P. Whetton, R. Little and M. Littleboy. Canberra: Modelling and Simulation Society of Australia and New Zealand.
Kaly, U. and C. Pratt. 2000. *Environmental vulnerability index: Development and provisional indices and profiles for Fiji, Samoa, Tuvalu and Vanuatu.* Phase II Report for NZODA. SOPAC Technical Report 306.
Kaly, U. L., L. Briguglio, H. McLeod, S. Schmall, C. Pratt and R. Pal. 1999. *Environmental vulnerability index (EVI) to summarise national environmental vulnerability profiles.* SOPAC Report to NZODA.
Saaty, T. L. 1980. *The analytic hierarchy process.* New York: McGraw-Hall.
Stratus Consulting Inc. 1999. *Compendium of decision tools to evaluate strategies for adaptation to climate change.* Final Report. FCCC/SBSTA/ 2000/Misc.5. Bonn, Germany: UNFCCC Secretariat.
USAID 1997. *FEWS Project: Vulnerability assessment*, Published for USAID, Bureau for Africa, Disaster Response Co-ordination, Arlington, VA.
WCED (World Commission on Environment and Development). 1987. *Our Common Future.* Report of the World Commission on Environment and Development. Oxford: Oxford University Press.
Yin, Y. 2001. *Designing an integrated approach for evaluating adaptation options to reduce climate change vulnerability in the Georgia basin.* Final report submitted to Adaptation Liaison Office, Climate Change Action Fund, Ottawa, Canada.
Yin, Y., G. Huang and K. W. Hipel. 1999. Fuzzy relation analysis for multicriteria water resources management. *Journal of Water Resources Management and Planning* 125:41–47.
Yin, Y. and S. Cohen. 1994. Identifying regional policy concerns associated with global climate change. *Global Environmental Change* 4:245–260.
Yin, Y. and G. Huang. 1990. Applying fuzzy sets to identify the spatial patterns of air pollution factors. Paper presented to The GLOBE'90 Conference. March 19–23. Vancouver, B.C.
Yin, Y. Y., Z. M. Xu and A. H. Long. 2007a. Chapter 12 Evaluation of adaptation options for the Heihe River Basin of China. In: Neil Leary et al., (eds.) *Climate Change and Adaptation.* Earthscan.
Yin, Y. Y., N. Clinton, B. Luo and L. C. Song. 2007b. Chapter 5 Resource system vulnerability to climate stresses in the Heihe River Basin of western China. In: Neil Leary et al., (eds.) *Climate Change and Vulnerability.* Earthscan.
Yohe, G. and R. S. J. Tol. 2002. Indicators for social and economic coping capacity — moving toward a working definition of adaptive capacity. *Global Environmental Change* 12:25–40.

Chapter 22

APPLICATION OF ASIA–PACIFIC INTEGRATED MODEL IN EAST ASIA

XIULIAN HU and KEJUN JIANG

1. Introduction

It has been acknowledged that climate change is one of the important issues facing humanity in the present century and it needs to be studied over a long-term horizon. The effects of climate change are already visible in different parts of the world, and the surface temperature over the globe has increased, on average, by ~0.6°C during the last century (IPCC 2001). However, some regions have cooled rather than warmed so it is necessary to study climate forcings at the regional scale in order to predict future climate response at regional and global scales.

The Asia–Pacific Integrated Model project started in 1990 and has been specifically developed using a collaborative approach with participation of governments and developing countries to get their own integrated assessment tools. These countries have already applied their own integrated assessment models to their policy making processes (Hu et al. 1996; Jiang et al. 1998; Kainuma et al. 2003; Shulka et al. 2004).

The research has been funded by the Ministry of Environment in Japan and the team includes the following collaborators: Asian Institute of Technology of Thailand, Energy Research Institute of China, Fuji Research Institute Corporation of Japan, Indian Institute of Management, Institute of Geographical Sciences and Natural Resources Research of China, Institute for Global Environmental Strategies of Japan, Korea Environment Institute, Kyoto University of Japan, Maulana Azad National Institute of Technology of India, National Institute for Environmental Studies of Japan, Ritsumeikan University of Japan, Seoul National University of Korea, University Putra Malaysia, and Winrock International, India.

2. The Model

The Asia–Pacific Integrated Model (AIM) is a set of computer simulation models used to assess policy options for sustainable development in the region. It started as a tool to evaluate policy options to mitigate climate change and its impacts, and its function was extended to analyze other environmental issues such as air pollution control, water resources management, land use management, and environmental industry encouragement.

More than 20 modules have been developed so far. They include AIM/Enduse, AIM/CGE, AIM/Impact, and AIM/Impact [policy]. They are separately or jointly used to simulate climate change impacts and mitigation potential in the future. AIM/Enduse is used to measure global and regional reduction potential and direct costs, AIM/CGE to estimate economic burdens of emission reduction to comply with the path, AIM/Impact to estimate country level climate impacts, and AIM/Impact [policy] to project allowable emission paths from the view point of climate stabilization. These four models follow the paths illustrated in Fig. 22.1.

3. Climate Change and its Impacts in Asia

3.1. Observed and simulated climate changes

3.1.1. Trends of annual mean temperature and precipitation

The simulated temperature increases during the 21st century are larger in China, Japan, and Korea, and smaller in Thailand than in India. The uncertainty of the temperature increases is also large where temperature increase is large. The simulated temperature changes by the end of 21st century are larger in the SRES-A2 scenario (based on high population growth and lower agricultural productivity) than in the A1B (balanced energy technology across all sources), B2 (medium population growth with environmental protection), and B1 (low population growth and limited emissions) scenarios. [see Intergovernmental Panel on Climate Change Special Report on Emission Scenarios (SRES; Nakicenovic et al. 2000) for a full description of the scenarios].

While future precipitation trends are quite uncertain, both models and emission scenarios suggest that increases in precipitation will be a common future trend for China.

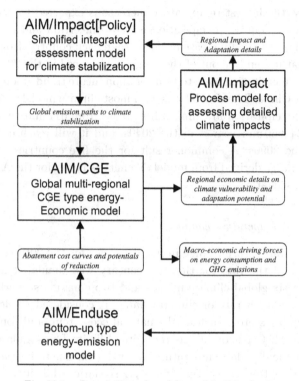

Fig. 22.1. Description of models used in this chapter.

In India, the simulated temperature change from 1990s to 2040s ranges between 1.04°C and 1.75°C, and the temperature change from 1990s to 2090s ranges from 2.1°C to 4.05°C due to the emission scenario diversity. However, the differences among climate models are quite large, and the temperature change from 1990s to 2090s with the SRES A2 scenario ranges between 2.83°C and 5.31°C. The uncertainty in range resulting from model diversity is larger than that resulting from emission scenarios when one model is used.

In China, the simulated temperature change from 1990s to 2090s ranges from 3.23°C to 5.48°C, and it is higher than that in India. The differences among climate models are also larger than those for India. On the other hand, the simulated temperature changes and the difference among models are smaller in Thailand than in India. In all five countries, the simulated temperature change by the end of the 21st century is larger in the A2 scenario, than in the A1B, B2, and B1 scenarios. In the first half of the

21st century, the largest temperature increase is expected to occur under the B2 scenario. Future precipitation is quite uncertain and it is difficult to tell whether it would increase or decrease in each country. However, it is expected that it could be out of the range experienced in the 20th century. Increases in precipitation seem to be a common future trend among emission scenarios for China. The simulations with most climate models suggest that the annual precipitation in China will increase from 611 mm year^{-1} in the 1990s to 624–660 mm year^{-1} in the 2040s, and it will reach 664–715 mm year^{-1} in the 2090s. A common result for the five countries is that the uncertainty range derived from model diversity is larger for the A2 scenario than the B2 scenario.

3.1.2. *Rate of temperature change*

The rate of temperature change in five countries for the 20th century was determined and the rate for the 21st century was simulated from those results using six global climate models and two emission scenarios (SRES A2 and B2). A high rate of climate change is considered to significantly affect vulnerable sectors such as the natural ecosystem, and some studies have defined 0.2°C per decade as the critical limit. The simulated results suggest that the decadal temperature rate will continue to increase even in the latter half of the 21st century in the A2 scenario, while it will be rather stable in the B2 scenario. Even in the B2 scenario, the simulation suggests that the rate would be more than 0.4°C per decade in the first several decades and it may not cross the critical line of 0.2°C per decade during the century in most countries. In order to fulfill the climate control target of 0.2°C per decade, more drastic mitigation of greenhouse gas emission is required.

3.1.3. *Fluctuation of monthly mean temperature during a year*

The simulated decadal mean of intra-annual temperature fluctuation index for scenarios A2 and B2 was determined as above with six global climate models. The intra-annual temperature fluctuation index is defined as the difference between monthly mean temperature and annual mean temperature viz;

$$\text{Intra-annual fluctuation index} = \sum_{m=1}^{12} |temp_m - temp_{\text{ave}}|/12.$$

Here, $temp_m$ denotes monthly mean temperature and $temp_{ave}$ denotes annual mean temperature. The simulated results for China and India show a clear trend of decrease in the index through both the observation and simulation periods. The results are interpreted to mean that future temperature increases in the cold season will be larger than those in the hot season and the fluctuation of monthly temperature during a year will decrease. In Korea and Japan, there was a decreasing tendency in the 20st century, but this may not continue in the 21st century. In Thailand, the index did not change significantly through the observation period and the simulated results suggest that it will not change significantly in the simulated period.

3.1.4. Fluctuation of annual mean temperature during a decade

The inter-annual temperature fluctuation index of each decade, defined as the difference between decadal mean temperature and annual mean temperature, was also determined. The index for 1990s is described as

$$\text{Inter-annual fluctuation index} = \sum_{y=1991}^{2000} |temp_{ave,y} - temp_{dec,1990s}|/10.$$

Here, $temp_{ave,y}$ denotes annual mean temperature in year y and $temp_{dec,1990s}$ denotes decadal mean temperature in the 1990s. Although the future trend of inter-annual fluctuation is quite different among the six models, the simulations suggest that it may increase significantly in India and China. The increasing tendency of the index implies that the extremely hot summers and disastrous climate events experienced in many regions will occur more often in the future.

3.2. Climate change impacts in Asian countries

Flooding, one of the main natural disasters in China, occurs frequently, not only in southern China where a humid monsoon climate prevails but also in arid and semiarid northern China. Changes in risk of flooding are considered to be one of the potential impacts of climate change, since some studies indicate an increase in frequency/intensity of heavy rain. On the other hand, the additional investment in infrastructure for preventing flood disasters in the early decades of this century have the potential to mitigate not only the additional flood disasters caused by future climate change but also those which currently occur because of climate variability.

We evaluated the optimal amount of investment in infrastructure for preventing flood disasters by estimating flood damage to cultivated land under four alternative scenarios;

(i) policy makers do not arrange adaptation investment for projected damage, and climate change does not occur (baseline),
(ii) policy makers do not arrange adaptation investment for projected damage, but climate change does occur,
(iii) policy makers arrange adaptation investment for projected damage, but climate change does not occur, and
(iv) policy makers arrange adaptation investment for projected flood damage, and climate change does occur.

The simulated results indicate that flooding will damage 1.13% of cultivated land in 2100 even if investment takes climate change into consideration and climate changes do occur (scenario iv). The damage will gradually increase to 3.11% by the end of the century when there is no investment in flood prevention infrastructure to combat projected climate change but climate change does occur (scenario ii). The consumer's utility per capita will decrease by 0.83% (compared with the baseline case; scenario i), when no adaptation investment is implemented, but climate change occurs (scenario ii). The decrease of utility can be reduced to 0.1% if adaptation investment is implemented (scenario iv). Even if investment is larger than the optimal level, consumer's utility per capita will not be completely lost. According to this analysis, it appears to be a good strategy to invest in flood prevention infrastructure even if the occurrence and magnitude of climate change is still uncertain.

Under the unsustainable scenario, which reflects a high rate of population growth and low rate of improvement in water-use efficiency, the water stress index (ratio of water withdrawal to renewable water resource in each basin) will significantly increase in India, leading to increased drought risk. On the other hand, under the sustainable scenario, water stress index will decrease slightly in most basins reflecting a high rate of improvement in water use efficiency. (There are a small number of exceptional basins where water stress index increases because of the regional pattern of climate change.)

Whether current forest vegetation will be damaged or not is identified by comparing potential velocity of forest moving with the velocity of vegetation zone shift that is estimated by considering climate change scenarios. The velocity of forest moving is estimated to be in the range of 0.25–2.0 km year^{-1} in our assessment. In the SRES A2 scenario, where the temperature increase is higher than that in the other SRES scenarios, the extinction area of the Korean Peninsula would be 2.08% if the velocity of forest moving is assumed to be 0.25 km year^{-1}.

4. Greenhouse Gas Emission and Reduction Potential in Asia

4.1. China

Using IPAC-AIM/China with various gross domestic products, population and technology assumptions, simulated results were obtained for energy demand and carbon dioxide emission for China in the 21st century. The results suggest there will be a significant increase in primary energy demand, which ranges from 2.3 to 3.7 billion tons of coal equivalent in 2020, and 2.7 to 4.4 billion tons of coal equivalent in 2030. The results also suggest that carbon dioxide emission will increase quickly with economic development. In 2000, 2010, and 2030, carbon dioxide emissions will be 1.7, 2.5, and 3.8 times those in the 1990 base year for the medium case, which can be regarded as a possible development case, and 1.6, 2.1, and 3.3 times those in 1990 for the policy case, which gives the lowest growth rate for carbon dioxide emission. The simulations suggest that in 2010 most of the emissions in China will emanate from the coast region. The results also indicate that there could be significant potential for saving energy and reducing carbon dioxide emission. The simulated savings in 2030 range from by 8% to 46%, and the cost of the reduction of carbon dioxide emission for the major sectors is presented in Fig. 22.2.

4.2. India

The preceding analysis highlights the growing trend in greenhouse gases and its impact on the regions natural resource. However, technological developments resulting from both autonomous investments in research and development, and policies from national governments has resulted in the development of technologies with the potential to abate greenhouse gas emissions.

Figures 22.3 and 22.4 present abatement potential and marginal costs for greenhouse gas abatement for some important sectors. India produced 27.1 million metric tons of finished steel from main and secondary producers in 2000. The Indian steel industry has grown above 7% (compounded) per

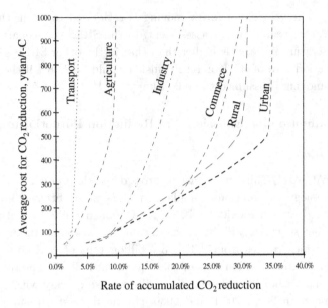

Fig. 22.2. Cost of carbon dioxide emission reduction by major sectors.

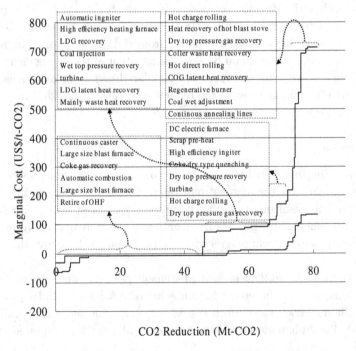

Fig. 22.3. Marginal abatement curve for carbon dioxide in the steel sector in 2020.

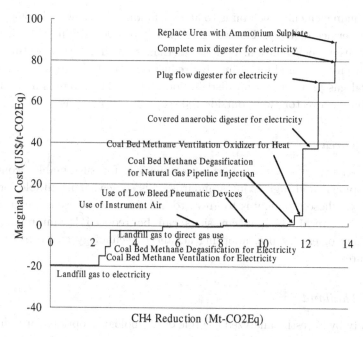

Fig. 22.4. Marginal abatement curve for methane across sectors in 2020.

annum during the 1990s and is poised to grow even faster during the present decade due to thrust on infrastructure development and housing by the Government. The total steel production is projected to reach 52.5 Mt in 2010, 81 Mt in 2020, and 103 Mt in 2030.

Presently, the Indian steel industry is about 1.5 times more energy intensive than the average international steel producers, both for the integrated steel plants and the medium and small steel-rolling mills. Therefore, it has considerable potential for energy conservation. Figure 22.3 shows the CO_2 reduction potential in the steel sector and the associated costs of the technologies for the year 2020. Also, the cumulative carbon emissions from the Indian steel sector for 2000–2030 are projected to be about 1.5 billion tons of coal equivalent. Energy conservation and carbon mitigation technologies may mitigate about one-third of these.

Investigation on coal bed methane for its commercial exploitation is a comparatively recent phenomenon in India (Fig. 22.4). The government announced in October 2003 that 14 bids for eight blocks had been received under the second round of bidding for coal bed methane exploitation. Also, Indian cities are introducing private sector participation in solid waste management. The aim is to attract funding and new technologies for better service delivery. The waste to energy technologies in India include incineration, pelletization, and bio-methanation. The other sectors that have been considered for estimating reduction potential are paddy rice and natural gas production and distribution sectors. The biomass and enteric fermentation sectors were outside the scope of this analysis.

4.3. Korea

Since energy is a very important component for sustainable economic development and high living standards, it is somewhat inevitable for one country, whose economy is to grow, to increase greenhouse gas emissions. However, greenhouse gas emissions can be reduced by improving the efficiency of energy use, adopting advanced technology and other policy measures.

4.4. Thailand

A study by Shrestha analyzed the role of technological options for reducing greenhouse gas emissions under the Clean Development Mechanism in the power sector of three Asian countries, including Thailand. This study showed that cleaner thermal power generation technologies involving fuel switching from coal to gas or oil would be the main mechanism for reducing carbon dioxide emission not only at the presently prevailing prices for certified emission reduction viz. 3–6 dollars, but also at significantly higher prices.

5. Conclusions

Impact studies have shown that far sighted investments in infrastructure in China in the early decades of the 20th century for preventing flood disaster have the potential to mitigate not only additional flood disasters caused by future climate change, but also flood disasters which currently occur because of climate variability. The impact analysis also covers the impact of climate change on the vegetation of the Korean Peninsula and the water resources in India. The Korean country study highlights the extent of vegetation damage caused by climate change while the Indian analysis indicates the acute risk of drought that India faces if the country grows in an unsustainable manner.

However, technological developments resulting from both market driven investments in research and development, and policies from national governments have resulted in the development of technologies with the potential to abate greenhouse gas emissions and thus mitigate adverse impacts on the environment. A section of this study using the AIM Global framework analyses the greenhouse gas emission trends, both past and future, and the reduction potentials in countries in the Asia-Pacific region.

The AIM Global modeling results show greenhouse gas reduction potentials in different sectors for world regions for the year 2020. The study illustrates the potential for mitigating greenhouse gas emissions and the costs involved. To achieve this reduction potential it is essential that a mechanism be set up to facilitate transfer of advanced technologies and financial aid to less developed nations of the world.

Also, an issue that has been much debated is that of burden sharing among countries, including mitigation and adaptation costs. Measuring contributions by countries to mitigate climate change has been difficult. However, an attempt has been made to simulate the contribution of each country to mitigate greenhouse gas emissions and the resulting impact on each country's gross domestic product. The analysis clearly indicates that international collaboration and establishment of mechanisms that facilitate technology transfer and investments from developed countries to developing countries can help achieve the desired level of emission mitigation, with much lower loss to world gross domestic product.

Literature Cited

Hu, X., K. Jiang and J. Liu. 1996. Application of AIM/Emission model in P. R. China and preliminary analysis on simulated results, AIM Interim Paper, IP-96-02, Tsukuba, Japan.

IPCC (Intergovernmental Panel on Climate Change). 2001. *Climate change 2001: The scientific basis. Contribution of working group I to the third assessment report of the Intergovernmental Panel on Climate Change*, edited by J. T. Houghton, Y. Ding, D. J. Griggs, M. Nogure, P. J. van der Linden and X. Dai. Cambridge, UK: Cambridge University Press.

Jiang, K., X. Hu, Y. Matsuoka and T. Morita. 1998. Energy technology changes and CO_2 emission scenarios in China. *Environment Economics and Policy Studies* 1:141–160.

Kainuma, M., Y. Matsuoka and T. Morita. 2003. *Climate policy assessment*. Tokyo: Springer.

Nakicenovic, N., J. Alcamo, G. Davis, B. de Vries, J. Fenhann, S. Gaffin, K. Gregory, A. Grübler, T. Y. Jung, T. Kram, E. L. La Rovere, L. Michaelis,

S. Mori, T. Morita, W. Pepper, H. Pitcher, L. Price, K. Riahi, A. Roehrl, H.-H. Rogner, A. Sankovski, M. Schlesinger, P. Shukla, S. Smith, R. Swart, S. van Rooijen, N. Victor and D. D. Zhou. 2000. *Special report on emissions scenarios for the intergovernmental panel on climate change*. Cambridge: Cambridge University Press.

Shukla, P. R., A. Rana, A. Garg, M. Kapshe and R. Nair. 2004. *Climate policy assessment for India*. New Delhi: University Press.

Chapter 23

HUMAN DRIVERS OF CHANGE IN THE EAST ASIAN MONSOON SYSTEM

KAREN SETO, DENNIS OJIMA, QINGYUAN SONG and ARVIN MOSIER

1. Introduction

The East Asian Human–Monsoon System is one of the most dynamic regions of the world. Over the past three decades, most East Asian countries underwent significant policy reforms and experienced rapid economic growth. Many of the most pressing economic, demographic, and political–institutional changes in the world and associated environmental impacts are occurring in this region: rapid development in China has led to the widespread conversion of natural and agricultural ecosystems to urban areas; the Asian financial crises during the 1990s highlighted the region's environmental vulnerability to external economic pressures; heavy foreign debt loads and new land use laws have led to intensive grazing practices in Mongolia; the collapse of the Soviet Union has left a power vacuum in Siberia where extensive deforestation is common; expansion of industrial activities and a greater reliance on personal vehicles throughout the region have dramatically increased energy consumption, especially the use of coal, and deteriorated local and regional environmental quality, increased carbon emissions and other aerosols; growing industrial, urban, and agricultural demands on water supply have resulted in new water development projects; and a growing "middle-class" has resulted in shifts in consumption patterns and livelihood expectations.

Socioeconomic and political changes in the region are also creating new opportunities for efficient resource use and sustainable development. Air quality in Tokyo is among the best of the cities in the world; water quality is improving in China's urban rivers; the end of forced relocation to the countryside has resulted in the regrowth of grasslands in Central Asia;

market forces are allowing biophysical comparative advantage to determine agricultural production and thereby reduce pressures to farm marginal land.

With more than one quarter of the global population and combined economies that will be the largest in the world within the next decade, the magnitude of changes in socioeconomic conditions in East Asia have the potential to affect the functioning of the Earth system. Therefore, an understanding of the major characteristics of the social systems in the region is critical to an integrated human–environmental study of the East Asian Monsoon System. This chapter discusses salient socioeconomic conditions and their potential roles in driving ecosystem change in the region.

2. Current Trends and Conditions

2.1. Demographics

One of the defining characteristics of the East Asian Monsoon System is the region's demographic dynamics. While the region's population size is impressive — more than one quarter of the global population — three other demographic trends are likely to drive significant ecosystem change.

First, the region is experiencing unprecedented rates of urbanization. Historically, the region's population was primarily rural and urban populations constituted a small portion of total population. With the end of forced frontier settlement, hundreds of thousands of farmers are relocating to cities and coastal regions. This large-scale rural to urban migration has profound implications for virtually every aspect of society and the environment. The infrastructure requirements will be huge: growth in demand for energy and water in urban areas will require extraordinary actions; providing public services and housing for such large numbers of people will present civic and environmental challenges; efficient land-use will be essential. During the period 1950–2000, Asia had the second highest urban population growth rates in the world (3.46%), second only to Africa (4.42%), and the highest urbanization rate in the world (1.23%). Because Monsoon East Asia is still considerably rural, it is expected to experience rapid rates of urbanization between 2000 and 2030. (United Nations 2002; Peng and Han, Chapter 20, this volume).

The direct and indirect effects of rapid urbanization in East Asia will be one of biggest socioeconomic and environmental challenges of the 21st Century. China will be a key player in this urban revolution. According to the United Nations, nearly 500 million people will be added to Chinese urban areas between 2000 and 2030 (United Nations 2002). Over this same period, hundreds of towns and small cities with fewer than 150,000 people will become major urban centers of 500,000 to over one million in population. Urbanization will affect not only East Asia, but also the world: increases in energy and grain prices, acid rain in Japan, and floods in South Asia can be attributed to changes brought about by rapid Chinese urbanization.

Second, the impact of the region's population size and growth rates on ecosystem functioning may be secondary relative to household dynamics. In Monsoon East Asia, household numbers are growing as a result of the concurrent increase in population size and the reduction in household size (Liu et al. 2003a). While these patterns mirror worldwide trends, they are particularly significant because of the region's sheer population size. The decline in household size will pose considerable challenges to natural resource use, consumption of fuel wood, alteration of habitats for expansion of housing stock, and greenhouse gas emissions. More households require more housing stock, energy, and durable goods. The decline in household size is particularly worrisome because many "hotspots" of biodiversity are located near regions with increasing numbers of households and associated rises in demand will place increasing pressure on local resources.

Third, urbanization and the growth of household numbers will drive urban land-use change. Over the last two decades, urban land-use change in Monsoon East Asia has occurred primarily along the coast and in productive farmlands (Seto et al. 2000; Liu et al. 2003b). Throughout the region, thousands of hectares of the most fertile agricultural lands have been taken out of production and converted into residential, commercial, and industrial land. The conversion of natural and agricultural land will affect the regional carbon cycle through reduction in both the annual rate of net primary production and the size of the carbon reservoir. The increase in impervious surface and associated changes in surface albedo will also alter surface energy balance. The urban heat island has been noted in many East Asian cities, but their aggregate effects on regional climatology are unknown.

A final concern related to demographic trends in the region needs special mention: the rapid growth in the population greater than 60 years of age, almost doubling from approximately 145 million in 1995 to an estimated 285 million in 2020 (UNESCAP 1999). The rapid aging of the current work force and the change in birth rates will have tremendous impacts on the ability of these countries to care for their elderly and to maintain a labor force within the region. Some national policies are attempting to make adjustments in health care provisioning; however, the cost of caring for the elderly and the social–economic implications have not been fully understood as life-expectancies continues to increase.

2.2. Transition economies

Reforms have transformed the economies of Monsoon East Asia during the past three decades. During the late 1970s, China launched an ambitious program of market-oriented reform and opening. The transition from a command-and-control to market-based economy fundamentally changed regional and global economic and political realities. For the next two decades, China's economy grew at average annual rates of 10% or more. In contrast, global average annual rates of growth ranged between 2% and 3% during this period. During the 1980s, Hong Kong, Singapore, Taiwan, and South Korea also transformed their economies by focusing on export-oriented industries. The overwhelming success of these countries served as a role model for reforms in Thailand, Indonesia, Philippines, and Malaysia. At the heart of reforms were four primary policy changes: (1) liberalization of trade and investment policies, (2) decentralization, (3) creation of land markets, and (4) rural and agricultural price reforms. These changes interact to generate enormous cultural and economic changes.

Liberalization of trade and investment policies has given rise to a large influx of foreign direct investment into the region. Foreign investments into five East Asian countries represent more than 40% of all investment inflows to developing countries and nearly 13% of the world total (Table 23.1). Foreign direct investment has been a direct and indirect driver of land-use change. It drives land-use change directly by funding infrastructure development and construction of factories and residential complexes. These in turn stimulate local economies and in-migration, further increasing demand for local resources.

Political and economic decentralization has occurred in most of Monsoon East Asia. In some countries, decentralization policies were initially instituted in response to mandates from international organizations such as the International Monetary Fund.

Decentralization was a mechanism to improve the delivery of services, reform the tax system, and increase the efficiency of resource allocation. The decentralization process involved the devolution of authority from

Table 23.1. Foreign direct investment inflows (millions US$) for selected East Asian countries and world (UNCTD 2005).

	1970	1980	1990	1995	2000	2001	2002	2003
World	13,032	54,986	208,646	335,734	1,387,953	817,574	678,751	559,576
Developed countries	9,477	46,530	171,109	204,426	1,107,987	571,483	489,907	366,573
Developing countries	3,555	8,421	36,897	115,953	252,459	219,721	157,612	172,033
China	0	57	3,487	37,521	40,715	46,878	52,743	53,505
Hong Kong	50	710	3,275	6,213	61,939	23,775	9,682	13,561
Korea	66	17	759	1,249	8,572	3,683	2,941	3,752
Mongolia	0	0	0	10	54	43	78	132
Taiwan	62	166	1,330	1,559	4,928	4,109	1,445	453
% of Developing countries	5.0	11.3	24.0	40.1	46.0	35.7	42.4	41.5
% of World	1.4	1.7	4.2	13.9	8.4	9.6	9.9	12.8

the central government to lower-tiered governments in the areas of trade, banking, and taxation. The net effect is that localities have more economic freedom to chart their development strategy.

A third pillar of neoliberal reforms was the establishment of land markets. This usually involved changes in the land tenure system with a shift toward increasing private ownership. In China, land reforms created a multi-tier real estate market where land could be purchased, leased, or rented, but the central government maintained ownership of the land.

In essence, state remains the landowner, but usufruct rights can be controlled by individuals. This land policy change has created a new pattern of land use and management in Northern China and Mongolia where livestock has been privatized and land is being leased for long-term usage. Taxation and the continued land use rights depend on the maintenance of the land holdings. In addition, these changes in land policy have resulted in the establishment of more permanent housing sites. It has been argued that without ownership rights, farmers and land-users are more likely to exploit the land for short-term gain, depending on the duration of their lease.

A final change associated with these transition economies is the elimination of collective farming in favor of household and individual farming and by instituting price and marketing reforms. Prior to reforms, agricultural production in command and control economies was managed by a production team and production brigade. Production team members were permitted to farm small plots for their personal use, and within the commune, farmers were allowed to engage in minimal trading. However, farmers were not permitted to sell the products generated from their personal farming activities, and teams were required to meet a production quota. Agricultural reforms allowed farmers to choose their crop and engage in individual farming. Price reforms had the effect of eliminating the emphasis on self-sufficiency in which each region was expected to be self-sufficient in grain production. Under the old self-sufficiency policy, any regional comparative advantage in grain production was eliminated. Because the state procurement prices were low, regions with favorable growing conditions were reluctant to grow more grain than required by the quota while areas with inhospitable growing conditions were still required to be self-sufficient. In order to meet the grain quota requirements, farmers often expanded grain production at the expense of other farm products. This policy led to inefficient uses of land, and regional shortages of grain.

2.3. Globalization

Integration into the world economy, liberalization in trade, and cooperation with international agreements has lead to unparalleled economic interdependence. This interdependency is driven and characterized by cross-boundary capital movements, technology transfers, information flows, and the relocation of mass manufacturing. The demand for intensive labor, energy, and raw material resources for manufacturing has fundamentally changed the global energy demand and balance, causing a potential shift in geo-political balance. Since the transfer of advanced technologies usually lags behind manufacturing transfers, the emission of air-borne pollutants, wastewater and carbon dioxide, and landfill contamination are all expected to increase in developing countries.

Another characteristic of globalization is that state authority is increasingly being challenged by new actors such as non-governmental organizations, multi- and trans-national corporations, and world commodity markets. This interdependency means that decision-making and social systems in one region of the world can impact land use, biodiversity, and ecosystem functioning in another region. For example, international demand for shrimp has driven an increase in shrimp production in China and other East Asian countries. The conversion of coastal and inland ecosystems for aquaculture has contributed to numerous changes in the environmental subsystem, including loss of biotic diversity, intensification of the biogeochemical cycles, and re-allocation of water.

Culturally, the combined impact of globalization, increased urbanization, and a growing "middle-class" is that the Asian culture has the appearance of being increasingly influenced by Western-style culture. This trend has raised concerns in the region about the erosion of cultural identity and traditional values. The widespread availability of Western consumer goods has influenced consumption patterns in East Asia.

The proliferation of Western lifestyles is evident in consumer choices for international fast food chains, single-family homes, modern technologies, and personal automobiles. In turn, these choices have significant impacts on local energy demand, land use, agricultural production patterns, and transportation infrastructure needs.

With globalization and policy reforms, rates of economic growth are now faster than that of the past. East Asian countries in particular are among the fastest developing, with growth rates higher than those of

Table 23.2. Growth of gross domestic product per capita (percent) (Maddison 1995).

Country	1940–1950	1950–1960	1960–1970	1970–1980	1980–1990
Argentina	1.83	1.09	2.76	1.22	−2.23
Australia	1.97	1.69	3.14	1.72	1.75
China	−2.34	3.64	2.21	2.96	6.33
France	2.69	3.65	4.46	2.63	1.73
Germany	−2.55	7.05	3.50	2.56	1.97
Japan	−3.82	7.55	9.31	3.33	3.53
Mexico	2.97	2.92	3.10	3.36	−0.50
South Korea	−6.13	4.04	5.42	6.39	8.14
Taiwan	−3.52	4.26	6.76	7.66	6.24
United Kingdom	0.45	2.27	2.24	1.80	2.47
United States	3.15	1.58	2.87	2.09	1.81

Table 23.3. Average annual trade export growth rates (percent) (UNCTD 1999, 2002).

Country	1960–1970	1970–1980	1980–1990	1990–2000
Argentina	4.8	18.0	2.1	10.1
Australia	7.7	15.9	6.3	5.0
China	1.3	20.0	12.9	14.5
France	9.7	20.3	7.5	4.2
Germany	11.2	19.1	9.2	3.9
Japan	17.5	20.8	8.9	4.1
Mexico	6.1	24.8	8.2	16.1
South Korea	39.8	37.2	15.1	10.1
Taiwan	23.2	28.6	14.8	7.2
United Kingdom	5.9	18.5	5.8	5.4
United States	8.1	18.5	5.7	7.3

the United States and European countries during similar periods of rapid expansion (Table 23.2).

Another result of globalization is the growth in trade. Growth in trade among East Asian countries has been enormous, both in term of total volume of trade and rates of increase (Table 23.3). The region is increasingly the engine of global production.

2.4. Consumption patterns

Rises in disposable income have dramatically changed food and energy consumption patterns. As with most countries, income gains in East Asian countries have been coupled with increased intake of luxury goods such as meat and fish products (Fig. 23.1). An intensification of both meat

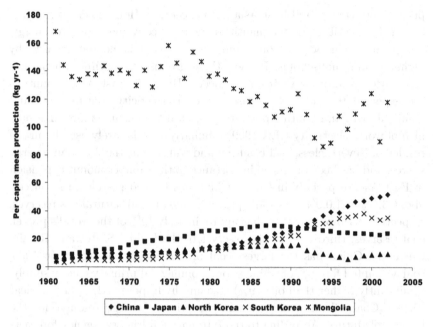

Fig. 23.1. Per capita meat production for selected East Asian countries (FAO 2001).

and fish production systems have increased in the region. Aquaculture is growing rapidly in this region due to demand for high quality protein sources and diminishing capture fisheries in the surrounding coastal areas (Bartley 1997). There is mounting evidence that the environmental impacts of aquaculture are significant (Naylor et al. 2000). In addition, although direct grain consumption has declined, a meat-based diet requires more grain. The increase in demand for grain and meat is satisfied by intensifying production on existing lands or by converting more land from forests to agricultural and pasturelands. Possible methods of intensification include shorter or no fallow periods, row cropping, and increased applications of fertilizers. Alternatively, an increase in demand for grain may be satisfied most easily by increasing agricultural land and pastureland, thereby placing further pressure on natural ecosystems. This expansion of agriculture into forested land was the dominant source of global carbon emissions until the 1940s, when emissions of carbon from fossil fuel combustion surpassed those from land conversion (Houghton and Skole 1990).

Energy demand is also on the rise and energy mix is changing. Household energy consumption usually increases with gross domestic product per capita, and East Asia is no exception. Households with higher income levels utilize greater quantities and more convenient forms of energy. For example, in South Korea and Taiwan, coal is no longer used by higher income households. Rather, these groups rely entirely on natural gas and electricity for their energy needs. By 2025, East Asian countries are expected to lead the increase in world electricity consumption, and China, Japan, and South Korea are projected to consume more than one-fifth of world electricity (IEA 1999). Similarly, coal is rarely used by urban residents. Nevertheless, coal has been and will stay as the dominant energy source, and has been of crucial importance to the rapid economic expansion in East Asia, especially in China. China has limited petroleum resources, about 3.5% and 0.9% of world's proven crude oil and natural gas reserves, respectively. However, China has approximately 12% of the world's proven coal reserves, third only to the reserves of the United States and Russia. Currently, China has the largest coal production in the world. In 1996, coal accounted for 76% of China's total commercial primary energy supply, significantly higher than oil (19%), gas and hydropower (2%), and nuclear (0.4%). Coal will inevitably remain the primary energy resource for the foreseeable future. According to the International Energy Agency forecasts (IEA 1999), in 2020, coal will still account for 67% of total primary energy demand.

This growing concern for energy self-reliance and the need to meet urban population needs across a broad spectrum of sectors has also given rise to the increased number of water projects including the Three Gorges Project on the Yangtze River. In order to provide an additional source of energy, the hydroelectric output of the Three Gorges Dam is estimated to replace approximately 40–50 million tones of coal per year. In addition, the project will enhance flood control and facilitate navigation of larger barges in the region. Given these economic benefits, there is still considerable concern regarding the environmental and social impact the project will have on the region during the coming years.

Advances in technologies could dramatically reduce the emission of carbon dioxide (CO_2) and other pollutants, especially in developing countries where the outdated combustion technologies have very low thermal efficiencies. China's CO_2 emissions declined 7.3% between 1996 and 2000, and CH_4 emissions declined by 2.2% between 1997 and 2000, despite economic growth of 36% (Streets et al. 2001). China was able to accomplish this feat by restructuring its economy, switching to cleaner energy sources and improving energy efficiency. Advanced technologies in coal gasification, clean-coal, bio-fuel, and electric, hybrid and fuel-cell vehicles will become increasingly important to East Asia in terms of energy security and independence, CO_2 emission reduction and environmental quality (Jiang and Wen, Chapter 22, this volume).

2.5. Land-use change

The demand for space for economic activity, rural–urban migration, and industrialization will drive changes in land cover. Increases in population and per capita income tend to promote urbanization, which expands cities and towns into forested and agricultural areas. Industrialization also increases pressure on forests and existing agricultural lands by converting these lands for commercial and residential uses. These land-use changes will result in changes in the biophysical attributes of the land surface and ecosystem functions. Such changes contribute to regional and global climate change by modifying the surface energy and water budgets, and the ecosystem storage and cycling of carbon. Extensive urban land-use change in South China has already affected precipitation patterns at the local scale (Kaufmann et al. 2007). Given the magnitude of new city developement throughout the region, the cumulative impacts on regional climatology are likely to be significant. Government policies also play vital roles in determining land use. China's afforestation and reforestation programs between 1949 and 1998 have been heralded as successful for increasing forest biomass carbon storage (Fang et al. 2001). In southern China where subtropical monsoon evergreen forests dominate, the conversion of forests to agriculture translates to a decrease in net mean primary productivity from 585 to $290\,\mathrm{g\,C\,m^{-2}\,year^{-1}}$ (Whittaker and Likens 1973). Land-use conversions also have dramatic effects on albedo, which is one of the key surface parameters influencing climate (Fu and Wang, Chapter 12, this volume).

3. Conclusions

Monsoon East Asia is a region undergoing tremendous socioeconomic, political, and institutional changes. The magnitude and rate of developmental pathways of this region are unprecedented. The continuous expansion of economies, demographic dynamics, globalization, policy reforms, and cultural and lifestyle changes will interact and affect the global Earth system at all spatial and temporal scales. Taken together, rapid and large-scale land-use change, urbanization, globalization, industrialization, and changes in consumption patterns in the region will affect the properties and resilience of ecosystem services and Earth system functioning. The region is currently on an upward path of change. In order to understand and predict changes in ecosystem dynamics, climate, atmospheric composition, biogeochemistry, or hydrology, it will be critical to understand the underlying human activities and social drivers of change.

Literature Cited

Bartley, D. 1997. East Asia, in FAO Fisheries Circular No 886 FIRI/C886 (Rev.1); *Review of the state of world aquaculture*, Rome.

Fang, J., A. Chen, C. Peng, S. Zhao and L. Ci. 2001. Changes in forest biomass carbon storage in China between 1949 and 1998. *Science* 292:2320–2322.

Houghton, R. A. and D. L. Skole. 1990. Carbon. Pp. 393–408 in *The Earth as transformed by human action*, edited by B. L. Turner II, W. C. Clark, R. W. Kates, J. F. Richards, J. T. Matthews and W. B. Meyer. New York: Cambridge University Press.

IEA (International Energy Agency). 1999. *Coal in the energy supply of China*. Report of the Coal Industry Advisory Board. Paris: Organisation for Economic Co-operation and Development/International Energy Agency.

FAO (Food and Agriculture Organization). 2001. *FAOSTAT database collections*. http://www.apps.fao.org. Rome, Italy: FAO.

Kaufmann, R. K., K. C. Seto, A. Schneider, L. Zhou, and Z. Liu. 2007. Climate Response to Rapid Urban Growth: Evidence of a Human-Induced Precipitation Deficit. *Journal of Climate* 20(10):2299–2306.

Liu, J., G. C. Daily, P. R .Ehrlich and G. W. Luck. 2003a. Effects of household dynamics on resource consumption and biodiversity. *Nature* 42:530–533.

Liu, J., M. Liu, D. Zhuang, Z. Zhang and X. Deng. 2003b. Study on spatial patterns of land use change in China during 1995–2000. *Science in China* 46:373–384.

Maddison, A. 1995. *Monitoring the World Economy 1820–1992*. Table D-1a. Pp. 194–206. Paris: Organisation for Economic Co-operation and Development.

Naylor, R. L., R. Goldburg, J. H. Primavera, J. Clay, N. Kautsky, M. C. M. Beveridge, J. Clay, C. Folke, J. Lubchenco, H. Mooney and M. Troell. 2000. Effect of Aquaculture on World Fish Supplies. *Nature* 405:1017–1024.

Seto, K. C., R. K. Kaufmann and C. E. Woodcock. 2000. Landsat reveals China's farmland reserves, but they're vanishing fast. *Nature* 406:121.

Streets, D. G., K. Jiang and J. Sinton. 2001. Recent reductions in China's greenhouse gas emissions. *Science* 294:1835–1837.

UNCTD (United Nations Conference on Trade and Development). 1999, 2002, 2005. *Handbook of international trade and development statistics*. New York: United Nations.

UNESCAP (United Nations Economic and Social Commission for Asia and the Pacific). 1999. *Population ageing and development: Implications for Asia and the Pacific.* Japan: United Nations Economic and Social Commission for Asia and the Pacific.
United Nations. 2002. *World urbanization prospects: The 2001 revision.* New York: United Nations.
Whittaker, R. H. and G. E. Likens. 1973. Carbon in the Biota. Pp. 281–302 in *Carbon and the Biosphere,* edited by G. M. Woodwell and E. V. Pecan. US Atomic Energy Commission Symposium Series 30. Springfield, Virginia: National Technical Information Service.

LIST OF CONTRIBUTORS

Borjiginte Ailikun
Institute of Atmospheric Physics
Chinese Academy of Sciences
P.O. Box 9804
Beijing 100029, China

Arthur Chen
Institute of Marine Geology
 and Chemistry
National Sun Yat-sen University
Kaohsiung 80424, Taiwan

Zhisheng An
State Key Laboratory of Loess
 and Quarternary Geology
Institute of Earth Environment
CAS, Xian 710075, China

Xing Chen
Department of Atmospheric
 Sciences
Nanjing University
Nanjing 210093, China

Joanne Linnerooth-Bayer
International Institute
 for Applied Systems
 Analysis (IIASA)
A-2361 Laxenburg
Austria

Yihui Ding
Sustainable Development
 Research Institute
University of British Columbia
2029 West Mall, Vancouver, BC
Canada V6T 1Z2

Wenjie Dong
Regional Center for Temperate
  East Asia (TEA)
Institute of Atmospheric
  Physics
Chinese Academy of Sciences
P.O. Box 9804
Beijing 100029, China

Anmin Duan
State Key Laboratory of
  Atmospheric Science and
  Geophysical Fluid Mechanics
Institute of Atmospheric Physics
Chinese Academy of Sciences
P.O. Box 9804
Beijing 100029, China

Nobuhiko Endo
Institute of Observational
  Research for Global Change
Japan Agency for Marine–Earth
  Science and Technology
Yokosuka, Japan

John Freney
CSIRO Plant Industry
62 Gellibrand Street
Campbell, ACT 2612
Australia

Congbin Fu
Institute of Atmospheric Physics
Chinese Academy of Science
P.O. Box 9804
Beijing 100029, China

Roland Fuchs
International START Secretariat
2000 Florida Avenue, N.W., Suite 200
Washington, DC 20009, USA

Yi Ge
State Key Laboratory of Pollution
  Control & Resource Reuse
School of the Environment
Center for Environmental
  Management and Policy
Nanjing University
Nanjing 210008, China

Peng Gong
Sustainable Development
  Research Institute
University of British Columbia
2029 West Mall, Vancouver
BC Canada V6T 1Z2

Xuehui Han
Institute of Population Research
Fudan University
Shanghai, China

Sandy Harrison
Department of Geography
University of Bristol, University Road
Bristol BS8 1SS, UK

Jeff Hicke
Natural Resource Ecology
  Laboratory
Colorado State University
Fort Collins
Colorado 80523, USA

List of Contributors

Dunxin Hu
Institute of Oceanography
Chinese Academy of Sciences
7 Nanhai Road
Qingdao, SD 266071, China

Xiulian Hu
Energy Research Institute
State Development Planning
  Commission
Guohong Building, No. (A) 11
Muxudi-beili, Xicheng District
Beijing 100038, China

Ronghui Huang
Institute of Atmospheric
  Physics
Chinese Academy of Sciences
P.O. Box 9804
Beijing 100029, China

Kejun Jiang
Energy Research Institute
State Development Planning
  Commission
Guohong Building, No. (A) 11
  Muxudi-beili
Xicheng District
Beijing 100038, China

Shipei Jun
College of Resources
  Sciences and Technology
Beijing Normal University
Beijing 100875, China

David Kicklighter
The Ecosystem Center
Marine Biological Laboratory
Woods Hole, MA 02543, USA

Young-Joon Kim
Advance Environmental
  Monitoring Research Center
Kwangju Institute of Science
  and Technology
Gwangju, Korea

Kanehiro Kitayama
Center for Ecological Research
Kyoto University, 509-3 Ohtsuka
Kamitanakami, Hirano-Cho
Otsu, Shiga 520-2113, Japan

Akio Kitoh
Meteorological Research Institute
Japan Meteorological Agency
Tsukuba, Japan

Dong Kyou Lee
Atmospheric Sciences Program
School of Earth and
  Environmental Sciences
Seoul National University
Seoul, Korea

Youn Lee
Department of Agricultural
  Environment
National Institute of Agricultural
  Science and Technology
Seodoon-dong 249
Suwon 441-707, Korea

Ruby Leung
Pacific Northwest National
　Laboratory
P.O. Box 999, Richland
Washington 99352
USA

Chongyin Li
Institute of Atmospheric Physics
Chinese Academy of Sciences
P.O. Box 9804
Beijing 100029, China

Li Li
State Key Laboratory of Loess
　and Quarternary Geology
Institute of Earth Environment
CAS, Xian 710075
China

Tong Ling
Yellow Sea Fisheries Research
　Institute
106 Nanjing Road
Qingdao 266071, China

Jiyuan Liu
Institute of Geographical
　Sciences and Natural
　Resources Research
Chinese Academy of Sciences
Beijing 100101, China

Mingliang Liu
Institute of Geographical
　Sciences and Natural
　Resources Research
Chinese Academy of Sciences
Beijing 100101, China

Shaw Liu
Environmental Change
　Research Project
Institute of Earth Sciences
Academia Sinica, P.O. Box 1-55
Nankang, Taipei 11529, Taiwan

Yunfeng Luo
National Natural Science
　Foundation of China
83 Shuangqing Road, Haidian District
Beijing 100085, China

Keping Ma
Laboratory of Quantitative Ecology
Institute of Botany
Chinese Academy of Sciences
20 Nanxin cun, Xiangshan
Beijing 100093, China

Zhuguo Ma
Institute of Atmospheric Physics
Chinese Academy of Sciences
P.O. Box 9804
Beijing 100029, China

List of Contributors

Mike McCracken
Climate Institute
1785 Massachusetts Avenue NW
Washington, DC 20036
USA

Jiangyu Mao
Institute of Atmospheric Physics
Chinese Academy of Sciences
P.O. Box 9804
Beijing 100029
China

Jerry Melillo
The Ecosystems Center
Marine Biological Laboratory
7 MBL St., Woods Hole
MA 02543, USA

Surabi Menon
The Atmospheric Sciences
    Department
Lawrence Berkeley
    National Laboratory
MS51R208, 1 Cyclotron Road
Berkeley, CA 94720, USA

Nobuo Mimura
Center for Water
    Environment Studies
Ibaraki University
4-12-1 Nakanarusawa
Hitachi, Ibaraki 316-8511, Japan

Arvin Mosier
1494 Oakhurst Dr.
Mt. Pleasant
South Carolina 29466, USA

Dennis Ojima
Natural Resource Ecology
    Laboratory
Colorado State University
Fort Collins, Colorado 80523, USA

The H John Heinz III Center
    for Science, Economics, and the
    Environment
900 17th Street, NW, Suite 700
Washington, DC 20006, USA

Norio Okada
Disaster Prevention
    Research Institute
Kyoto University
Kyoto 611-0011, Japan

Shufen Pan
Institute of Atmospheric Physics
Chinese Academy of Sciences
P.O. Box 9804
Beijing 100029, China

Chongguang Pang
Institute of Oceanology
Chinese Academy of Sciences
7 Nanhai Road
Qingdao, SD 266071
China

List of Contributors

Graeme Pearman
School of Geography and
 Environmental Science
Monash University
Clayton, Victoria 3800
Australia

Xizhe Peng
Institute of Population Research
Fudan University
220 Handan Road
Shanghai 200433, China

Yun Qian
Pacific Northwest National
 Laboratory
P.O. Box 999/K9-24
Richland, Washington
99352, USA

Tom Riley
Natural Resource Ecology
 Laboratory
Colorado State University
Fort Collins
Colorado 80523, USA

Karen Seto
Department of Geological
 Environmental Sciences
Stanford University
Stanford, CA 94305-2115, USA

Guangyu Shi
Nansen-Zhu International
 Research Centre
Institute of Atmospheric Physics
Chinese Academy of Sciences
P.O. Box 9804
Beijing 100029, China

Peijun Shi
Beijing Normal University
19 Xinjiekou Wai Street
Haiden District
Beijing 100029, China

Anond Snidvongs
Old SWU Pathumwan 5 Building
5th Floor
Chulalongkorn University
Henri Dunant Rd.
Bangkok 10330, Thailand

Qingyuan Song
Ford Scientific Research Laboratory
Physical and Environmental
 Sciences Department
Dearborn, Michigan
48188, USA

John Stewart
118 Epron Road
Salt Spring Island, BC
Canada V8K 1C7

List of Contributors

Qisheng Tang
Yellow Sea Fisheries Research
  Institute
Chinese Academy of Fishery
  Sciences
106 Nanjing Road
Qingdao 266071, China

Hanqin Tian
School of Forestry and Wildlife
  Sciences
Auburn University
Auburn, AL 36849, USA

Peter Tyson
Climatology Research Group
University of Witwatersrand
P.O. Box 130110
Bryanston 2074, South Africa

Hassan Virji
International START Secretariat
2000 Florida Avenue
N.W., Suite 200
Washington, DC 20009, USA

Pinxian Wang
Department of Marine Geology
Tongji University
Shanghai 200092, China

Qingye Wang
Institute of Oceanology
Chinese Academy of Sciences
7 Nanhai Road, Qingdao
SD 266071, China

Shuyu Wang
Institute of Atmospheric Physics
Chinese Academy of Science
P.O. Box 9804, Beijing 100029
China

Zifa Wang
Nansen-Zhu International
  Research Centre
Institute of Atmospheric Physics
Chinese Academy of Sciences
P.O. Box 9804, Beijing 100029
China

Gang Wen
State Development Planning
  Commission
Guohong Building, No. (A) 11
Muxudi-beili, Xicheng District
Beijing 100038, China

Guoxiong Wu
State Key Laboratory of
  Atmospheric Science and
  Geophysical Fluid Mechanics
Institute of Atmospheric Physics
Chinese Academy of Sciences
P.O. Box 9804, Beijing 100029
China

Li Xie
Institute of Atmospheric Physics
Chinese Academy of Sciences
P.O. Box 9804
Beijing 100029, China

Zhengqin Xiong
Institute of Soil Science
Chinese Academy of Sciences
P.O. Box 821, Nanjing, China

Jiongxin Xu
Institute of Geographic
  Sciences and Natural
  Resources Research
Chinese Academy of Sciences
A11 Datun Road
Beijing 100101, China

Kazuyuki Yagi
National Institute for
  Agro-Environmental Sciences
3-1-3 Kannondai
Tsukuba, Ibaraki 305-8604
Japan

Weijin Yan
Institute of Geographic
  Sciences and Natural
  Resources Research
Chinese Academy of Sciences
A11 Datun Road
Beijing 100101, China

Xiaodong Yan
Institute of Atmospheric Physics
Chinese Academy of Sciences
Beijing 100029, China

Hongwei Yang
Energy Research Institute
State Development Planning
  Commission
Guohong Building, No. (A) 11
  Muxudi-beili, Xicheng District
Beijing 100038
China

Ying Yang
Institute of Atmospheric
  Physics
Chinese Academy of Sciences
P.O. Box 9804
Beijing 100029, China

Tetsuzo Yasunari
Frontier Research Center for
  Global Change
JAMSTEC
Yokohama, Kanagawa
Japan

Yongyuan Yin
Adaptation & Impacts
  Research Division (AIRD)
Environment Canada, and
  Department of Forest
  Resources Management
University of British Columbia
4619-2424 Main Mall, Vancouver
BC Canada V6T 1Z4

List of Contributors 357

Hiromune Yokoki
Center for Water
   Environment Studies
Ibaraki University
4-12-1 Nakanarusawa
Hitachi, Ibaraki 316-8511
Japan

Ge Yu
Nanjing Institute of Geography
   and Limnology
Chinese Academy of Sciences
Nanjing 210008, China

Shaocai Yu
Environmental Protection Agency
USA

Qiong Zhang
Institute of Atmospheric Physics
Chinese Academy of Sciences
P.O. Box 9804
Beijing 100029, China

Zongci Zhao
National Climate Center
China Meteorological
   Administration
No. 46, S. Road, Zhongguancun
Beijing 100081, China

Yingqun Zheng
Institute of Meteorology
PLA University of Sciences and
   Technology
Nanjing 210000, China

Dadi Zhou
Energy Research Institute
State Development Planning
   Commission
Guohong Building, No. (A) 11
Muxudi-beili
Xicheng District
Beijing 100038, China

Guangsheng Zhou
Laboratory of Quantitative
   Vegetation Ecology
Institute of Botany
Chinese Academy of Sciences
20 Nanxincun, Xiangshan
Beijing 100093, China

Zhaoliang Zhu
Institute of Soil Science
Chinese Academy of Sciences
P.O. Box 821, Nanjing, China

INDEX

Acer mono, 203
Betula costata, 203
Betula dahurica, 203
Betula platyphylla, 203
Fraxinus rhynchophylla, 203
Juglans mandshurica, 203
Larix olgensis, 203
Phellodendron amurense, 203
Populus davidiana, 203
Quercus mongolica, 203
Stipa krylovii, 204
Tilia spp., 203
Ulmus spp., 203
ex ante financing, 301
ex post financing, 301

abandoned croplands, 214
abatement, 329–331
above-ground biomass, 204–206
abundance, 240–244, 246, 247
adaptation, 257, 262
adaptation costs, 333
adaptation investment, 328
adaptation option, 309, 311, 312, 315, 316, 319–321
adaptation strategies, 5
adaptive capacity, 261, 262
advanced gas turbines, 280
advanced nuclear power, 280
advanced technology, 279, 280, 332, 333
advection of ammonia, 234
aeolian mass accumulation rates, 63
aeolian sequences, 27, 29
aerosol particles, 287

aerosols, 105, 106, 111, 115–120, 122–127, 185
agricultural activity, 227
agricultural infrastructure, 216
agricultural production, 214
agricultural productivity, 324
agronomic activity, 215
AIM Global, 333
air pollution, 3, 187, 279, 287, 290, 324
air temperature, 240
albedo, 3
algal blooms, 186
altimeter data, 252
ammonia volatilization, 184, 185, 190, 232
Amur, 253
analytic hierarchy process, 318, 321
anchovy, 242–246, 248
angler, 243
animal population, 179, 182
animal production, 184
animal wastes, 179
anomaly, 76–78, 80, 82, 83
anthropogenic, 105, 109
anthropogenic emissions, 2, 105, 109
anthropogenic impact, 232
anticyclonic circulation, 79, 82
aquifers, 253
arable land, 181–183, 189
arid regions, 211, 213, 214
aridity, 204, 212, 214
artificial coast, 258
artificial islands, 258
Asian dust, 133–138, 141

359

Asian monsoon, 9, 10, 15, 20, 21, 153, 155, 157
Asian summer monsoon, 76, 77
atmosphere–land cover change interaction, 158
atmosphere-ocean model, 83, 85
atmospheric aerosols, 1
atmospheric composition, 1, 2, 6
avalanches, 293, 298

balanced energy technology, 324
basements, 258
beach erosion, 255
benthic community biomass, 4
benthic systems, 249
benthophagic, 243
benthos, 240
bio-methanation, 332
biodiversity, 3, 195, 199, 208
biodiversity conservation, 298
biogeochemical processes, 207, 208
biological fixation, 179
biological productivity, 255
biomass, 195, 204–206, 240–244, 246–248
biomass burning, 185
biomass fuels, 2
biome, 196–199, 207, 217
BIOME 6000 data sets, 62
biotic communities, 241
black carbon, 115, 116, 119–122, 124–127
black carbon particles, 2
Bohai Sea, 226, 227, 240, 241, 244, 246–248
boreal forests, 206
boreal species, 243, 244
bottom trawl surveys, 248
Boyang Lake, 79
Brahmaputra Rivers, 261
breakwaters, 255, 256
broad-leaved mixed forest, 202
burden sharing, 333
burning of coal, 2

business as usual, 106
Buyatia, 216

carbon balance, 3
carbon biogeochemistry, 211, 215
carbon cycle, 165, 171, 174, 212, 214, 215
carbon dioxide, 3, 105, 329, 330, 332
carbon dioxide exchange, 211
carbon mitigation technologies, 331
carbon monoxide, 287
carbon sequestration, 217, 218, 281
carbon stores, 211, 217
carbon tax system, 283
carnivorous fish, 246
casualties, 293
catastrophe bond, 301
catastrophe reserve, 301
catch, 241–246, 248
Cenozoic, 25
Chinese Lake Status Database, 61
Chinese Loess database, 61
Chinese Loess Plateau, 24, 26, 27, 29
Chita Oblast, 216
chlorophyll, 246, 247
circulation, 79, 80, 82–84
clean coal technologies, 280, 282
clean technology, 5, 279, 282, 290
climate change, 3, 5, 150–152, 159, 240, 249, 251, 254, 256, 257, 261, 262, 280, 282, 283, 293–296, 298, 303–305, 323, 324, 326–328, 332, 333
climate models, 262, 325, 326
climate stabilization, 324
climate variability, 327, 332
climate vulnerability, 5
climate warming, 212, 214, 217, 293
climate-related disaster, 5
climatic effects, 124, 126, 127
cloud formation, 3
cloud microphysics, 3
cloud nucleation, 3

coal, 279, 280, 282, 284–287, 289, 329, 331, 332
coal bed methane, 331
coastal dyke, 258
coastal erosion, 225, 253
coastal flood, 296
coastal systems, 239
coastal wetland, 254
coastal zone, 4, 225, 232, 236, 251, 256–258
cod, 241, 243
cold surges, 75, 76
cold temperate coniferous forest, 202
collective loss sharing, 300
community structure, 241, 246, 248
conifer broad-leaved mixed forest, 202
conservation, 225, 229, 230, 234–236
consumption, 335, 337, 341–344, 346
continental margins, 248
convection, 75
coral, 4, 254, 255
coral bleaching, 254
coral mining, 254
crop residue, 179
cropland abandonment, 211, 212, 214
cropland conversion, 212, 218
cross-equatorial flow, 76
crustacean, 241
cultivation, 4
Cultural scenery, 199

dam construction, 225, 230, 234, 236
damage analyses, 302
dark conifer forest, 202
deaths, 295, 304
decadal temperature, 326
deciduous broad-leaved forests, 197, 198, 202, 205, 207
decreased water and sediment fluxes, 227, 229
decreasing precipitation, 107
deep-sea records, 53
deforestation, 3, 4, 188, 225, 287, 296, 298

deforestation observed, 214
degradation, 211, 212, 214–216, 218
degraded rangelands, 214
delta growth, 226
deltas, 251–254, 256, 257, 261
demersal species, 241–243, 246, 248
demographic dynamics, 6
demographic patterns, 216
denitrification, 184, 190, 232
deposition, 133, 136–138, 141, 143, 234
desertification, 3, 188, 211, 214, 215
destocking, 211, 214
detritus, 195
dietary preferences, 179
direct cost, 324
disaster damage, 293, 295
disaster management, 297–300, 302
Disaster Monitoring and Forecasting Satellites, 302
disaster prevention, 297, 298, 303, 304
disaster prevention facilities, 256
disaster prone regions, 296
disaster reduction policies, 298
disaster reduction strategy, 297
disaster relief, 297, 304
disasters, 283, 287
dissolved inorganic nitrogen, 232, 234
diversity, 240, 242, 248
dominant species, 242, 243, 247, 248
Dongting Lake, 79
doubled CO_2 concentration, 106
driving forces, 1, 5, 202
drought, 76, 77, 79, 107, 109, 110, 293, 296, 298, 328, 332
dust aerosol fluxes, 211
dust aerosols, 133, 135, 137–139
dust bowl, 186
dust emissions, 217
dust export, 212
dust inputs, 214
dust particles, 3
dust plumes, 214, 217
dust storms, 3, 187

dykes, 256, 257
dynamical impact, 82

early warning systems, 297
Earth system function, 211
earth's long-wave radiation, 3
Earth's orbital parameters, 111
East Asia, 115, 116, 120, 123, 124, 126, 127, 163, 164, 167–174
East Asian monsoon, 1, 2, 4, 23–25, 27–29, 31, 32, 34, 35, 76, 82–85, 105, 107, 108, 110
East Asian winter monsoon, 75, 79, 80, 82, 83, 85
East China Seas, 4
eastern plains, 198
Eastern Russia, 216
eco-city, 283, 290
eco-city development, 5
eco-city policies, 5
ecological footprints, 273, 274
ecological replacement, 243
economic burden, 324
economic development, 5, 282, 288–290
economic loss, 257, 293–296
ecosystem change, 199, 207
ecosystem dynamics, 6
ecosystem model, 167–170, 172, 173
ecosystem structure, 195
ecosystems, 225, 232, 236
ecotypes, 240, 243, 244
El Niño, 2, 75, 77–83, 239, 241, 242, 254, 296
El Niño events, 231
El Niño Southern Oscillation, 243
elemental carbon, 218
elevation, 204
emission, 115, 116, 119–121, 123, 127
emission mitigation, 283
emission paths, 324
emission reduction, 324, 330, 332
emission scenarios, 288
emission standards, 284–287

emission trading, 283
emissions, 105–109
energy, 5
energy and water budgets, 155
energy consumption, 275–277
energy demand, 329
energy efficiency, 279, 280, 284, 288, 289
energy tax system, 283
ENSO, 75–77, 79, 80, 82–85
Entisols, 197
environment policies, 282, 289
environmental change, 3, 251, 261
environmental gradients, 199, 203, 206
environmental issues, 324
environmental modeling, 283
environmental policies, 5
environmental problems, 185, 186, 188
environmental protection, 324
environmental quality, 5
environmental trading systems, 283
epidemic, 296
erosion, 225, 227, 229
Eurasia, 75, 85
eutrophication, 186, 225, 232, 236, 249
evergreen broad-leaved forests, 197, 198, 202, 205
evergreen coniferous forests, 197
evergreen conifers, 197, 198
extreme events, 105, 251
extreme precipitation, 296
extreme temperature, 296
extreme weather, 293

Fagus crenata, 208
famine, 296
fatalities, 295
fencing, 204, 205
fertility, 267–270
fertilizer nitrogen, 3, 179–184, 186–190, 232

fertilizer use efficiency, 190
fertilizers, 3, 4
financial burdens, 299
fish resources, 4, 239, 240, 248
fisheries, 239, 244, 248
fishing, 254, 258
flatfish, 241, 243
flood control, 225
flood disasters, 327, 328, 332
flood hazard, 297
flood prevention, 328
flooding, 327, 328
floods, 76, 107–110, 225, 226, 230, 293, 295–300, 302–304
floods control systems, 297
flow-cut-off, 227, 229
fluctuation index, 326, 327
flux, 3, 227, 229, 230, 232, 234, 236
food consumption patterns, 3
food production, 179, 187, 189
food resources, 195
food security, 287
food supply, 3
food webs, 239, 246, 248
fossil energy, 279
frequency of floods, 108, 110
freshwater outflow, 4
fuel switching, 332
functioning of marine ecosystems, 239

Ganges, 261
Ganges River, 231
gaseous emissions, 184
general circulation models, 5, 105, 106, 108
geographical distribution of the precipitation, 107
geophysical disasters, 295, 296
glacial boundary, 69
glacial cycles, 29, 34
glaciers, 214, 252
global carbon cycle, 200, 206
global change, 200, 202, 205, 206, 239, 249

global datasets, 259
global emissions, 2, 105
global warming, 1, 2, 105, 106, 239, 240, 251, 252, 254, 255
globalization, 6, 341, 342, 345, 346
Gobi, 215, 217
government, 5
grain production, 177, 181–183, 188
grasslands, 215–218
grazing, 195, 203–205, 214–218
greenhouse gas emission, 279–281, 326, 329, 332, 333
greenhouse gases, 2, 5, 105, 106, 109, 111, 293, 329
ground subsidence, 252

Hadley, 79
Hanjiang, 235
heart disease, 290
heat source, 10, 14, 19
heat wave, 297
heavy rain, 91, 101
Heihe Basin, 313, 314, 317
Heilongjiang, 253
herders, 216
herring, 240, 242–246
high fertilizer inputs, 232
high waves, 258
high-latitude subtropical evergreen/deciduous broad-leaved forest, 202
Himalayas, 196
historical climate changes, 109
Hokkaido, 206, 207
Holdridge life zone classification scheme, 206
Holocene, 226
Holocene maximum, 109, 111
Huaihe River, 77, 79
human health, 279, 290
human induced emissions, 107, 109
human responses, 211
human-induced global warming, 1
hydroelectric power, 225

hydrologic conditions, 1
hydrological cycle, 116, 119, 123, 124, 126, 214
hydrology, 6

ice sheet forcing, 70
icecaps, 252
impact of nitrogen, 177
impacts of agriculture, 177
improved pasture, 216
Inceptisols, 197
inchthyophagic, 243
increased sediment discharge, 225, 226
increased snow cover, 85
Indian Ocean, 76, 85
indicators, 287, 288
industrial development, 214
industrial pollution, 2
industrial sector emissions, 280
infrastructure, 327, 328, 331, 332
Inner Mongolia, 204, 216
input of nitrogen, 179
insect infestation, 296
institutional changes, 5
insurance, 300, 301, 303
integrated assessment, 309, 311, 321
integrated assessment model, 323
integrated gasification, 281, 282, 285
intensification of cropping, 179
inter-annual climate variability, 239
inter-annual fluctuation, 327
inter-tidal zone, 255
interannual, 1, 3
interannual variability, 92, 97
interannual variation, 75, 77, 215
interdecadal variation, 1, 84
Intergovernmental Panel on Climate Change, 106
intraseasonal oscillation, 80, 85
intrusion of seawater, 253
inundation, 253, 258, 259
investment, 327-329, 332, 333
irradiation, 241, 242, 246

irrigated cropland, 216
irrigation, 254

Japan, 79
Jialingjiang, 234, 235
Jinshajiang, 234

key technologies, 280, 282
Kuroshio, 108, 109

La Niña, 79, 80, 83
land cover, 1-3
land cover change, 149-153, 155, 157-159
land reclamation, 302
land reform policy, 218
land subsidence, 4
land surface processes, 85
land tenure policies, 214, 216
land terracing, 229, 230
land use, 149-153, 157-159
land use change, 1-4, 195, 199, 200, 206, 208, 211, 215, 217, 233, 296, 302, 337, 338, 345, 346
land use management, 324
Last Glacial Maximum, 59, 111
Last Glaciation, 30
Last Glaciation Maximum, 31-33
last interstadial, 60
latitudinal plane, 197
leaching, 184, 185, 190, 232
lead, 281-283, 290
legal arrangements, 298, 299
legislation, 298, 302, 303
levees, 297, 302, 303
Lijin station, 227, 231
Little Ice Age, 32, 111
livestock, 179, 182-185, 187
livestock production, 214, 216
living resources, 239-241, 249
loess hills, 227
loess plateau, 24-27, 29, 31-33, 226, 227, 229
loess-paleosol sequence, 23, 24, 27, 29

long-range transport, 133–136, 138
loss sharing, 300, 305
low frequency variation, 85
low hills, 198
low noise level, 287
low pollutant emissions, 279
low subtropical monsoon evergreen broad-leaved forest, 202
low-lying areas, 253

mackerel, 242–246
macro-topographic regions, 198
man-made systems, 253, 256, 261
management policies, 5
managing risks, 5
mangroves, 255
manure nitrogen, 233, 234
marine biota, 4
market accessibility, 216
market availability, 214
meat consumption, 182–184
meat production, 177, 182
Medieval Warm Period, 109, 111
mega cities, 253, 254
Mei-yu front, 76
Meiyu, 91, 100, 101
Mekong River, 261
meridional wind, 76, 81
mesopelagic zones, 248
methane, 105
mid-Holocene, 59
mid-latitude subtropical evergreen broad-leaved forest, 202
migration, 5, 203, 207, 208
mineral recycling, 195
mitigation, 297, 299, 302, 303
mitigation potential, 324
mobile pastoralism, 216
model simulations, 2
mollisols, 212
Mongolia, 211, 212, 215–218
Mongolian plateau, 211
monsoon, 327
monsoon circulation, 155–158, 197, 198

monsoon–ENSO relationship, 83–85
montane drawf suffruticose desert, 198
montane fruticose desert, 198
montane shrub–grass tussock, 198
mortality, 267–269, 290
mountain tropical rain forest, 202
mountain waves, 9, 15
mowing, 204, 205
mud flats, 255
muddy coasts, 254, 256
multi-criteria, 311, 313, 318, 320, 321
multinominal logit model, 205, 207

National Disaster Reduction Plan, 302
natural beaches, 258
natural disasters, 293–296, 299, 303
natural gas, 280, 281, 284, 286, 287
natural vegetation, 1
necromass, 195
net carbon exchange, 3
net primary productivity, 214, 217
net primary productivity model, 206
neurological damage, 290
new energy, 279, 282, 290
nitric acid, 3
nitrogen, 232–234, 236
nitrogen cycle, 177, 185
nitrogen fertilizer application, 232, 234, 236
nitrogen fixation, 232
nitrogen oxides, 287
nitrogen pollution, 187
nitrogen use efficiency, 3, 189, 190
nitrous oxide, 105, 184, 185
nomadic herding, 199
nomadic pastoralism, 215
North China Plain, 215
Northern Hemisphere glaciation, 27, 28
numerical simulation, 83
nutrient cycling, 240
nutrient delivery, 4

nutrient transport, 225
nutrients, 207, 225, 232, 236

ocean ecosystem, 240
ocean warming, 239
optical depth, 3
orbital cycles, 46
orbital-forcing, 67
orographic effects, 198
orographic uplift, 196
over-cultivating, 287
overexploitation, 241, 243, 246
overgrazing, 186, 188, 202, 212, 214, 215, 287
overploughing, 186
oxygen equivalent, 5

Pacific Decadal Oscillation, 84
Pacific Ocean, 76, 85
Pacific Walker circulation, 83
paleo-monsoon, 23, 34, 39, 42, 44, 45, 49, 53
paleoclimate change, 109, 111
paleoclimate modeling, 59
paleoenvironmental databases, 61
Papua New Guinea, 261
pastoral systems, 4
pasture degradation, 216
Pearl River, 253
pelagic fish, 240, 243–246
pelagic species, 242–244, 246, 248
permafrost, 197, 206
permafrost model, 206
Philippine Sea, 82
phosphorus, 207
photosynthesis, 207, 208
phytoplankton, 240, 241, 246, 247
pilchard, 243–245
Pingshan, 234
plankton, 239
planktonic larvae, 240
plant productivity, 212, 217
plant species number, 204, 205

policies, 5, 211, 214, 216, 218, 279, 280, 282, 283, 287–290
policy framework, 282
policy implementation, 290
policy makers, 328
policy making, 323
policy option, 324
policy reforms, 6
political reforms, 4
pollination, 195
pollutants, 248
pomfret, 242–246
population aging, 270
population dynamics, 241
population growth, 232, 233, 288, 324, 328
population trends, 5
port facilities, 256, 258
Poseidon, 252
post-glacial rebound, 252
post-glacial sea-level change, 254
potential evapotranspiration, 198
potential evapotranspiration rate, 198
prawn, 240, 241
precipitation, 2, 3, 77, 78, 84, 105, 107–111, 202, 204–206, 212, 213, 215–218, 231, 232, 234, 251, 324, 326
precipitation change, 107
precipitation characteristics, 93, 94
prevention, 297–299, 302–304
primary production, 241, 246, 247, 249
primary productivity, 195, 204, 206, 207
proactive "risk management", 293, 297, 305
proactive disaster financing, 297
process-based model, 207
productivity, 240, 243, 246, 247
projections, 105–110
projections of precipitation, 107
protection facilities, 256

Index

proxies, 43–45, 47, 48, 53
proxy index, 24

quality indices, 288
quasi-biennial oscillation, 84, 85, 231
quasi-periodic variation, 84
Quaternary, 23, 28, 29, 34

radiative forcing, 105, 116, 122, 123, 126, 218
rain forest, 202, 206, 207
rainfall, 76, 77, 79, 83–85, 212, 218, 219, 241, 242, 246
rainfall anomalies, 77, 79
rainfall gradient, 202
rangelands, 211, 212, 214–218
rangelands recovery, 217
receding, 214
reconstruction, 297, 298, 300, 301, 303, 304
recruitment, 240–242, 246, 249
red-clay, 24, 25, 27
reduced upwelling, 4
reefs, 4
reforestation, 287, 302
regional climate, 2, 3
regional climate and environment, 150, 152
regional climate model, 105, 108, 109, 155
reinsurance, 300, 301, 303
relocating population, 302
renewable energy, 279, 280, 283, 285–287
reservoir construction, 236
reservoirs, 226, 230, 236, 254
resilience, 254, 255, 261
respiratory diseases, 290
response strategies, 261, 262
risk assessment, 297
risk management, 293, 297–302, 305
risk transfer, 300, 301
river runoff, 241
run-off, 184, 185, 232

salinity, 240, 241
salinization, 211, 214
sand mining, 255, 256
sandstorms, 287
sanitation, 287
Sanmenxia Reservoir, 230
sardine, 243, 244
satellite altimeter, 252
scad, 243–245
scenarios, 5, 205, 206, 324–326, 328, 329
scenarios studied, 106
sea level rise, 4, 225, 251–262
sea surface temperature, 240
sea-level, 251–262
searobin, 243
seasnail, 243
seasonal variation, 241, 247
seawalls, 255, 256, 258
seawater temperature, 251, 252
securities transfer, 300
sedentary cultivation, 199
sediment, 4, 253–256
sediment discharge, 225–228, 231, 235, 236
sediment flux, 4, 227, 229, 230, 234
sediment load, 254
sediment transport, 4
sedimentation, 230
semiarid ecosystems, 211, 215
semiarid lands, 211
sequestering atmospheric CO_2, 214
sewage outfall, 256
Shannon index, 203
sharing the disaster burdens, 297
shocks, 218
shrimp, 255
shrub–steppe, 198
shrublands, 202
Siberia, 206
Siberian, 80
Siberian cold air mass, 198
simulated temperature, 324, 325
simulation, 205, 289, 290
simulation model, 324

sink-source relationship, 214
Skates, 243
slides, 296
small islands, 251
snow cover, 85, 216
snow depth, 85
socio-economic scenarios, 5
socioeconomic, 5
socioeconomic factors, 211
soil carbon, 215, 216
soil erosion, 187, 188
soil organic carbon, 204, 205
soil organic matter, 212, 218
solar activity, 111
solar short-wave radiation, 3
Songnen Plain, 203
soot, 218
South Asian (Indian) monsoon, 76
South China Sea, 23, 31
South China Sea monsoon, 76
South Korea, 79
Southern Hemisphere monsoon, 75
southern oscillation, 75
spatial distribution, 241
spawning, 239, 241, 242, 246, 248
spawning habitats, 239
species composition, 241
species diversity, 4, 240
spectral analyses, 231
squid, 243
steel, 329-331
steppe, 198, 202-204, 206, 212, 215-217
steppe soils, 212
storm surges, 253, 255-259, 261
structural mitigation, 297, 302
structural mitigation projects, 297
suborbital variations, 50
subsistence tillage, 199
substrate structure, 240
subsurface ocean temperature, 80, 82, 83
sulfate, 115, 116, 119-127
sulfate aerosols, 106, 111
sulfur dioxide, 3, 5, 283, 287, 290
sulfur dioxide emissions, 5
summer monsoon, 1, 2, 23-34

surface air temperature, 2, 105, 106, 110
surface climate, 157
surface roughness, 3
sustainability indicator, 5
sustainable development, 251, 262, 287, 301, 324
symbiotic algae, 254
systematic replacement, 243

Taklimakan, 217
technological strategies, 280
technologies, 279-282, 284, 290
technology advancement, 5
technology transfer, 333
teleconnection process, 79
temperate coniferous, 202
temperature, 105-111, 211, 212, 214, 251, 252, 254, 255, 323-327, 329
terrestrial carbon stocks, 3
terrestrial ecosystems, 163-165, 167-173
terrestrial transects, 202
thermal expansion, 252
Tibetan Plateau, 1, 9, 10, 12, 14-21, 23-25, 28, 34, 85, 197, 198, 206
tide gauge, 252
Topex, 252
topographic gravity waves, 76
total nitrogen, 204
tourism, 254, 256
transition economies, 338, 340
transitional mountain area, 198
trend, 91-99, 101, 293-296
tropical cyclones, 85
tropical Pacific Ocean, 108, 109
tropical seasonal rain forest, 202
tropospheric biennial oscillation, 84, 85
tsunamis, 258, 259
Tuva, 216
typhoon, 251, 257, 258, 262, 293, 296, 304
typhoon numbers, 108-110

Udults, 197
Ulaanbaatar, 217

upwelling regions, 248
urbanization, 5, 271, 272, 274–277, 336, 337, 341, 345, 346

variability, 1
variation, 226, 234
vascular plants, 195
vegetation belts, 197
vegetation feedbacks, 60
vegetation zone shift, 328
volcanic activity, 111
volcano, 296
vulnerability, 5, 251, 256–259, 261, 262
vulnerability assessment, 258, 262, 309–311, 321
vulnerable regions, 293

Wang and Hao, 158
Wang et al., 157
warm temperate deciduous broad-leaved forest, 202
warmth index, 197
waste management, 332
water consumption, 228, 229
water discharge, 254
water diversion, 225, 236
water erosion, 184
water gates, 256
water policy, 228
water pollution, 254
water regime, 211, 214
water resources, 298, 302
water resources management, 324
water stress index, 328
water use efficiency, 205
water vapor exchange, 211

water-use efficiency, 328
water/soil conservation, 225, 229, 230, 234–236
wave/surge, 296
waves, 251, 253, 255, 256, 258
weather-related disasters, 293–297, 305
westerly-easterly boundary, 10–12
Western China, 309, 311, 313–317
wild fires, 296
wind erosion, 184, 186, 187
windstorms, 293, 296
winter monsoon, 1, 2, 23–32
woodlands, 202
Wujiang, 235

Xiaolangdi reservoir, 230
Xilin River, 204
Xinjiang, 216

Yalongjiang, 234
Yangtze River, 4, 77, 79, 253, 254, 257, 261
Yangtze River (Changjiang), 225, 231–236
yellow croaker, 241–243
Yellow River, 4, 79, 252–254, 256, 257
Yellow River (Huanghe), 226
Yellow Sea, 240–244, 246, 248, 249
Younger Dryas, 30, 31

Zheng et al., 157
zonal pressure, 82
zonal wind, 76, 81, 82
Zonobiomes, 196
zooplankton, 240, 241, 246, 247, 249